Control of Reproduction in the Cow

Current Topics in Veterinary Medicine

Volume 1

Control of Reproduction
in the Cow

**A Seminar in the EEC Programme of Coordination of
Research on Beef Production held at Galway,
September 27-30, 1977**

Sponsored by The Commission of the European Communities,
Directorate-General for Agriculture, Coordination of Agricultural Research

Edited by

J. R. Sreenan

The Agricultural Institute
Belclare, Tuam
Co. Galway

Martinus Nijhoff - The Hague/Boston/London   1978
for
The Commission of the European Communities

Publication arranged by

Commission of the European Communities,
Directorate-General Scientific and Technical Information
and Information Management, Luxembourg.

EUR 6014 EN

ISBN 90 247 2135 0

Table of contents

PREFACE

This publication contains the proceedings of a Seminar held in Galway, Ireland on September 27 - 30, 1977 under the asupices of the Commission of the European Communities, as part of the EEC programme of co-ordination of research on beef production.

The programme was drawn up by a scientific working group on PHYSIOLOGY OF REPRODUCTION on behalf of the beef production committee. The working group comprised: Dr. J.M. Sreenan (Chairman), Ireland; Mr. L.E.A. Rowson, United Kingdom; Professor C. Thibault, France; Dr. B. Hoffman, Germany (Fed. Rep.); Professor L. Henriet, Belgium; Dr. F. du Mesnil du Buisson, France; Dr. J. Riemensberger, Germany (Fed. Rep.); Dr. P. Mauleon, France; and, in the planning of the seminar, they were joined by Mr. P. L'Hermite, CEC and Dr. J.C. Tayler (temporarily seconded to the CEC, 1975).

The subject chosen for this seminar was drawn from the list of priorities in research objectives drawn up in 1973 by members of a committee (now the Standing Committee on Agricultural Research, CPRA) given in Appendix 1. One of the functions of this series of seminars was to summarise and update the information available on the selected subjects and to discuss future needs for research, so as to assist the Commission in evaluating the probable impact of research on agricultural production within the community.

The Commission wishes to thank those representatives of Ireland who took responsibitility in the organisation and running of this seminar; notably Dr. J.M. Sreenan (Chairman and local organiser) and Dr. M.T. Kane and Dr. J.P. Gosling, both of University College, Galway.

Thanks are also due to the Chairmen of sessions, Dr. D. Schams, Professor P. Mauleon, Mr. L.E.A. Rowson, Professor N. Rasbech, Mr. R. Newcomb and Dr. J.M. Sreenan.

OBJECTIVE OF SEMINAR

To review recent research, including research projects funded by the Commission, on techniques employed in the control of reproduction in the cow and to formulate objectives and priorities for future research.

BACKGROUND

Oestrous cycle synchronisation techniques could lead to increased efficiency because they enable a greater number of cows, particularly beef cows, to be bred by artificial insemination. Using proven bull semen, the rate of genetic gain for economically important traits could be improved.

Induction of twin-calving could also lead to increased efficiency by increasing the size of the calf crop.

The development of such techniques requires a study of many factors relating to the oestrous cycle and its control.

SESSION I

FOLLICULAR GROWTH AND OVULATION

Session 1a - Normal Physiological Processes

Session 1b - Nature of PMSG

Session 1c - PMSG - Superovulation Studies

Chairman:
(Session 1a and 1b)

D. Schams

Chairman:
(Session 1c)

P. Mauleon

SOME OBSERVATIONS ON THE DEVELOPMENT AND FUNCTION OF OVARIAN FOLLICLES[*]

S.C.S. Chang,* R.J. Ryan,* Y.H. Kang and W.A. Anderson
*Department of Molecular Medicine,
Mayo Medical School, Rochester, MN 55901, USA
Department of Zoology & Howard University Cancer Research Center
Howard University, Washington, DC 20059, USA

INTRODUCTION

This article is not intended to be a review of the literature as several excellent reviews are available (see, for example, Greenwald, 1974). Rather, it is an effort to 1) put together a number of observations that we have made over the past three years with respect to the structure and function of ovarian follicles, and 2) express some thoughts concerning follicular development and point out areas that require further study. Some of these observations have been published previously, while others are presented here for the first time.

PREANTRAL FOLLICLES

The fine structure of oocyte-follicle relationships has been studied extensively by standard transmission electron microscopy in a variety of species (for example, Motta et al., 1971; Zamboni, 1974; Albertini and Anderson, 1974). It is our intent here to add to that literature some observations made with high voltage transmission EM.

A primordial follicle is illustrated in Figure 1 which is a high voltage (1 000 kv) electron micrograph of a thick (0.5 - 1 mu) section of rat ovary. The follicle is separated from the surrounding interstitium by a thick basement lamina which is composed of amorphous material and collagen fibrils. The granulosa cells are spindle-shaped and 2 to 3 in number. Granulosa cell mitochondria are tubular in shape. There is intimate contact between the granulosa cells and the oval oocyte and demosomes or macula adherens can be noted. The

[*] Supported by funds from the National Institutes of Health (NICHD 9140) and the Mayo Foundation.

4

dictyate oocyte nucleus is spherical and its diameter is approximately three-fourths the diameter of the cell. The nuclear
membrane is distinct, without visible pores. Masses of hetero-
chromatin and a reticular nucleolus can be noted. The oocyte
mitochondria are globular, highly electron dense and associated
with elongated cisternae of the endoplasmic reticulum. This
association has not been noted previously.

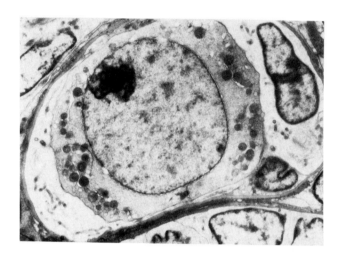

Fig. 1   High voltage (1 000 kv) transmission electron micrograph (TEM) of a
         thick (0.5 - 1 mu) section of rat ovary showing a primordial follicle.
         Note the highly dense spherical oocyte mitochondria associated with
         cisternae of the endoplasmic reticulum. Ovaries were fixed in 2.5%
         glutaraldehyde-formaldehyde in 0.1 M cacodylate buffer, pH 7.2, for
         1 - 2 hours and post fixed for 60 min in 2% OS_2O_4. They were de-
         hydrated in graded alcohol solutions, embedded in Epon, cut with
         glass knives, mounted on formvar-coated grids and stained with uranyl
         acetate-lead citrate.

The intial phase of follicular growth is concerned with enlargement of the oocyte, particularly the ooplasm. This is
followed by an increase in the number of granulosa cells as
illustrated in Figures 2A and B. There is an increase in the
number of spindle-shaped granulosa cells (Figure 2A) followed
by the development of a more cuboidal shape of the cells (Figure
2B). More macula adherens can be noted (Figure 2B) but otherwise there is little change in cell structure. It should be
noted (Figure 2A) that the interstitial cells at this stage

are largely fibroblastic in nature and are not yet organised
into a theca.

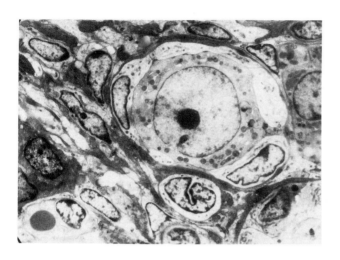

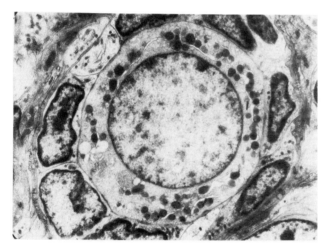

Fig. 2 High voltage TEM of rat ovary: A (top): a primordial follicle
        without a defined theca. Note the ER cisternae associated oocyte
        mitochondria, the contact of oocyte plasma membrane with the base-
        ment lamina of the follicle and macula adherens between the oocyte
        and granulosa cells. B (bottom): a slightly later stage of follicular
        development in which the number of granulosa cells has increased
        and the granulosa cells have hypertrophied.

Figure 3 shows a follicle with a single layer of 20 cuboidal to low columnar granulosa cells encircling the oocyte. The tubular mitochondria of the granulosa cells begin to accumulate on the oocyte side of the cell. Distinct cytoplasmic processes of the granulosa cells begin to appear in the distended portions of the peri-oocyte space. The oocyte mitochondria are not as clearly associated with cisternal spaces of the endoplasmic reticulum.

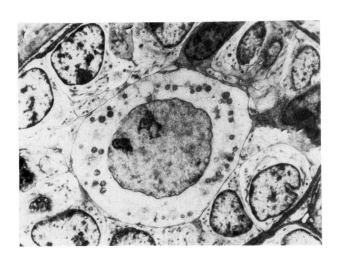

Fig. 3 High voltage TEM of rat ovary. The follicle is composed of a single layer of 20 cuboidal to low columnar granulosa cells surrounding the oocyte. Granulosa cell mitochondria are tubular and towards the oocyte surface. The association of oocyte mitochondria with ER cisternae is less apparent.

In Figure 4A the oocyte is encircled by two layers of granulosa cells. The peri-oocytic space contains thick, long and short granulosa cell microvilli, some of which parallel the oocyte surface (Figure 4B), and fewer, smaller oocyte microvilli. Electron opaque amorphous zona pellucida material begins to form between the villous processes and macula adherens develop at the attachments of granulosa cell microvilli to the oocyte surface. At this stage in the oocyte there is the first appearance of lamellar complexes, the mitochondria are clearly dissociated from the ER cisternal spaces, multivesicular bodies and nacent cortical granules appear in the ooplasm (4B). The theca

surrounding the follicle begins to develop into a clear entity
and the cells take on a more epithelioid character.

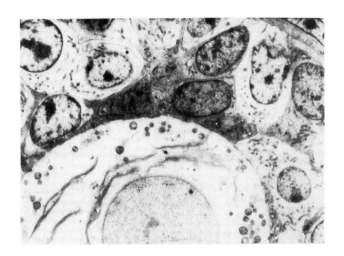

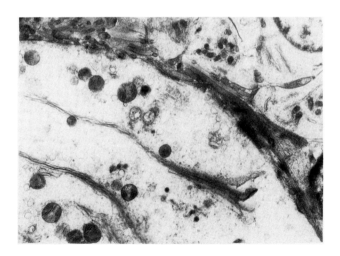

Fig. 4 A high voltage TEM of rat ovary.   A (top): a follicle with two
layers of granulosa cells. In the oocyte lamellar complexes have
appeared and the mitochondria are located peripherally. Amorphous
material is beginning to appear in the peri-oocytic space. B (bottom):
a higher magnification of the same follicle illustrated in A showing
the appearance of long granulosa cell microvilli and short oocyte
microvilli in the peri-oocytic space. A few nacent cortical granules
can be noted in the ooplasma.

The follicle in Figure 5 is developed to the point where the granulosa is 2 - 3 layers thick. The zona pellucida region is now clearly defined and the organisation of the theca and its capillary network is apparent.

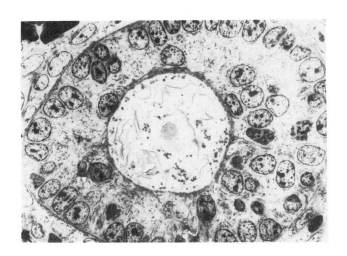

Fig. 5 A high voltage TEM of a rat ovarian follicle with 2 - 3 layers of granulosa cells. Note the increase in thickness of the zona pellucida and the formation of a defined theca.

With further development of the follicle, there is an increase in the number and peripheral localisation of oocyte mitochondria and cortical granules and the Golgi enlarges and becomes more saccular. The zona pellucida develops into a meshwork of long granulosa microvilli and shorter oocyte microvilli with frequent macula adherens at contact points (Figure 6). The granulosa cells increase to 8 or more layers in depth with the appearance of fluid containing clefts between the cells. These clefts ultimately coalesce to form the antrum. The granulosa cells show an increasing number of tubular mitochondria and an increasing amount of smooth endoplasmic reticulum with cisternae.

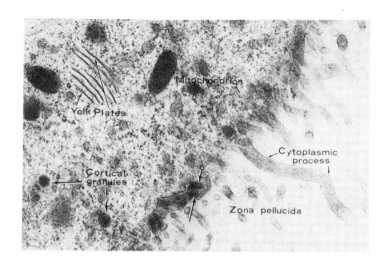

Fig. 6   A high voltage TEM of the zona pellucida region of a large preantral
follicle of the rat ovary. Granulosa cell cytoplasmic process or
microvilli are noted to transverse the zona and make contact with
the oocyte plasma membrane. The arrows indicate macula adherens
junctions.

The initiation of growth of primordial follicles and their
development to the stage of antrum formation occurs in the neo-
natal (Ben-Or, 1963; Peters, 1969) and prepubertal periods
(Pedersen, 1969; Peters, 1969), following hypophysectomy (Smith,
1930; Greep et al., 1942), in anencephalic humans (Ross, 1974),
in women with hypogonadotrophic hypogonadism (Ross, 1974), in
women taking sex steroids (Ross, 1974), and in rats injected with
antigonadotrophic serum (Lunenfeld and Eshkol, 1968). Since all
these conditions are characterised by diminished or absent concentrations of gonadotrophins in blood, it appears safe to
conclude that follicle development to the stage of antrum formation does not have an absolute requirement for gonadotrophins.
Furthermore, there are data to indicate that in the first week
following birth the mouse ovary contains primarily primordial
follicles and follicles with only 1 - 2 layers of granulosa
cells and an undeveloped theca. These follicles are refractory
to small doses (2 - 4 U) of PMSG (Ben-Or, 1963). This may imply
that the follicle at these early stages lacks gonadotrophin receptors since ^{125}I-HCG binding, at least, cannot be demonstrated

in rat ovaries of less than 5 days of age (Presl et al., 1971, 1972).

Even though follicle development up to antrum formation does not require gonadotrophins, and even though follicles with less than two layers of granulosa cells and an undeveloped theca do not respond to gonadotrophins, there is still evidence for hormonal influences on preantral follicle development. Perhaps the clearest evidence is the effect of pharmacological doses of oestrogen. In intact (Paesi, 1952; Smith, 1961) and hypophysect-omised rats (Croes-Buth et al., 1959), large doses of oestrogen cause an increase in the number of large (4+ layers of granulosa cells) preantral follicles to a point where they constitute a significant portion of the ovarian mass. Not only do oestrogens increase the number of granulosa cells but they also increase the sensitivity of the ovary to FSH (Bradbury, 1961; Paesi, 1952) and increase ^{125}I-FSH binding to granulosa cells (Richards, 1977). This is an appropriate sequence in a physiologic sense since the preantral follicle must become responsive to FSH for the induction of an antrum (Greep et al., 1942).

This effect of oestrogen gives rise to several questions: Is the effect of oestrogen a physiologic response or merely a pharmacologic phenomenon? If physiological, what is the source of the oestrogen? How is the oestrogen mediated? The fact that stimulation by injected oestrogen occurs in hypophysectomised rats, and that oestrogen application to one ovary increases fol-licular development in that ovary to a greater extent than the counterlateral one (Bradbury, 1961), indicates that the stim-ulation is due to direct effect on ovarian structures. Thus, injections of pharmacologic doses may be required to achieve physiologic concentrations in the ovary. However, conclusive evidence is lacking.

The source of ovarian oestrogen for the growing follicle is not clear. It could, of course, arise from the interstitium or larger vesicular follicles. There is no evidence that it arises from the theca-granulosa complex of the particular growing pre-antral follicle. The fact that FSH (but not LH) binding sites appear on granulosa cells of larger preantral follicles and that LH (but not FSH) binding sites occur on theca cells of larger preantral follicles (Richards, 1977) does not necessarily prove that oestrogen is being secreted.

The mechanism by which oestrogen increases the number of large preantral follicles could be stimulation of smaller pre-antral follicles to further develop or the prevention of atresia. The recent data of Harman et al. (1975) suggest that oestrogens inhibit atresia.

ANTRAL FOLLICLES

The antral follicle is arbitrarily classified as small, medium or large but the absolute dimensions for these sizes vary with the species under study since as a general rule the size of the largest follicle is proportional to the size of the ani-mal. It should also be pointed out that the greatest increase in follicular size occurs in the period immediately prior to ovulation.

1. Morphologic considerations

The morphologic features of the development of antral fol-licles have been described previously (Hertig and Adams, 1967; Crisp and Channing, 1972; Albertini and Anderson, 1974; Chang et al., 1977). With progression from small to large follicles, there is an increase in the number of layers of granulosa cells, an increase in follicular fluid and a differentiation of the granulosa into two regions, one zone surrounding the egg, the cumulus, and the other lining the remainder of the follicle. The granulosa cells themselves increase in size, predominantly because of an increase in cytoplasm. The nucleus becomes more rounded, the chromatin becomes less clumped, the nuclear mem-brane becomes more distinct and multiple discrete nucleoli appear. Mitochondria increase in number and become lamelliform, the endoplasmic reticulum increases and numerous lipid droplets and lysosomes appear. All of these features are characteristic of steroid secreting cells. In the period immediately before

ovulation, the granulosa cells show a decrease in the number of
mitosis, nuclei, particularly in the cumulus region become pyc-
notic and there is a marked increase in lipid droplets, smooth
ER and mitochondria.

Concomitant with these changes, there is hypertrophy of
the theca cells and increased vascularisation of the theca layer.
In the preovulatory period the theca cells also become luteinised
and shortly following ovulation the theca vessels penetrate the
basement lamina and come in contact with the luteinising granulosa cells.

A more recently recognised feature of granulosa cell development is shown in Figure 7. Granulosa cells lining small (Figure
7A) and medium (Figure 7B) follicles in pig ovaries are rather
smooth while those from large follicles are covered with microvilli (Figures 7C and D). Transmission EM indicates that these
villi contain microfilaments. A study by Chang and Ryan (1976)
also indicated that microvilli are present on granulosa cells
from large follicles of the rat ovary and that they appear between 24 and 36 hours after injection of PMSG and persist up to
72 hours.

Also recognised in recent years is the development of extensive gap junctions among adjacent granulosa cells throughout
the follicular wall (Albertini and Anderson, 1974). These
junctions probably serve to synchronise granulosa cell function
since there is evidence that the cells are electrically coupled.

2. Properties of follicular fluid

Follicular fluid is not a simple solution but rather a gel
with an organised structure. It contains mucopolysaccharides,
rich in hyaluronic acid and chondroitin sulfate, that are secreted into the antral space by the granulosa cells (Odebald,
1954; Zachariae, 1957; Motta, 1965[1,2]). A transmission EM
picture of rat follicular fluid stained for mucopolysaccharides
(Figure 8) indicates a complex network of coarse fibres. These
fibres tend to coalesce and adhere to the surface of granulosa

cells (Kang et al., 1977). It is known that the gelatinous properties of follicular fluid lessen as the time of ovulation approaches.

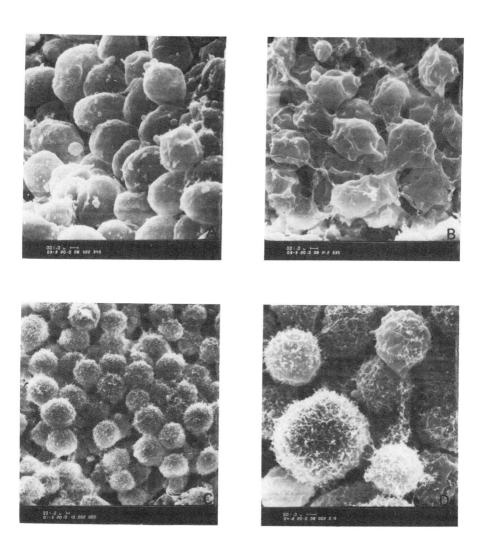

Fig. 7   Scanning electron micrographs of granulosa cells lining the antral surfaces of pig ovarian follicles. A: a large (6 - 12 mm) follicle. B: a medium (3 - 5 mm) follicle. C: a large (6 - 12 mm) follicle. D: a higher magnification of C. The degree of magnification is indicated by the micron scale at the bottom of each photograph. Taken from Chang et al. (1977).

Follicular fluid is in part a production of granulosa cell secretions and a transudate of serum. The basement lamina, however, acts as a molecular sieve and excludes some large molecules. Thus, molecules smaller than serum albumin ($\sim$ 65 000 daltons) enter freely while those larger than $\sim$ 800 000 daltons are excluded. Molecules between 65 000 and 800 000 daltons are present in follicular fluid in concentrations inversely proportional to their size (Shalgi et al., 1973).

Edwards (1974) has extensively reviewed the properties of follicular fluid and summarised the available data concerning pH, ionic composition and concentrations of such metabolites as glucose, lactate and pyruvate. We shall summarise here only our own contributions to this subject.

a) Lipid composition. Data in Table 1 indicate that porcine follicular fluid contains less triglycerides, phospholipids and both total and free cholesterol than porcine serum. All of the lipid is protein bound, but it is contained only in high density lipoprotein (HDL) as low (LDL) and very low density lipoproteins (VLDL) are excluded from follicular fluid (Chang et al., 1976). It should be noted that the major lipids in HDL are cholesterol and phospholipids such as lecithin.

TABLE 1

LIPID COMPOSITION OF FOLLICULAR FLUID FROM PORCINE OVARIES.

	Follicular fluid mg/dl			Serum mg/dl
	Small	Medium	Large	
Triglycerides	18	16	13	50
Phospholipids	49	47	48	132
Total Cholesterol	32	29	29	115
Cholesterol Esters	23	21	22	100

Freshly collected fluid from small (1 - 2 mm), medium (3 - 5 mm) and large (6 - 12 mm) follicles was analysed. Data from Chang et al., 1976.

Follicular fluid contains a lecithin-cholesterol acyltransferase comparable in its kinetics to that found in serum but in

lower absolute activity and without significant variation between follicles of different sizes (Yao et al., 1977). This
study led to an analysis of the fatty acid composition of lecithin in follicular fluid as compared to serum and these data
are presented in Table 2. The general similarity of plasma
and follicular fluid lecithin should be noted. The predominance
of palmitic (16 : 0) and stearic (18 : 0) acids at position C1
was noted while arachidic (20 : 4), oleic (18 : 1) and linoleic
(18 : 2) acids predominated at position C2 of lecithin. The
high content of arachidic acid (20 : 4) at the C2 position
of lecithin may play an important role in providing the
precursor for prostaglandin synthesis (Demers et al., 1973;
Ramwell et al., 1977).

TABLE 2

CONTRIBUTION AND POSITIONAL DISTRIBUTION OF FATTY ACIDS (FA) IN THE C1
AND C2 POSITIONS OF LECITHIN IN PORCINE OVARIAN FOLLICULAR FLUID (POFF)
AND IN PORCINE SERUM (PS)

FA	POFF				PS			
	Total	C-1	C-2	Reconst.	Total	C-1	C-2	Reconst.
	%				%			
16:0	20.4	33.1	3.1	18.1	19.4	28.8	1.3	15.1
18:0	24.8	53.7	1.1	27.4	26.0	58.7	0.7	29.7
18:1	15.0	6.9	23.5	15.2	14.5	7.5	19.3	13.4
18:2	12.9	1.6	21.2	11.4	15.7	1.4	26.4	13.9
20:3	1.9	-	4.4	2.2	1.4	0.1	3.2	1.6
20:4	18.0	0.7	33.9	17.3	17.4	0.7	36.4	18.6
22:4	1.8	0.2	3.9	2.0	1.4	-	3.1	1.5
22:5	2.0	-	4.8	2.4	1.6	-	3.8	1.9
22:6	0.6	-	1.5	0.8	0.6	-	1.9	0.9

Data from Yao et al., 1971.

LDL has been implicated, through a receptor, in the transfer of cholesterol and cholesterol esters into fibroblasts
(Goldstein and Brown, 1974). The absence of LDL from follicular
fluid suggested to us that HDL might serve a similar role for

TABLE 3

CONCENTRATIONS OF AMINO ACIDS AND RELATED COMPOUNDS IN FRESHLY COLLECTED
PORCINE FOLLICULAR FLUID AND SERUM

Amino compound	SFF	MFF	LFF	Serum
		(μM/dl)		
Phosphoserine	0.7	0.8	0.9	0.9
Phosphoethanolamine	4.2	2.5	2.2	0.7
Taurine	9.6	8.9	7.3	18.0
Aspartic Acid	3.7	2.0	1.5	2.0
Threonine	21.0	17.0	16.0	13.0
Serine	16.0	13.0	12.0	13.0
Aspargine + Glutamine	77.0	62.0	60.0	62.0
Sarcosine	13.0	10.0	11.0	12.0
Proline	27.0	19.0	17.0	22.0
Glutamic Acid	46.0	30.0	25.0	20.0
Citrulline	5.5	4.9	5.2	6.9
Glycine	83.0	68.0	64.0	72.0
Alanine	79.0	68.0	56.0	61.0
α-Aminoadipic Acid	3.7	3.4	2.7	3.0
α-Amino-n-butyric Acid	5.0	4.5	5.4	2.8
Valine	27.0	25.0	28.0	26.0
Half-Cystine	18.0	16.0	15.0	1.6
Methionine	4.5	3.4	3.1	4.0
Isoleucine	11.0	9.2	10.0	10.0
Leucine	20.0	17.0	18.0	18.0
Tyrosine	11.0	8.5	8.0	8.1
Phenylalanine	13.0	10.0	9.1	8.5
β-Alanine	1.5	2.1	1.7	0.9
Ornithine	3.1	2.7	2.6	6.9
Lysine	37.0	25.0	22.0	20.0
Histidine	13.0	9.4	9.1	10.0
3-Methylhistidine	5.0	4.6	4.4	4.6
Carnosine	7.6	10.0	6.8	11.0
Arginine	23.0	18.0	16.0	26.0

Data from Chang et al., 1976.

the granulosa cell. A study of ^{125}I-HDL binding to porcine
granulosa cells has been made and indicates the existence of
saturable high affinity (10^{-9}M) binding sites that vary in
number with granulosa cell maturation (Chang and Ryan, 1977).
The number of sites found were 2.2 x 10^5 for cells from small
(1 - 2 mm) follicles, 3.5 x 10^5 for medium (3 - 5 mm) follicles,
and 5 x 10^5 for cells from large (6 - 12 mm) follicles. Nothing,
as yet, is known concerning the regulation of HDL binding sites
but it is tempting to speculate that prolactin may play a role
since it is known to increase the cholesterol content of the
ovary and testis (Armstrong et al., 1969).

b) Amino acids and related compounds. Data presented
in Table 3 indicate the concentrations of amino acids and related compounds in freshly collected porcine follicular fluid
in comparison with serum. Significant differences from serum
are found with respect to phosphoethanolamine, threonine, aspargine + glutamine, glutamic acid, alanine, α-amino-η butyric acid,
half-cystine and lysine. These amino acids tend to show higher
concentrations in fluid from small follicles than in fluid from
large follicles (Chang et al., 1976). Furthermore, exposure of
follicular fluid to granulosa cells in intact follicles for
several hours at 4° - 8^OC results in significant changes in the
concentrations of these amino acids, particularly decreases in
the concentrations of half-cystine and aspargine + glutamine and
increases in taurine, glycine, aspartic acid and glutamic acid.
These changes in follicular fluid composition should be of interest to those wishing to culture granulosa cells in defined
media and should provide some clues for studying the detailed
metabolism of granulosa cells.

c) Hormones and related materials. The data presented
in Table 4 and illustrated in Figure 9 represent our results
concerning the concentrations of cAMP, progesterone, oestrogen
(oestradiol + oestrone) and testosterone in porcine follicular
fluid. They indicate a progressive increase in concentration
and content of all of these constituents with follicular maturation. One point of interest in these data is the higher

TABLE 4

CONCENTRATIONS OF cAMP OESTROGEN, PROGESTERONE AND TESTOSTERONE IN FOLLICULAR FLUID FROM ICE-STORED OVARIES

Compound	SFF	MFF (Mean $\pm$ S.E.).	LFF
cAMP			
p mole/ml	1.5 $\pm$ 0.1	4.4 $\pm$ 0.1	5.6 $\pm$ 0.2
f mole/follicle	22	183	875
Progesterone			
ng/ml	100 $\pm$ 2	115 $\pm$ 25	215 $\pm$ 55
ng/follicle	1.5	4.8	33.6
Oestrogen (Oestradiol + Oestrone)			
ng/ml	9.0 $\pm$ 1	75 $\pm$ 5	250 $\pm$ 10
ng/follicle	0.13	3.12	39.1
Testosterone			
ng/ml	14 $\pm$ 1	–	39 $\pm$ 1
ng/follicle	0.20	–	6.1

concentration of oestrogen than testosterone in fluid from large follicles. If the two-cell hypothesis of Dorrington et al. (1975) is correct, namely that androgen of thecal origin is converted to oestrogen by granulosa cells, then one might expect the concentration of testosterone to be higher than that of oestrogen. This two-cell theory is also challenged by the data of Younglai and Short (1970) and Channing and Coudet (1976).

McNatty and co-workers (1975) have reported on the concentrations of FSH, LH and prolactin in human ovarian follicles. Surprisingly, the concentrations of all three protein hormones are lower in follicle fluid than in serum. They also show remarkable small variations in concentrations among follicles of different sizes and at different times during the menstrual cycle. LH does, however, tend to be highest in follicles larger than 8 mm during the late follicular phase and prolactin is at its lowest concentration in these follicles.

3. Studies on isolated granulosa cells and granulosa cells in short-term culture:

a) Hormone binding to receptors. Binding of ^{125}I-HCG to porcine granulosa cells from small, medium and large ovarian follicles has been studied by Lee (1977). The specificity and affinity (K_d 2 X 10^{-10}M) of binding to the 3 classes of antral follicles is identical. The number of binding sites per cell vary, however, with the stage of development of the follicle. As illustrated in Figure 8, the number of LH-HCG binding sites per granulosa cell increases from an average of 350 for small follicles to 9 200 for large follicles. These data are very similar to those reported by several other investigators (see Channing and Tsafriri, 1977, for a review).

Autoradiographic localisation of radioiodinated HCG, FSH and prolactin have been reported by Midgley (1973). These studies indicate localisation of ^{125}I-HCG to interstitial cells, theca cells, granulosa cells of some large antral follicles and corpora lutea. FSH localises primarily to granulosa cells of larger preantral follicles and smaller antral follicles but not to theca

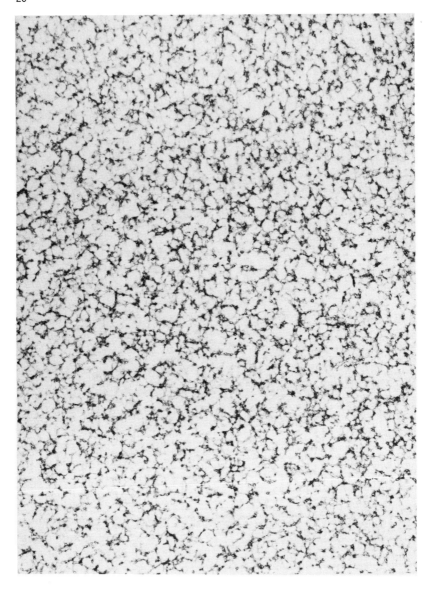

Fig. 8 High voltage TEM of rat ovarian follicular fluid. Ovaries were
        fixed in situ by perfusion with glutaraldehyde-paraformaldehyde
        in 0.1 cacodylate buffer, pH 7.2, embedded in glycol methacrylate,
        sectioned and stained for mucopolysaccharides using chromic acid-
        phosphotungstic acid. Note the complex reticular network of muco-
        polysaccharide. Taken from Kang et al. (1977).

cells or corpora lutea. Prolactin binds only to granulosa cells
of some large antral follicles, luteal cells and interstitial
cells.

Nimrod et al. (1976) reported quantitative data on hFSH
binding to rat and rabbit granulosa cells. The number of binding sites on granulosa cells from large preantral follicles (DES
treated hyposectomised rats) was approximately 1 200/cell,
decreased to approximately 600 per cell in medium sized antral
follicles on the day of proestrus and increased again to about
1 200/cell in large antral follicles on the day of proestrus.
The rabbit showed differences in binding affinity when follicles
of varying sizes were studied. The number of FSH sites, therefore, appear to be low and show surprisingly little variation in
contrast to the LH-HCG binding sites.

The regulation of ovarian receptors has been studied most
extensively in the rat by Richards (1977). Oestradiol receptors
are present in granulosa cells and their numbers are increased
by administration of oestradiol or FSH, the latter acting probably by stimulation of endogenous oestrogen. Oestradiol alone,
and FSH alone, increased the binding of FSH to granulosa cells,
but the combination of oestradiol and FSH is more effective.
FSH induces the appearance of LH receptors in granulosa cells.
LH induces the loss of its own receptors and the appearance of
prolactin receptors in luteinising granulosa cells of large
follicles. Prolactin induces the appearance of LH receptors in
luteal cells. The LH induced loss of its own receptor is related to the desensitisation of LH response adenylate cyclase,
also induced by LH (Ryan et al., 1977).

b) Activation of adenylate cyclase. Figure 10 illustrates
dose-response curves for the activation of adenylate cyclase
activity in granulosa cells from small porcine ovarian follicles.
In this system FSH stimulates adenylate cyclase activity at doses
between 10 and 1 000 ng/ml. The maximum activity achieved is
10-fold greater than basal activity. HCG, at a maximum effective
dose of 1 000 ng/ml produces only a 1.8-fold stimulation over

basal. The effect of HCG combined with FSH is only slightly
greater than the effect of FSH alone. The combined effect may
be additive, but it is certainly not synergistic. PMSG also
stimulates cyclase activity and the maximum achieved is comparable to that seen with FSH alone or FSH + HCG.

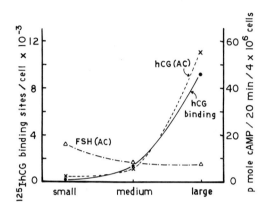

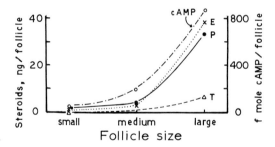

Fig. 9 Top: ^{125}I-HCG binding (●——●),HCG stimulatable adenylate cyclase
(X----X) and FSH stimulatable adenylate cyclase (Δ -.-.Δ) in
granulosa cells from small (1 - 2 mm), medium (3 - 5 mm) and large
(6 - 12 mm) porcine ovarian follicles.

Bottom: The content of cAMP (O-.-.O), oestrone + oestradiol (X....X),
progesterone (●——●) and testosterone (Δ---- Δ) in small, medium
and large porcine ovarian follicles.

The data is from Chang et al. (1976), Lee (1977) and unpublished
observations.

As illustrated in Figure 9, the effects of FSH and HCG

on adenylate cyclase activation are reversed when granulosa cells
from large follicles are studied. Here HCG produces a 10-fold
stimulation over basal activity while FSH produces only a 2-fold
stimulation. PMSG produces at least a 10-fold stimulation of
cyclase activity in granulosa cells from large follicles (not
illustrated).

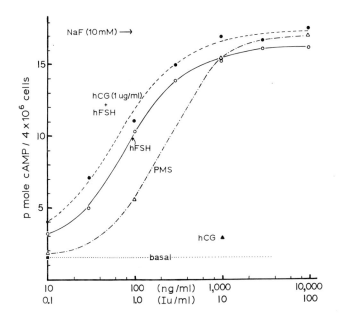

Fig. 10 Basal, hormone and sodium fluoride stimulatable adenylate cyclase
activity in granulosa cells obtained from small porcine ovarian
follicles. HCG (Cr 121) and HFSH (100 NIH-FSH-S1 U/mg) were highly
purified materials and doses are expressed in ng/ml. PMS was a
commercial preparation and dosages are expressed in iu/ml. The
cyclase assays were performed as previously described (Lee, 1977;
Ryan et al., 1977).

Although both FSH and HCG produce at least a 10-fold increase in cyclase activity relative to basal activity in their
most responsive cell types, the absolute quantities of cAMP produced at maximum stimulation are grossly different. As
illustrated in Figure 8, the maximum achieved with FSH in cells
from small follicles is 18 pm cAMP/20 min/4 x 10^6 cells, while
that achieved by HCG with cells from large follicles is 55 pm/

20 min/4 X 10^6 cells. This difference correlates with the cAMP
concentration and content of follicular fluid (Figure 9) and the
gross differences in numbers of FSH and HCG binding sites (see
above). The difference between fold-increase and absolute quantities of cAMP produced relate, at least in part, to the lower
basal cyclase activity in granulosa cells from small follicles
as compared to large follicles.

The relatively meagre adenylate cyclase response to FSH
raises the question as to whether the effects of this hormone
are mediated by cAMP. One of the effects of FSH is to induce
granulosa cell proliferation (see below). In general, cAMP
levels are lower in rapidly dividing cells than in differentiated
slowly dividing cells.

For further details concerning adenylate cyclase and hormone binding in ovarian tissues, see the reviews by Ryan et al.
(1977) and Channing and Tsafriri (1977).

c) Effects of FSH, LH and PMSG on DNA synthesis by
cultured porcine granulosa cells. Granulosa cells from
small (1 - 2 mm) and large (6 - 12 mm) follicles were cultured
in Lumbro multidish plates in medium 199 with 10% foetal calf
serum. After 48 hours monolayers were formed and the media was
changed to medium 199 with 5% foetal calf serum and hormones or
buffer were added. Incubations were then continued for 4, 6,
10 or 20 hours with the addition of 0.2 µCi of (^{3}H) thymidine
(250 mCi/M) during the last hour. The incubation was then ter-
minated with cold 10% trichloroacetic acid and the washed pre-
cipitable material was then analysed for DNA and (^{3}H).

Basal DNA synthesis by granulosa cells from small follicles (51 $\pm$ 10 cpm (^{3}H)/µg DNA) was approximately 5 times
greater than that by cells from large follicles (9 $\pm$ 4 cmp/µg
DNA). FSH and LH did not stimulate DNA synthesis by granulosa
cells from large follicles. As illustrated in Figure 11, FSH
and LH stimulated DNA synthesis by granulosa cells from small
follicles. This stimulation occurred only if the cells were

exposed to the gonadotrophins for at least 10 hours. LH caused
a maximum stimulation of 6575% over basal at a dose of 1 - 2 µg/ml.
On the other hand, the response to FSH was of a greater magnitude
and a maximum response was not achieved with a dose of 2 µg/ml.
Since the FSH preparation was approximately 1% pure and the LH
preparation approximately 50% pure, the data would suggest that
the sensitivity to FSH is much greater than the sensitivity to LH.

PMSG also stimulated DNA synthesis by granulosa cells from
small follicles. In contrast to LH and FSH, a stimulatory effect
was noted with only 4 hours exposure to the gonadotrophin. This
accelerated effect could not be mimicked by mixtures of FSH and
LH. The PMSG effect differed in another regard. Maximum stimulation (135% increase in DNA synthesis) was seen with 2 U/ml
while higher doses gave responses less than this.

CONCLUSION

During the past decade a great deal has been learned
about the patterns of gonadotrophin and sex steroid concentrations
in blood throughout the menstrual or oestrous cycle in many species.
LH receptors have been identified, characterised, correlated with
various physiologic states and some of the mechanisms of action
of the hormone have been defined. Although FSH, prolactin and
oestradiol binding to ovarian tissues has been shown, much remains to be done with respect to characterisation of their
receptors and understanding their mode of action. The effects
and mechanisms of action of androgens and progestins on ovarian
cell function have been grossly neglected.

A considerable amount of data is available concerning
steroid synthesis and metabolism by ovarian cells. There is a
lack, however, of information concerning the general metabolism
of granulosa, theca and luteal cells and how this metabolism is
affected by hormones and, conversely, how this metabolism affects
responses to hormones.

Most of the data available concerning hormonal regulation

of ovarian cell function has been acquired using luteal cells
and granulosa cells from antral follicles. There is a real need
to study granulosa cells from small preantral follicles and to
study further theca and interstitial cells. In particular, there
is a need to study the factor(s) that initiate differentiation
of the primordial follicle.

Perhaps the largest and most difficult problem related to
our understanding of follicular development and function is
atresia. It is the fate of most follicles, yet we have only a
few clues with respect to its regulation as a consequence of
varying concentrations of gonadotrophins, oestrogens and androgens. It is a difficult problem to study because the phenomenon
is recognised on a statistical basis or the application of histologic criteria after the process has already occurred. Indeed,
many of the histologic criteria are not absolute since they are
also discernible in the ovulating follicle. As an example of
the problem presented by atresia, many of the changes noted above
in the composition of follicular fluid or the surface features
of granulosa cells in antral follicles could be consequences of
atresia rather than follicular maturation.

Since this conference deals with superovulation, a few
thoughts concerned with this phenomenon will be given. The
general procedure to induce superovulation has been to administer
PMSG and a fraction of a cycle length later to inject HCG or
allow a spontaneous LH release to induce ovulation. In the mouse,
19 days or 4 - 5 cycle lengths are required for a follicle to
develop from the primordial stage to the ovulatory stage (Pedersen,
1970). Most of this time (about 14 days) is required to reach
the stage of antrum formation. It would thus appear that PMSG
is not acting by recruiting primordial or early preantral follicles. Rather, it suggests that PMSG is salvaging from atresia
a pool of follicles that are already in the antral or late preantral stages. Greenwald (1962, 1963, 1972, 1974) in studying
superovulation in hamsters referred to this pool as 'reserve'
follicles. The data illustrated in Figure 12 suggest that the
size of this reserve pool varies from species to species and is

proportional to the normal litter size of the species. The data
suggest that there are 5 to 7 reserve follicles for each follicle
that is normally ovulated.

Whether PMSG affects the size of the 'reserve' pool in
successive cycles has not been studied, to our knowledge. It
would be of interest to administer PMSG in cycle one and then
induce superovulation at 2, 3, 4 or 5 cycle-lengths later.

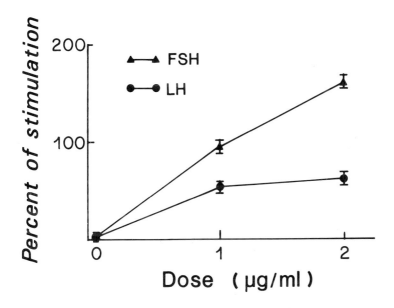

Fig. 11 FSH and LH stimulatable incorporation of ^{3}H thymidine into DNA of
cultured granulosa cells from small porcine follicles. Cells were
cultured for 20 hours in the presence of FSH (NIH-FSH-S10) or LH
(NIH-LH-S18) with the addition of labelled thymidine during the
final hour.

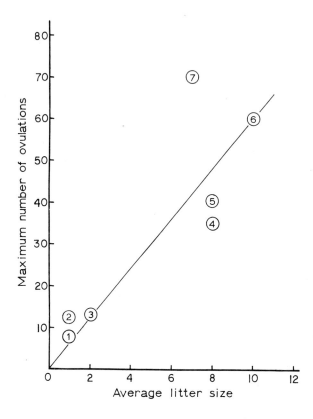

Fig. 12 A relationship between the maximum number of eggs obtained by superovulation and the average litter size of seven species.

(1) Human, Gemzell and Roos (1966); (2) Macaca mulatta, Simpson and
Van Wagenen (1962); (3) Ewe, Casada et al. (1944); (4) Rat, Evans
and Simpson (1940); (5) Rabbit, Parkes (1943); (6) Mouse, Zarrow
and Wilson (1961); (7) Hamster, Greenwald (1962).

REFERENCES

Albertini, D.F. and Anderson, E. 1974 The appearance and structure of intra-
cellular connections during the autogeny of the rabbit ovarian follicle
with particular reference to gap junctions. J. Cell Biol. 63:234-250

Armstrong, D.T., Miller, L.S. and Knudsen, K.A. 1969 Regulation of lipid
metabolism and progesterone in rat corpora lutea and ovarian interstitial elements by prolactin and luteinising hormone. Endocrinology
85:393-401

Ben-Or, S. 1963 Morphological and functional development of the ovary of
the mouse. I. Morphology and histochemistry of the developing ovary
in normal conditions and after FSH treatment. J. Embryol. Exptl.
Morphol. 11:1-11

Bradbury, J.T. 1961 Direct action of oestrogen on the ovary of the immature
rat. Endocrinology 68:115-120

Casida, L.E., Warwick, E.J. and Meyer, R.K. 1944 Survival of multiple pregnancies induced in the ewe following treatment with pituitary gonadotrophins. J. Animal Sci. 3:22-28

Chang, S.C.S., Jones, J.D., Ellefson, R.D. and Ryan, R.J. 1976 The porcine
ovarian follicle: I. Selected chemical analysis of follicular fluid
at different developmental stages. Biol. Reprod. 15:321-328

Chang, S.C.S. and Ryan, R.J. 1976 Time study of effects of pregnant mare's
serum gonadotrophin on surface characteristics of granulosa cells in
rat ovaries. Mayo Clinic Proc. 51:621-623

Chang, S.C.S. and Ryan, R.J. 1977 Binding of high density lipoprotein to
porcine granulosa cells. Submitted for publication.

Chang, S.C., Anderson, W., Lewis, J.C., Ryan, R.J. and Kang, Y.H. 1977 The
porcine ovarian follicle. II. Electron microscopic study of surface
features of granulosa cells at different stages of development. Biol.
Reprod. 16:349-357

Channing, C.P. and Coudert, S.P. 1976 Contribution of granulosa cells and
follicular fluid to ovarian oestrogen secretion in the Rhesus monkey
in vivo. Endocrinology 98:590-597

Channing, C.P. and Tsafriri, A. 1977 Mechanism of action of luteinising
hormone and follicle stimulating hormone in the ovary in vitro. Metabolism 26:413-468

Crisp, T.M. and Channing, C.P. 1972 Structural events correlated with progestin secretion during luteinisation of Rhesus monkey granulosa cells
in culture. Biol. Reprod. 7:55-72

Croes-Buth, S.F., Paesi, J.A. and de Jongh, S.E. 1959 Stimulation of ovarian
follicles in hypophysectomised rats by low doses of oestradiol benzoate.
Acta Endocrinol. 32:399-410

Demers, L.M., Behrman, H.R. and Greep, R.O. 1973 Effects of prostaglandins
and gonadotrophins on luteal prostaglandin steroid biosynthesis. Adv.
Biosciences 9:701-707

Dorrington. J.H., Moon, Y.S. and Armstrong, D.T. 1975 Oestradiol 17β biosynthesis in cultured granulosa cells from hypophysectomised immature
rats; stimulation by follicle-stimulating hormone. Endocrinol. 97:
1328-1331

Edwards, R.G. 1974 Follicular fluid. J. Reprod. Fertil. 37:189-219

Evans, H.M. and Simpson, M.E. 1940 Experimental superfecundity with pituitary
gonadotrophins. Endocrinol. 27:305-308

Gemzell, G. and Roos, P. 1966 Pregnancies following treatments with human
gonadotrophins with special reference to the problem of multiple births.
Am. J. Ob. & Gyn. 94:490-496

Goldstein, J.L. and Brown, M.S. 1974 Binding and degradation of low density
lipoproteins by cultured human fibroblasts. Comparison of cells from
a normal subject and from a patient with homozygous familial hyper-
cholesterolaemia. J. Biol. Chem. 249: 5153-5162

Greenwald, G.S. 1962 Analysis of superovulation in the adult hamster.
Endocrinology 71:378-389

Greenwald, G.S. 1963 Effect of an anti PMS serum on superovulation in the
hamster. Endocrinology 73:436-441

Greenwald, G.S. 1972 Experimental manipulation of follicular development
in the cycling hamster, in Velardo, J.T. (ed). Biology of Reproduction
Basic and Clinical Studies, pp 155-166

Greenwald, G.S. 1974 Role of follicle-stimulating hormone and luteinising
hormone in follicular development and ovulation. In Handbook of
Physiology, Sec. 7, Vol. 4, Part 2, pp 293-324, Amer. Physiol. Soc.
Washington

Greep, R.O., Van Dyke, H.B. and Chow, B.F. 1942 Gonadotrophins of the swine
pituitary. I. Various biological effects of purified thylakentrin
(FSH) and pure metakintrin (ICSH). Endocrinology 30:635-649

Harman, S.M., Louvet, J.P. and Ross, G.T. 1975 Interaction of oestrogen
and gonadotrophins on follicular atresia. Endocrinol. 96:1145-1152

Hertig, A.T. and Adams, E.C. 1967 Studies on the human oocyte and its
follicle. I. Ultra structural and histochemical observations on the
primordial follicle stages. J. Cell Biol. 34:647-675

Kang, Y.H., Anderson, W.A., Chang, S.C. and Ryan, R.J. 1977 Studies on the structure of the extracellular matrices of the mammalian follicles as revealed by high voltage electron microscopy and cytochemistry. In press.

Lee, C.Y. 1977 The porcine ovarian follicle. III. Development of chorionic gonadotrophin receptors associated with increase in adenyl cyclase activity during follicle maturation. Endocrinology 99:42-48

Lunenfeld, B. and Eshkol, A. 1968 The role of gonad stimulating hormone on the development of the infantile ovary. In Gonadotrophins, 1968, ed. E. Rosemberg, Geron-X, Los Altos, pp 197-204

McNatty, K.P., Hunter, W.M., McNeilly, A.S. and Sawers, R.S. 1975 Changes in the concentration of pituitary and steroid hormones in the follicular fluid of human graafian follicles throughout the menstrual cycle. J. Endocr. 64:555-571

Midgley, A.R., Jr. 1973 Autoradiographic analysis of gonadotrophin binding to rat ovarian tissue sections. In O'Malley and Means (eds), Receptors for Reproductive Hormones, Plenum Press, New York, pp 365-378

Motta, P 1965 Sulla formazione del liquor folliculi nee'ovajo dells coniglia. Biol. Lat. 18:252-271

Motta, P., 1965 Sur l'ultrastructure des "corpus d call et Exner" dans l'ovaire du lapin. Z. Zellforsch. 68:308-319

Motta, P., Tahev, S. and Nesci, E. 1971 Etude ultrastructurale et histochimique des rapports entre les cellules folliculaires et l'ovocyte pendant le developpement du follicle ovarien chez les mammifers. Acta anat. 80:537-562

Odeblad, E. 1954 Studies on the physiology of the follicular fluid. Acta Endocr. Copenh. 15:313-316

Nimrod, A., Erickson, G.F. and Ryan, K.J. 1976 A specific FSH receptor in rat granulosa cells: Properties of binding in vitro. Endocrinology 98:56-64

Paesi, E.J. 1952 The effect of small doses of oestrogen on the ovary of the immature rat. Acta Endocrinol. 11:251-268

Parkes, A.S. 1943 Induction of superovulation and super fecundation in rabbits. J. Endocrinol. 3:268-279

Pedersen, T. 1969 Follicle growth in the immature mouse ovary. Acta Endocrinol. 62:117-132

Pedersen, T. 1970 Follicle kinetics in the ovary of the cyclic mouse. Acta Endocr. 64:304-323

32

Peters, H. 1969 The development of the mouse from birth to maturity. Acta Endocrinol. 62:98-116

Presl, J., Figarova, V., Pospisil, J., Wagner, V. and Horsky, J. 1971 Evidence for human chorionic gonadotrophin binding sites in some organs of the immature rat. Effect of non labelled chorionic gonadotrophin and follicle stimulating hormone on the distribution of ^{125}I labelled human chorionic gonadotrophin. Folia Morphol. 19:171-176

Presl, J., Pospisil, J., Figarova, V. and Wagner, V. 1972 Developmental changes in the uptake of radioactivity of the ovaries, pituitary and uterus after ^{125}I labelled human chorionic gonadotrophin administration in rats. J. Endocr. 52:585-586

Ramwell, P.W., Leovey, E.M.K. and Sinetos, A.L. 1977 Regulation of the arachidonic acid cascade. Biol. Reprod. 16:70-87

Richards, J.S. Hormonal control of follicular growth and maturation-mammals. In Jones, R.E. (ed), Evolution of the Vertebrate Ovary. Plenum Press, New York, 1977.

Ross, G.T. 1974 Gonadotrophins and preantral follicular maturation in women. Fertil. & Steril. 25:522-543

Ryan, R.J., Birnbaumer, L., Lee, C.Y. and Hunzicker-Dunn, M. 1977 Gonadotrophin interactions with the ovary and testis as assessed by receptor binding and adenylate cyclase activity. In R.O. Greep (ed), International Rev. Physiol. Reproductive Physiology II, Vol. 13, Univ. Park Press, Baltimore, pp 85-152

Shalgi, R., Kraicer, P., Rimon, A., Pinto, M. and Soferman, N. 1973 Proteins of human follicular fluid: the blood-follicular barrier. Fertil. & Steril. 24:429-434

Simpson, M.E. and Van Wagenen, G. 1962 Induction of ovulation with human urinary gonadotrophins in the monkey. Fertil. & Steril. 13:140-152

Smith, B.D. 1961 The effect of diethylstilboestrol on the immature rat ovary. Endocrinology 69:238-245

Smith, P.E. 1930 Hypophysectomy and a replacement therapy in the rat. Amer. J. Anat. 45:205-274

Yao, J.K., Chang, S.C.S., Ryan, R.J. and Dyck, P.J. 1977 A model for studying LCAT reaction. I. In vitro cholesterol esterification in pig ovarian follicular fluid. Lipids, in press (1978).

Younglai, E.V. and Short, R.V. 1970 Pathways of steroid biosynthesis in intact graafian follicles of mares in oestrus. J. Endocrinol. 47:321-331

Zachariae, F. 1957 Studies on the mechanism of ovulation; autoradiographic
    observations on the uptake of radioactive sulfate (35S) into the
    ovarian follicular mucopolysaccharides. Acta Endocr. Copenh. 26:215-223

Zamboni, L. 1974 Fine morphology of the follicle wall and follicle cell-
    oocyte association. Biol. Reprod. 10:125-149

Zarrow, M. and Wilson, 1961 The influence of age on superovulation in
    the immature rat and mouse. Endocrinology 69:851-855

OOGENESIS AND FOLLICULAR GROWTH IN THE COW: IMPLICATIONS FOR SUPEROVULATION

T.G. Baker and R.H.F. Hunter*

Department of Obstetrics & Gynaecology and
*School of Agriculture University of Edinburgh

INTRODUCTION

In view of the importance of cattle, at least in the so-called developed world, in providing meat and dairy products, it is perhaps surprising that detailed and systematic studies of the basic reproductive physiology of this important animal have not been carried out. For example, the few published reports on cellular aspects of ovarian physiology in this species were derived largely from studies in which the primary objective was not related to reproduction. Thus our knowledge of oogenesis stems largely from studies designed to assess the radiosensitivity of female germ cells (Erickson, 1965; 1966[2]), while information on the embryology of the gonad and its associated ducts was obtained via studies of the bovine freemartin (Jost, Vigier and Prepin, 1972; Jost, Vigier, Prepin and Perchellet, 1973[1,2]). Bovine embryos and foetuses have also been used to study sex differentiation (Ohno and Gropp, 1965; Jost et al. 1972), to examine the role of granulosa cells in preventing the degeneration of germ cells (Ohno and Smith, 1964), and in studies of the configuration of chromosomes during the diplotene (arrested) period of oocyte maturation (Baker and Franchi, 1967[3]: see below). In most of these studies, it would seem that bovine tissues were selected because of freely available material at the slaughter-house, rather than to yield information on this important domestic species in its own right.

The purpose of this communication is twofold: firstly to provide a review of the available information on oogenesis and ovulation in the cow; and secondly to provide the impetus for further basic and applied research. Such studies would be important not only in terms of animal production. It will become apparent later that the process of oogenesis in cow and human ovaries is remarkably similar and thus the cow might provide a useful model in which to test the effects of drugs, chemical mutagens and ionising radiations on ovarian tissues (Baker and Neal, 1977; Erickson et al. 1976).

EARLY EMBRYOLOGY OF THE OVARY

The precise origin of the germ cells in cattle has seemingly not been determined, but there can be little doubt that such cells arise early in embryonic differentiation, as they do in other vertebrates (Baker, 1972; Zucherman and Baker, 1977). The only complete account of the migration of germ cells in the mammalian embryo is that of Witschi (1948) although Ohno and Gropp (1965) have identified some of the stages in sections of calf embryos. These authors used the alkaline phosphatase reaction to identify primordial germ cells, some 20 of which were found just beneath the aorta in embryos with a crown-rump length of 10-12 mm (ca 30th day of gestation). Other primordial germ cells were found within the gonad anlagen, which at this stage of embryology consist of a slight thickening of the surface of the mesonephroi (primitive kidneys) adjacent to the coelomic 'angle'. By the time the embryo has attained a crown-rump length of about 15 mm the primordial germ cells have largely entered the gonad, their route of migration being either via the blood stream, by amoeboid movements through the tissues, or a combination of these routes (Ohno and Gropp, 1965). Germ cell migration is completed by Day 35 post conception (pc), just a few days prior to the onset of gonadal sexual differentiation (Erickson, 1965).

Some further valuable information on gonadal development has been obtained from the classical studies of Jost and his collaborators on normal and freemartin embryos. For example, Jost et al. (1972) have shown that the sexual differentiation of the bovine testis occurs in the 25 mm embryo, which corresponds to an age of about 39 days pc. At this stage of embryonic development, the female gonad is recognised (as in the human

36

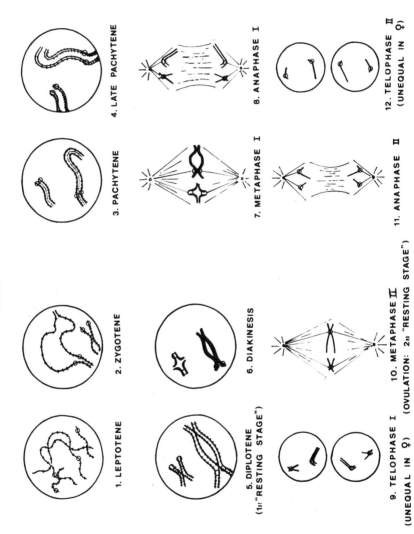

MEIOSIS

1. LEPTOTENE

2. ZYGOTENE

3. PACHYTENE

4. LATE PACHYTENE

5. DIPLOTENE
(1ₜₜ"RESTING STAGE")

6. DIAKINESIS

7. METAPHASE I

8. ANAPHASE I

9. TELOPHASE I
(UNEQUAL IN ♀)

10. METAPHASE II
(OVULATION: 2ₙ "RESTING STAGE")

11. ANAPHASE II

12. TELOPHASE II
(UNEQUAL IN ♀)

Fig. 1

embryo) by the fact that it does not resemble a testis (Gillman, 1948). In the female the gonad remains in the indifferent stage until about Day 48 to 49 pc when ovarian differentiation occurs (Erickson, 1966$_2$; Jost et al. 1972). This process consists of a slow prolonged thickening of the superficial ovarian layers. With the completion of this process, the ovary develops a leaf-shaped appearance and its germ cells are now correctly termed oogonia.

OOGENESIS IN FOETAL CALVES

The only complete study of ovarian development in the cow is that of Erickson (1966$_{1,2}$), although some qualitative observations have been added by other authors (Mauleon, 1969; Mauléon and Mariana, 1977. According to Erickson (1966$_{1,2}$), there is little change in the internal organisation of the ovary between the time of sexual differentiation on Days 48 - 49 and the 70th day pc, apart from a high rate of mitotic divisions affecting primordial germ cells and oogonia. His quantitative studies show an enormous increase in the population of germ cells from about 16 000 on Day 50, to 332 000 on Day 60, compared with 1 073 000 on Day 70 pc. (Table 1). This pattern of increasing numbers of germ cells is strikingly similar to that occurring in foetal human ovaries at a comparable stage of gestation (Baker, 1963, 1972).

The onset of meiotic prophase (and thus the first detectable primary oocytes) occurs between the 75th and 80th days pc (Erickson, 1966$_2$), which is slightly later in gestation than in human foetuses (Baker, 1963). From this time until Day 150 pc oogonia continue to proliferate by mitosis, and they co-exist within the ovarian cortex with primary oocytes at the leptotene, zygotene, pachytene and diplotene stages of first meiotic prophase (Figure 1). Studies on other mammalian species (although not on the cow) have shown that a meiotic inductor (seemingly produced by the rete ovarii: Byskov, 1975) is required to induce the onset of meiosis. The chemical nature of this inductor remains unknown, but its effect is short-lived: in its absence the germ cells

TABLE 1

DEVELOPMENT OF PRENATAL BOVINE OVARY

Stage of gestation (Days)	No. of mitotic figures	Total germ cells x 10^3	Oogonia %	% Oocytes in		
				Leptotene	Zygotene	Pachytene
50	13	15	100	–	–	–
60	304	322	100	–	–	–
70	142	1073	100	–	–	–
80	214	2183	38	21	41	0
90	182	2043	36	17	45	2
110	223	2739	36	21	39	4
130	69	2448	30	9	36	25
150	3	1528	9	4	24	63
170	0	107	3	2	6	89
190	0	81	3	0	0	97
210	0	81	12	0	0	88
230	0	116	7	0	0	93
250	0	51	2	0	0	98
270	0	68	2	0	0	98

Adapted from Erickson, BH (1966)

(oogonia) rapidly degenerate, but if the 'trigger' is present
for a mere day or so meiosis proceeds without interruption to
the diplotene stage (O and Baker, 1976, 1977).

By mid-gestation, the bovine ovary shows marked stratific-
ation, with a thin outer cortical zone containing mainly oogonia,
while the deeper layers contain progressively more advanced
stages of meiotic prophase. Oocytes at diplotene are found only
at the cortico-medullary boundary (Ohno and Smith, 1964; Baker
and Franchi, 1967[3]). The number of germ cells increases to its
peak value of 2 739 000 at Day 110 pc, but thereafter declines
to a value of only 68 000 by the time of birth (Erickson, 1966[1,2];
Table 1). This reduction in the population of germ cells is
brought about firstly by the cessation of mitotic activity in
oogonia (at about Day 150), and secondly by 'waves' of degeneration affecting germ cells at all stages in their development
(Baker, 1972). Thus the proportion of germ cells which have
reached the pachytene and diplotene stages of first meiotic
prophase increases sharply between Day 170 pc and the time of
birth (ca Day 240), and the pre-diplotene stages disappear around
this time (Hofliger, 1948; Ohno and Smith, 1964; Erickson,
1966[1,2]; Baker and Franchi, 1967[3]).

The follicle (granulosa) cells, which are initially far less
numerous than the germ cells (Erickson 1966[2]), proliferate late
during foetal life and surround the oocytes to form primordial
follicles. Such follicles first appear near the cortico-medullary
junction in foetuses with a crown-rump length of 300 mm (ca 140
days pc), and it has been suggested that the oocytes which fail
to become enclosed in primoridal follicles are pre-destined to
degenerate, although definitive proof of this situation is not
yet available (Ohno and Smith, 1964).

This section on the development of the ovary would not be
complete without at least a note on the bovine freemartin. It
has long been known that the female foetus, when co-twin to a
bull, becomes masculinised (e.g. Lillie, 1916), Jost et al.
(1972) have shown that ovarian development in the female is

initially retarded, but that structures resembling seminiferous
tubules develop in some freemartins around 90 days pc. The oogonia
fail to complete their mitotic proliferation and meiosis does not
occur (Jost et al. 1973[1,2]). The possible causes and consequences
of the masculinisation of the freemartin are beyond the scope of
the present review but are discussed fully elsewhere (Jost et al.
1972, 1973[1,2]). In the context of the theme of this Symposium,
it is important to remember that the use of techniques for
superovulating cattle will greatly increase the incidence of
freemartins, as well as normal offspring.

'ARRESTED' STAGE OF OOCYTE DEVELOPMENT

One of the most characteristic features of gametogenesis in
female mammals is that meiosis is interruped during first
prophase, and the process does not resume until shortly before
ovulation (see below). There has been some controversy as to
the precise stage of meiotic prophase at which arrested development occurs. On the basis of histological observations,
Henricson and Rajakoski (1959) and Rajakoski (1965) believed
that arrest occurs at the pachytene stage, diplotene appearing
relatively late during the process of follicular growth. However,
Baker and Franchi (1967[3]) using both light and electron microscopy were able to prove conclusively that the oocytes in
primordial follicles have reached the diplotene stage of meiotic
prophase at which meiotic arrest occurred. Baker and Franchi
(1967[1,2,3]) have also shown that oocytes at the arrested
(diplotene) stage of meiotic prophase contain structures similar
to the lampbrush chromosomes found in lower vertebrates. The
lampbrush chromosomes actively synthesise ribonucleoprotein
(Baker, Beaumont and Franchi, 1969), which may be utilised as
the organiser of early embryonic development, as well as during
follicular growth (Baker, 1971).

REPRODUCTIVE THRESHOLD

Little is known of the changes in ovarian structure
occurring between birth and puberty. It must therefore be

assumed, at least as a working hypothesis, that the events are similar to those occurring in other mammalian species where follicular growth up to and including the antral stage is commonly encountered prior to puberty (Baker, 1972; Mauléon, 1969). In rats and mice, follicular growth has to be 'triggered' at about the time of birth by the influence of both FSH and LH (Eshkol, Lunenfeld.and Peters, 1971; Baker and Neal, 1973). This process lasts only 1 to 3 days, after which follicular growth is independent of gonadotrophins until the oocyte is enclosed within a multilayered follicle (see below).

The age at which heifers reach puberty is highly variable, ranging from 8 to 18 months in Holstein cattle (Hunter, 1976). This would imply that the ovary is potentially ready for functional activity in quite young animals but awaits the maturation of the hypothalamic-hypophyseal-ovarian axis (Ramírez, 1973). This would be similar to the situation in young girls (especially those with virilising adrenal tumours) where menstruation, and possibly ovulation, can occur many years before the normal onset of puberty.

Useful summaries of the available data on puberty in cattle have been provided by Sorensen, Hansel, Hough, Armstrong, McEntee and Batton (1959) and Joubert (1963), although these relate mainly to the topic of nutrition rather than to ovarian changes. However, Hofliger (1948) and Erickson (1966[1]) have provided some information, particularly on changes in the population of oocytes with increasing age.

In terms of the present discussion on ovarian potential, it is probably more important to consider the number of available antral (Graafian) follicles. This remains fairly constant from 2 months of age until 8 to 10 years, after which it slowly declines (Erickson, 1966[1]). Tertiary follicles are found in the ovary before birth (Mauléon and Mariana, 1977; Robinson, 1977), but there is some doubt as to when antral follicles first appear. In terms of purely macroscopic studies, neither Desjardins and Hafs (1969) nor Black and Hansel (1972) could find such

follicles at the time of birth, although antral follicles were
detected as early as one month of age. On the other hand, small
Graafian follicles are readily seen in sections of pre-natal human
ovaries but would be difficult, if not impossible, to detect
macroscopically (Baker, 1972; van Wagenen and Simpson, 1973).
In either event, the presence of Graafian follicles prior to
puberty should provide an opportunity for utilising the occytes
(after appropriate treatment of calves with gonadotrophins) to
increase the reproductive potential of the animal (Hunter, 1976).
This is beyond the scope of this review but will be considered
further elsewhere in this Symposium (Ryan, 1977; Seidel et al.
1977).

CONTROL OF FOLLICULAR GROWTH IN ADULT ANIMALS

This is another area of reproductive biology which still
requires systematic study in cows and hence a number of
assumptions based on studies in other mammals will have to be
made. Figure 2 shows diagramatically the principal events of
follicular growth as they occur in the mammalian ovary. Studies
of hypophysectomised rats, hamsters, sheep, etc together with
the results from treatment with suitable antisera to FSH and LH,
show that the control of follicular growth up to stage 4 in
Figure 2 is independent of gonadotrophic hormones, although the
precise internal regulatory system within the ovary remains
obscure. The number of follicles in each particular size-group
appears to be constant, implying that follicles only develop to
fill a 'gap' left by one that has moved forward in its growth
phase (Pedersen, 1972; Mauléon and Mariana, 1977). There can be
little doubt, however, that FSH is required for the final
maturation of the pre-antral (multilayered) follicle and that
FSH, and possibly also LH, are required for the development of
the antrum. Injections of FSH or PMSG(which has both FSH-like
and LH-like activity) into cows causes a rapid change in
'growing' follicles which subsequently develop an antrum: such
hormones also bring about the expansion of the antrum in Graafian
follicles (Mauléon and Mariana, 1977). However, usually only
one of the Graafian follicles completes its maturation to the

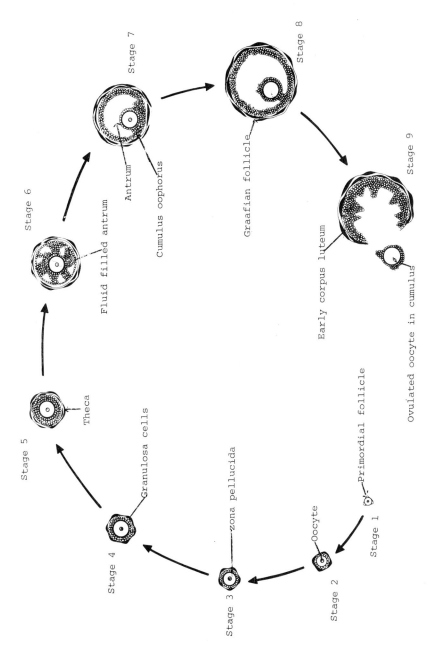

Fig. 2 Diagrammatic representation of follicular growth as it is believed to occur in the human ovary.

point of ovulation in the spontaneous oestrous cycle, the
remainder undergoing degeneration (atresia), probably in a
fashion similar to that described for the rhesus monkey (see
Koering, 1969). There can be no doubt, however, that such
degenerative changes are not pre-destined since multiple births
(especially following treatment with gonadotrophic hormones) are
well documented for the cow and for other monotocous mammals
(Mauléon and Mariana, 1977). Indeed, Hay and Moor (1977) have
shown that atretic sheep follicles maintained in organ culture
revert to seemingly normal follicles, and their oocytes, after
transfer to suitably prepared recipients, can produce viable
lambs. It must be recalled, however, that 'superovulating' doses
of hormones (usually PMSG followed by HCG) have given very
variable results in cows, pigs and sheep (Hunter, 1976). It is
to be hoped that these results would be greatly improved (as
they have been for the human female) by the use of gonadotrophins
which may act in a more physiological manner - such as bovine FSH
and bovine LH in the cow.

PRE-OVULATORY MATURATION OF OOCYTES

There can be little doubt that in vivo the so-called 'LH-
surge' (in reality an increase above tonic levels of both FSH
and LH in plasma) is responsible for the resumption of meiosis
beyond the diplotene stage. The timing of the post-diplotene
oocyte stages and of the interval between LH-surge and time of
ovulation is predictable for each species (Baker and Hunter,
1977). The oocyte nucleus (germinal vesicle) breaks down in a
matter of hours after the 'trigger' provided by LH and the stages
of metaphase I to metaphase II rapidly ensue. Thus an important
concept for animal production studies is that the number of
oocytes is fixed and finite once the last oogonium has been
eliminated from the foetal ovary, and the population of oocytes
will gradually fall to zero if the animal lives long enough.
However, as Robinson (1977) has pointed out, it is unlikely that
cows would be permitted to survive long enough for the population
of oocytes to fall to such a low level.

The actual mechanism of action of LH in inducing oocyte maturation and ovulation in Graafian follicles remains obscure, but it is improbable that gonadotrophins have a direct effect on oocytes. Oestrogens may play a permissive role, (Hunter, Cook and Baker, 1976), be the actual intermediates in hormone action, or may augment the effects of FSH and LH, but more defined experiments are required before this important problem can be solved (review by Baker and Hunter, 1977). Nevertheless, cow oocytes removed from Graafian follicles can be matured in a simple culture system (Sreenan, 1970), and when transplanted to the fallopian tubes of inseminated recipients, normal fertilis- ation can ensue (Hunter, Lawson and Rowson, 1972). A recent study has further shown that oocytes matured in culture can be fertilised by spermatozoa capacitated in vitro or in the uterus of an oestrous rabbit (Iritani and Niwa, 1977). In conclusion, therefore, the future holds the twin prospects of exploitation of the oocyte population in selected animals either by more controlled procedures of superovulation or by in vitro maturation of follicular oocytes; both approaches will be used in con- junction with the technique of egg or embryo transfer.

ACKNOWLEDGEMENTS

The authors are grateful to the Wellcome Trust and the Ford Foundation for financial support of their own studies mentioned in this review.

46

REFERENCES

Baker, T.G. 1963. A quantitative and cytological study of germ cells in
human ovaries. Proc. R. Soc. B, 158, 417-433

Baker, T.G. 1972. Oogenesis and ovarian development. In: Reproductive
Biology, ed. H. Balin and S.R. Glasser, pp. 398-437. Excerpta Medica
Amsterdam.

Baker, T.G. 1971. Electron microscopy of the primary and secondary oocyte.
In: Advances in the Biosciences, ed. G. Raspé, vol 6, pp. 7-23
Pergamon Press, Oxford.

Baker, T.G., Beaumont, H.M. and Franchi, L.L. 1969. The uptake of tritiated
uridine and phenylalanine by the ovaries of rats and monkeys.
J. Cell Sci. 4, 655-675

Baker, T.G. and Franchi, L.L. 1967$_1$. The fine structure of oogonia and
oocytes in human ovaries. J. Cell Sci. 2, 213-224

Baker, T.G. and Franchi, L.L. 1967$_2$. The structure of the chromosomes in
human primordial oocytes. Chromosoma, 22, 358-377

Baker, T.G. and Franchi, L.L. 1967$_3$. The fine structure of chromosomes in
bovine primordial oocytes. J. Reprod. Fert. 14, 511-513

Baker T.G. and Hunter, R.H.F. 1977. Interrelationships between the oocyte
and somatic cells within the Graafian follicle of mammals. Ann. Biol.
anim. Bioch. Biophys. in press.

Baker, T.G. and Neal, P. 1973. Initiation and control of meiosis and
follicular growth in ovaries of the mouse. Ann. Biol. anim. Bioch.
Biophys. 13, (Hors serie), 137-144.

Baker, T.G. and Neal, P. 1977. Action of ionizing radiations on the
mammalian ovary. In: The Ovary, ed. S. Zucherman and B.J. Weir, 2nd
edn, vol 3, p.1. Academic Press, New York.

Black, B.L. and Hansel, W. 1972. Endocrine factors affecting reproduction
in the bovine female. Univ. Mass. Agri. Expt Station Res. Bull. 596

Byskov, A.G.S. 1975. The role of the rete ovarii in meiosis and follicle
formation in different mammalian species. J. Reprod. Fert. 45, 201-209

Desjardins, C. and Hafs, H.D. 1969. Maturation of bovine female genitalia
from birth through puberty. J. Anim. Sci. 28, 502-507

Erickson, B.H. 1965. Symposium on atomic energy in animal science:
Radiation effects on gonadal development in farm animals. J. Anim Sci.
24, 568-583

Erickson, B.H. 1966$_1$. Development and senescence of the post-natal bovine
ovary. J. Anim. Sci. 25. 800-805

Erickson, B.H. 1966$_2$. Development and radio-response of the pre-natal
    bovine ovary. J. Reprod. Fert. 11, 97-105

Erickson, B.H., Reynclds, R.A. and Murphree, R.L. 1976. Late effects of ^{60}Co
    radiation in the bovine oocyte as reflected by oocyte survival,
    follicular development and reproductive performance. Radiat. Res. 68,
    132-137

Eshkol, A., Lunenfeld, B. and Peters, H. 1971. Ovarian development in infant
    mice. In: Gonadotrophins and Ovarian Development, ed. W.R. Butt, A.C.
    Crooke, and M. Ryle, pp. 249-258. Livingstone, Edinburgh.

Gillman, J. 1948. The development of the gonads in man, with a consideration
    of the role of foetal endocrines and the histogenesis of ovarian tumours.
    Contr. Embryol. Carneg, Instn, 32, 81-131

Hay, M.F. and Moor, R.M. 1977. Differential response of components of Graafian
    follicles to atresia. Ann, Biol. Anim. Bioch. Biophys. In Press.

Henricson, B and Rajakoski, E. 1959. Studies of oocytogenesis in cattle.
    Cornell Vet. 49, 494-503

Hofliger, H. 1948. Das Ovar des Rindes in den verschiedenen Lebensperioden
    unter besonderer Berücksichtigung seiner functionellen Feinstructur.
    Acta Anat. Suppl. 5, 1-196.

Hunter, R.H.F. 1976. Reproductive physiology in the bovine female: a review.
    pp. 79-104. In: Beef Cattle Production in Developing Countries, ed.
    A.J. Smith. Proc. 1974 Edinburgh Conf. University of Edinburgh.

Hunter, R.H.F., Cook, B and Baker, T.G. 1976. Dissociation of response to
    injected gonadotropin between the Graafian follicle and oocyte in pigs.
    Nature 260, 156-158

Hunter, R.H.F., Lawson, R.A.S. and Rowson, L.E.A. 1972. Maturation, trans-
    plantation and fertilisation of ovarian oocytes in cattle. J. Reprod.
    Fert. 30, 325-328

Iritani, A and Niwa, K. 1977. Capacitation of bull spermatozoa and
    fertilisation in vitro of cattle follicular oocytes matured in culture.
    J. Reprod. Fert. 50, 119-121

Jost, A., Vigier, B and Prepin, J. 1972. Freemartins in cattle: the first
    steps of sexual organogenesis. J. Reprod. Fert. 29, 349-379

Jost, A., Vigier, B., Prepin, J. and Perchellet, J.P. 1973$_1$. Le développement
    de la gonade des freemartins. Ann, Biol. Anim. Bioch. Biophys. 13,
    (Hors serie) 103-114

Jost, A., Vigier, B., Prepin, J. and Perchellet, J.P. 1973$_2$. Studies on sex
    differentiation in mammals. Recent Progr. Horm. Res. 29, 1-41

Joubert, D.M. 1963. Puberty in female farm animals. Anim. Breeding Abstr. 31, 295-306

Koering, M.J. 1969. Cyclic changes in ovarian morphology during the menstrual cycle in *Macaca mulatta*. Am. J. Anat. 126. 73-101

Lillie, F.R. 1916. Theory of the freemartin. Science, N.Y. 43, 611-612

Mauléon, P. 1969. Oogenesis and folliculogenesis. In: Reproduction in Domestic Animals, ed. H.H. Cole and P.T. Cupps, 2nd edn, pp. 187- Academic Press, New York.

Mauléon, P. and Mariana, J.C. 1977. Oogenesis and folliculogenesis. In: Reproduction in Domestic Animals, ed. H.H. Cole and P.T. Cupps, 3rd edn, pp. 175-202

O, Wai Sum and Baker, T.G. 1976. Initiation and control of meiosis in hamster gonads in vitro. J. Reprod. Fert. 48, 399-401

O, Wai Sum and Baker, T.G. 1977. Germinal and somatic cell interrelationships in gonadal sex differentiation. Ann. Biol. Anim. Biochem. Biophys. In Press.

Ohno, S. and Gropp, A. 1965. Embryological basis for germ cell chimerism in mammals. Cytogenetics, 4, 251-261

Ohno, S. and Smith, J.B. 1964. Role of foetal follicular cells in meiosis of mammalian oocytes. Cytogenetics, 3, 324-333

Pedersen, T. 1972. Follicle growth in the mouse ovary. In: Oogenesis, ed. J.D. Biggers and A.W. Schuetz, pp. 361-376

Rajakoski, E. 1965. Some views on oogenesis in cattle. Nord. Vet Med. 17, 285-

Ramírez, V.D. 1973. Endocrinology of puberty. Handbook of Physiology, Section on Endocrinology, vol. II part I, 1-28. American Physiological Society, Washington D.C.

Robinson, T.J. 1977. Reproduction in cattle. In: Reproduction in Domestic Animals, ed. H.H. Cole and P.T. Cupps, 3rd edn, pp. 433-454.

Ryan, R.J. 1977. Hormonal requirements for ovulation and implications for superovulation. (This Symposium)

Seidel, G.E. Jr., Elsden, R.P., Nelson, L.D. and Bowen, R.A. 1977. Super-ovulation in cattle with PMSG and FSH. (This Symposium)

Sreenan, J. 1970. In vitro maturation and attempted fertilisation of cattle follicular oocytes. J. Agric. Sci. 75, 393-396

van Wagenen, G. and Simpson, M.E. 1973. Post-natal Development of the Ovary in *Homo sapiens* and *Macaca mulatta* and Induction of Ovulation in the Macaque. Yale University Press, New Haven.

Witschi, E. 1948. Migration of the germ cells of human embryos from the
    yolk sac to the primitive gonadal folds. Contr. Embryol. Carneg.
    Instn, <u>32</u>, 67-80

Zuckerman, S. and Baker, T.G. 1977. The development of the ovary and the
    process of oogenesis. In: The Ovary, ed. S. Zuckerman and B.J. Weir,
    2nd edn, Vol <u>1</u>, pp. 1-   Academic Press, New York.

THE BIOLOGY OF PREGNANT MARE SERUM GONADOTROPHIN (PMSG)

W.R. Allen and Francesca Stewart

ARC, Institute of Animal Physiology,
307, Huntingdon Road, Cambridge, UK

INTRODUCTION

Pregnant mare serum gonadotrophin (PMSG), a glycoprotein ho
mone present in high concentration in the blood of pregnant equi
between Days 40 and 130 of gestation (Cole and Hart, 1930), is
unique amongst the gonadotrophic hormones in possessing both
follicle stimulating hormone (FSH) and luteinising hormone (LH)
biological activities within the one molecule (Gospodarowicz,197;
Papkoff, 1974). Despite the extensive use of PMSG in recent yeaı
in cattle, sheep and pigs for increasing ovulation rates for the
purposes of embryo recovery and transfer, relatively little is
known about the precise chemical and biological properties of the
molecule. Even less is known and understood about the biologica]
function of PMSG in the animal which produces it, and this paper
addresses the questions of how and why the pregnant mare secretes
such large amounts of a potent gonadotrophic hormone during the
first half of pregnancy.

THE SOURCE OF PMSG IN THE MARE

The classical studies of Cole and Goss (1943) first demonstrated that PMSG is secreted by the endometrial cups. These
structures are found only in equids and comprise a series of smal
cup-shaped, ulcer-like, endometrial outgrowths in the pregnant
horn of the uterus (Schauder, 1912). They first become visible
between Days 38 and 40 of gestation and are usually arranged in
a circle around the circumference of the uterine lumen so as to
surround the developing conceptus (Plate 1, Figure 1). The cups
grow rapidly to reach maximum dimensions of 2 to 10 cm in length
and 1 to 3 cm in width by Day 70 of gestation and thereafter under
a steady regression and necrosis while at the same time liberatinç
increasing amounts of a sticky, honey-coloured exocrine secretion

which adheres to the surface of the overlying foetal membranes
(Allen, 1975$_a$). This material is extremely rich in PMSG and may
contain up to 1 000 000 iu PMSG/g fresh tissue (Clegg et al.,
1954; Rowlands, 1963). The coagulum of necrotic cup tissue and
exocrine secretion is eventually sloughed off the surface of the
endometrium between 130 and 180 days of gestation where it may
give rise to the pedunculated allantochorionic pouches described
by Clegg et al. (1954).

Histologically the cups are composed of a discrete and
densely packed mass of large, epithelioid, decidual-like cells
(Plate 1, Figure 2). These possess a pale-staining foamy cytoplasm and two large ovoid nuclei with one or more prominent
nucleoli. The cytoplasm is filled with short profiles of endoplasmic reticulum which resemble granules with uniform density
and a prominent, densely staining halo surrounds the nuclei (Hamilton et al., 1973). Few blood vessels are present within the
cup tissue itself but a complicated network of small veins and
large lymph sinuses develops in the endometrial stroma beneath
the cup. Most of the apical portions and luminal openings of the
endometrial glands are obliterated during early cup development,
leaving the fundic portions of the glands at the base of the cup
swollen and distended with accumulated exocrine secretion. Liber-
ation of this accumulated material when the surface regions of
the cup tissue begin to degenerate gives rise to the PMSG-rich
endometrial cup secretion. Commencing immediately after development of the cups at Day 40, increasing numbers of small lymphocytes, esinophils and plasma cells accumulate in the endometrial
stroma at the periphery of the cup tissue. This aggregation of
leucocytes gives the appearance of a maternal immunological
attempt to 'wall off' the cup from the surrounding normal endometrial stroma and as necrosis of the large cup cells commences
from around Day 70, the accumulated leucocytes begin to invade the
cup and phagocytose the dying cells (Plate 1, Figure 3).

There is no physical connection between the fully developed
cups and the overlying foetal membranes and it was generally held
for many years that the cups were entirely maternal in origin,

developing as a form of maternal decidual response to the presence of the conceptus in the uterus (Clegg et al., 1954, Amoroso,
1955; Allen, 1970$_a$). More recently, however, Allen et al. (1973)
showed that the large, binucleate PMSG-secreting cells which comprise the bulk of the cup tissue are foetal in origin and develop
following a rapid invasion of the endometrium between Days 36 and
38 of gestation by specialised trophoblast cells from a region of
the foetal membranes known as the chorionic girdle. This discrete
and annulate band of trophoblast tissue was first described by
Ewart (1897) and forms at the junction of the developing allantoic
and regressing yolk sac membranes. It begins to develop from
about Day 25 as a simple folding of the chorion and by Day 36 consists of tightly packed finger-like projections of rapidly multiplying trophoblast cells interspersed with gland-like structures
which secrete a densely staining amorphous substance. This material lies between the chorionic girdle and underlying endometrial
epithelium and appears to act as an adhesive to bind the two tissues together. Between Days 36 and 38, the entire chorionic
girdle separates from the rest of the foetal membranes and the
girdle cells by means of pseudopodia-like projections, rapidly
invade and phagocytose the endometrial epithelial layer and pass
through the basement membranes. Once in the endometrial stroma
the cells quickly enlarge, become binucleate and begin to secrete
PMSG (Allen et al., 1973).

FACTORS INFLUENCING PMSG LEVELS IN SERUM

Following its initial appearance in the blood between Days
37 and 40 of gestation, PMSG concentrations rise rapidly to a
well-defined peak between Days 55 and 75 and thereafter decline
steadily to become undetectable again between Days 120 and 150.
Marked individual variations exist between mares in the total
amount of PMSG secreted (Text-figure 1) and peak concentrations
in the serum may range from as low as 10 to as high as 250 iu/ml
in any randomly selected group of animals (Allen, 1969$_a$).

The amount of PMSG produced by any particular mare will depend primarily on the total amount or volume of endometrial cup

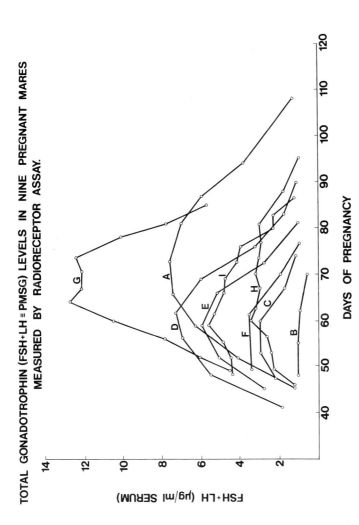

TOTAL GONADOTROPHIN (FSH+LH ≡ PMSG) LEVELS IN NINE PREGNANT MARES MEASURED BY RADIORECEPTOR ASSAY.

Fig. 1.   Gonadotrophin (PMSG) concentrations in μg/ml serum measured by radioreceptor assays at weekly intervals in the serum of nine pony mares (designated A to I) during early pregnancy. The FSH and LH activities have been added together for each serum sample to obtain a 'total' gonadotrophin concentration. Good correlation exists between results obtained by this assay method and those obtained by immunoassay which measures PMSG concentration in international units (Stewart et al., 1976). Both assays show major differences in serum PMSG concentrations between mares.

tissue which develops in the uterus, so that mares carrying twin conceptuses have two sets of endometrial cups and therefore double the 'normal' concentrations of PMSG in their blood (Rowlands, 1949). Marked variations exist between mares carrying a single conceptus in both the number and size of the endometrial cups they exhibit (Allen, 1970[a]). Since the cups develop as a result of the invasion of chorionic girdle cells into the endometrium between Days 36 and 38 of gestation the degree of cup development will depend upon the extent of the initial invasion process. This will be governed by such factors as the intimacy and area of contact between the chorionic girdle and endometrium at the time of invasion which will in turn depend upon such variables as the degree of folding or convolution of the endometrium in the area of contact.

Day and Rowlands (1947) reported that increasing age and parity of the mare may lead to a decline in total PMSG production with each successive pregnancy although the studies of Aylward and Ottway (1945) and Allen (1970[a]) did not support these findings. Similarly, Cole (1938), Day and Rowlands (1940), Rowlands (1963) and others have shown an inverse relationship between the size of the mare and peak PMSG concentrations in her serum. This is probably due to a dilution effect with both large and small mares producing approximately similar quantities of PMSG from equivalent amounts of endometrial cup tissue that is then subject to greater dilution in the blood and body tissues of the larger animals.

The single factor having by far the most pronounced effect on PMSG concentrations in the blood is foetal genotype. Mares mated to a male donkey and therefore carrying a mule conceptus, show greatly reduced serum PMSG levels and the hormone has generally disappeared from the blood as early as Day 80 of pregnancy (Bielanski et al., 1956; Clegg et al., 1962; Allen, 1969[b]). Conversely, the peak PMSG concentrations of 200 - 250 iu/ml which occur in female donkeys mated to a horse stallion and therefore carrying a hinny conceptus are up to 8 times higher than the gonadotrophin levels normally found in donkeys carrying a donkey conceptus (Allen, 1969[b]; Text-figure 2).

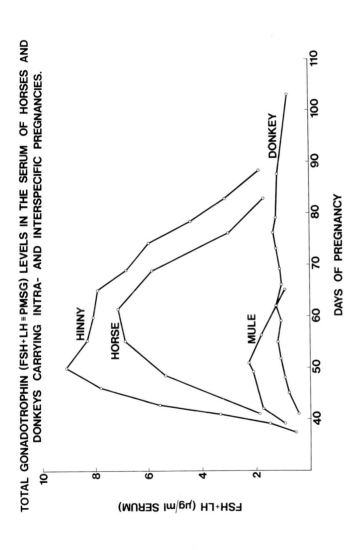

TOTAL GONADOTROPHIN (FSH+LH=PMSG) LEVELS IN THE SERUM OF HORSES AND DONKEYS CARRYING INTRA- AND INTERSPECIFIC PREGNANCIES.

Fig. 2.   Gonadotrophin (PMSG) concentrations measured by radioreceptor assays for FSH and LH in the serum of a mare carrying a horse conceptus, a mare carrying a mule conceptus, a donkey carrying a donkey conceptus and a donkey carrying a hinny conceptus. These curves are typical for the 4 different types of pregnancy and clearly illustrate the effect of foetal genotype on maternal PMSG concentrations.

The great reduction in PMSG production in mares carrying a mule conceptus results from a much more rapid than normal death and destruction of the endometrial cups in this hybrid situation. The steady accumulation of leucocytes around the cup tissue which occurs in a normal horse x horse pregnancy is greatly enhanced and speeded up in the mare carrying the mule and large numbers of lymphocytes have already invaded and completely destroyed the cup tissue by as early as Day 60 of gestation (Plate 2, Figures 1 and 2). In the donkey carrying the hinny conceptus on the other hand, many large and apparently quite healthy endometrial cups are still present at Day 60 of gestation (Plate 2, Figure 3) and these subsequently regress and die at about the same rate as the endometrial cups in mares carrying normal horse conceptuses and donkeys carrying normal donkey conceptuses. Although the accumulation of leucocytes in the endometrial stroma around the cup tissue is, as in the mare carrying the male foetus, considerably enhanced compared with the situation in the donkey carrying the normal donkey conceptus, the lymphocytes tend to remain aggregated in the endometrial stroma and seem unable or unwilling actively to invade the cup tissue and destroy the cup cells (Allen, 1975_a; Plate 2, Figure 4).

IN VITRO PRODUCTION OF PMSG

In vitro culture of chorionic girdle tissue recovered from equine conceptuses between 30 and 36 days of gestation, gives rise to discrete and stable monolayers of large epithelioid binucleate cells which are histologically indistinguishable from normal endometrial cup cells. These monolayers are remarkably stable and may continue to secrete large amounts of PMSG into the culture medium for up to 200 days, considerably longer than the normal lifespan of endometrial cups in the mare (Allen and Moor, 1972).

During the first 2 or 3 days in culture, the smaller mononucleate girdle cells multiply rapidly and spread out in the culture dish concentrically from each explant of tissue. The cells then uniformly undergo similar morphological changes to the girdle cells which invade the endometrium of the mare to form the

endometrial cups becoming rounded in outline, enlarging greatly
and acquiring a second nucleus. Having become binucleate the
cells apparently lose the ability to divide and no further development and expansion of the monolayers occurs. This high degree
of specialisation acquired by the trophoblast cells which comprise
the chorionic girdle, together with their apparent loss of mitotic
potential both in vivo and in vitro, prevents the sub-culture of
already cultivated girdle cells and so denies the establishment
of self-reproducing cell lines. This in turn severely limits the
potential for in vitro secretion of PMSG by cultured chorionic
girdle cells as a possible method for commercial production of
PMSG.

THE GONADOTROPHIC PROPERTIES OF PMSG

The original experiments of Cole and Hart (1930), Cole (1936)
and others, in rats clearly demonstrated the dual FSH-like and
LH-like biological activities of PMSG and the more recent studies
of Legault-Demare et al. (1961); Gospodarowicz and Papkoff (1967)
and others, established that PMSG is a single molecule. Moreover,
Papkoff (1974) has shown that the PMSG molecule, like the pituitary
gonadotrophins and the other major placental gonadotrophin, human
chorionic gonadotrophin (HCG), consists of an α and β sub-unit with
the β sub-unit being responsible for both the FSH-like and LH-like
activities of the intact molecule.

PMSG has a longer half-life in vivo than other pituitary and
placental gonadotrophins (Parlow and Ward, 1961; Cole et al.,1967;
Gay et al., 1970; McIntosh et al., 1975, and others) and this
property, along with the dual FSH and LH activities and ready
availability of PMSG has led to its wide-scale use for inducing
superovulation in experimental and domestic animals. There appears
to be considerable variations in sheep, cattle and pigs in the
degree and type of ovarian response obtained with standardised
doses of PMSG (Bellows and Short, 1972; Newcomb, 1976) which has
led to speculation about possible variations in the ratio of FSH
and LH activity in different extracts and commercially available
batches of the hormones (Polge and Rowson, 1975). Measurement of

58

TABLE 1

FSH : LH RATIOS MEASURED IN THE SERUM OF NINE PREGNANT PONIES

Mare	6	7	8	9	10	11	12	13	Mean ± SEM*
				Stage of gestation (weeks)					
A	–	1.37	1.34	1.36	1.57	1.34	1.50	1.27	1.39 ± 0.039
B	–	1.45	1.25	1.53	–	–	–	–	1.41 ± 0.038
C	–	1.30	1.16	1.57	1.62	1.42	–	–	1.41 ± 0.085
D	1.69	1.29	1.50	1.66	1.46	1.82	–	–	1.57 ± 0.078
E	–	1.50	1.68	1.65	1.82	1.85	1.31	–	1.64 ± 0.083
F	–	1.75	1.60	1.16	1.71	–	–	–	1.56 ± 0.135
G	–	1.92	1.73 1.81	1.31	1.24 1.10	1.31 1.13	1.31 1.11	1.24	1.38 ± 0.089
H	–	1.39	1.75 1.79	1.40 1.44	1.22 1.39	1.18 1.20	1.30 1.38	1.30 1.49	1.40 ± 0.052
I	–	1.36	1.38 1.15	1.42 1.63	1.18 1.40	1.25 1.50	1.47 1.50	1.74	1.42 ± 0.050
Mean **	–	1.48	1.51	1.47	1.43	1.40	1.36	1.41	
SEM	–	0.072	0.072	0.048	0.070	0.081	0.047	0.094	

Analysis of variance showed no significant difference between *mares (F = 1.355 P >0.05) or between **weeks of gestation (F = 0.570 P >0.05).

this ratio has proved difficult in the past because conventional in vivo bioassays and radioimmunoassays for FSH and LH do not readily lend themselves to the measurement of those activities in PMSG. However, the more recently developed radioreceptor assays for FSH and LH have been shown to be suitable for this purpose and were used to measure the FSH : LH ratio in PMSG from various sources (Stewart et al., 1976). These assays ultilise a preparation of rat testicular seminiferous tubules for measurement of FSH concentrations and rat testicular interstitial tissue for measurement of LH levels, with highly purified preparations of human pituitary FSH and LH as labelled ligands and reference standards.

The mean FSH : LH ratio of PMSG in mare's serum was found to be 1.45 $\pm$ 0.03 (SEM n = 67) and in a group of 9 pony mares bled weekly during the first 120 days of pregnancy no significant difference in ratio was observed either between mares or within any individual mare during pregnancy (Table 1). The FSH : LH ratio of PMSG in the medium recovered from cultures of chorionic girdle cells was lower than that in serum and a mean ratio of 0.74 $\pm$ 0.03 (SEM n = 21) was measured in serialmedium changes from cultures of chorionic girdle tissue recovered from five different mares (Text-figure 3). As in serum, the ratio remained remarkably constant, both within and between cultures from the individual mares despite major fluctuations in total gonadotrophin concentrations between them and during culture. The difference in the FSH : LH ratio of PMSG secreted in vivo (1.45) to that secreted in vitro (0.74) is thought to be due to the action of metabolic processes in vivo and/or a specific factor in horse serum which depresses the LH activity of PMSG (Stewart et al., 1976).

A mean FSH : LH ratio of 1.08 was measured in partially purified commercial preparations of PMSG ('Folligon', Intervet Laboratories, Bar Hill, Cambridge) and no statistically significant difference could be demonstrated in the FSH : LH ratios of 0.87, 0.98, 1.11, 1.16 and 1.30 which were measured in 5 different batches of Folligon, batch numbers 1327, 1491, 1108, 2101, 8727,

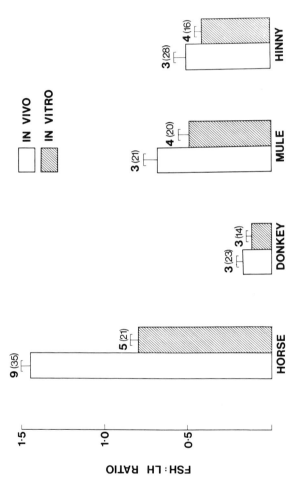

Fig. 3. The mean FSH : LH ratios measured by radioreceptor assays in PMSG produced in vivo (open bars) and in vitro (hatched bars) by horse, donkey, mule and hinny conceptuses. Bars represent SEM. Numbers in heavy type indicate the number of animals used in each group. The number of samples measured in each case are in parentheses.

respectively. It was of interest to note that when this material
was subsequently tested in 99 heifers by treating each animal
with 2 000 iu PMSG from one or other of the five batches, although
the number of ovulations in individual animals varied from 3 to
45, the mean ovulation rates in the five groups were found to be
insignificantly different (Newcomb and Rowson, personal communication). Similarly, when two groups of 47 and 13 sheep were
treated respectively with 1 000 iu PMSG of Folligon batch nos.
1108 or 1491, individual ovulation rates of the two groups were
not significantly different (Willadsen, Trounson and Moor, personal
communication). These results are in close agreement with the
observations of Schams et al. (1977) who found no significant
difference in the relative FSH-like and LH-like biological
activities in 4 different batches of commercially available PMSG.

As well as having a profound influence on total gonadotrophin
production in vivo foetal genotype markedly affects the FSH : LH
ratio of PMSG secreted both in vivo and in vitro (Stewart et al.,
1977). A ratio of 0.17 $\pm$ 0.01 (SEM n = 23) measured in the serum
of donkeys carrying normal donkey conceptuses, was considerably
lower than the ratio in the serum of mares carrying normal horse
conceptuses (1.45 $\pm$ 0.03 (SEM n = 67) with the FSH : LH ratios
of 0.64 $\pm$ 0.03 (SEM n = 21) in the mare carrying the mule conceptus
and 0.50 $\pm$ 0.02 (SEM n = 28) in the donkey carrying the hinny
conceptus falling approximately midway between those of the two
parental species (Text-figure 3). Similarly, in vitro the FSH :
LH ratio of PMSG secreted by cultured donkey girdle cells (0.12
$\pm$ 0.01 (SEM n = 14)) was appreciably less than that of horse
girdle cells (0.74 $\pm$ 0.03 (SEM n = 21)) with the ratios of the
PMSG secreted by mule and hinny girdle cells (0.45 $\pm$ 0.03 (SEM
n = 20) and 0.37 $\pm$ 0.02 (SEM n = 16) respectively) lying between
the two parental extremes (Text-figure 3; Stewart et al., 1977).

It is not yet known whether these major differences in FSH :
LH ratio, as measured by in vitro radioreceptor assay, in horse,
donkey and 'hybrid' PMSGs, are reflected by similar differences
in gonadotrophic activity between them when injected into laboratory or farm animals. Also, whether the apparently 'midway'
FSH : LH ratio of mule and hinny PMSG results from the production

of a 'hybrid' type of PMSG molecule by these conceptuses or
merely reflects the secretion of approximately equal amounts of
the donkey and horse types of PMSG.

BIOLOGICAL FUNCTION OF PMSG IN THE MARE

The primary corpus luteum of pregnancy in the mare begins
to decline in secretory activity from as early as Day 20 of gestation (van Rensburg and van Niekerk, 1968; Allen, 1970_a; Allen
and Rossdale, 1976) although it may remain present in the ovaries
and continue to secrete progesterone until as late as Day 160
(Squires and Ginther, 1975). Commencing on or around Day 40, and
in conjunction with the first appearance of PMSG in the blood, a
series of secondary ovulations and/or luteneised unruptured follicles develop in the ovaries (Cole et al., 1931; Amoroso et al.,
1948; Allen, 1970_a). These accessory corpora lutea remain present in the ovaries and actively secrete progesterone until around
Day 130 to 180 when they too regress following the disappearance
of PMSG from the blood. From then until term the mare's ovaries
contain neither follicles nor corpora lutea (Squires and Ginther,
1975) and during this time all of the progesterone required to
maintain pregnancy is secreted by the foeto-placental unit (Short,
1959). The coincidental appearance and disappearance of PMSG in
the blood with the development and regression of the secondary
corpora lutea has, for many years, led to the assumption that
PMSG plays a major role in stimulating follicular growth and
maintaining luteal function during early pregnancy in the mare.
There are, however, a number of recent findings which seem to cast
doubt on this hypothesis.

Although ovulation or luteinisation of follicles does not
normally occur until after the production of PMSG around Day 40,
the follicles which give rise to these secondary corpora lutea
undergo marked enlargement and maturation as early as Days 17 to
23, some 15 - 20 days before development of the endometrial cups
(Rowlands, 1949; Bain, 1967; van Rensburg and van Niekerk, 1968;
van Niekerk, 1973). Using a heterologous radioimmunoassay for
FSH which does not cross-react with PMSG, Evans and Irvine (1975)

have demonstrated that peripheral plasma FSH levels in the mare
show two pronounced peaks during the oestrous cycle, one around
the time of ovulation and coincident with a rise in plasma LH levels
and the other 10 - 11 days later in mid-dioestrus and quite independent of LH. Moreover, these apparently episodic 10-day pituitary
FSH releases which occur during the cycle continue unchanged
during pregnancy, only declining and eventually disappearing during
late autumn and early winter when non-pregnant mares are normally
beginning to pass into anoestrus. LH release does not follow this
pattern and peripheral plasma LH concentrations remain at base-
line values during at least the first 40 days of pregnancy (Allen,
W.R., unpublished data). The peaks of pituitary FSH release described by Evans and Irvine (1975) coincide closely with the waves
of follicular growth reported previously to occur during early
pregnancy in the mare by van Rensburg and van Niekerk (1968) suggesting perhaps that pituitary FSH, and not PMSG, is responsible
for stimulating follicular growth during early pregnancy, with
PMSG acting only to ovulate or luteinise the already mature follicles.

Additional evidence to support this theory is provided by the
finding that although secondary corpora lutea do not develop in
mares hysterectomised in dioestrus, ovarian follicular activity
in these animals is very similar to that in pregnant mares which
are being exposed to high concentrations of PMSG (Squires and
Ginther, 1975). Further, Allen, W.E. (1975) has shown that ovarian
activity in mares during early pregnancy is influenced more by
the season of mating than by serum PMSG concentration.

Other findings seem to minimise the luteotrophic role of PMSG
in the mare. If the entire conceptus is removed from the uterus
of mares after Day 38, either surgically via a laparotomy incision
or per vaginum following the administration of prostaglandins,
the chorionic girdle cells have already invaded the endometrium
to form the endometrial cups which continue to develop and regress
normally just as if the mare was still pregnant.

Accordingly, high concentrations of PMSG remain in the blood

of such mares for many weeks after the loss of the pregnancy
(Allen, 1970$_b$). Luteal function, as judged by peripheral plasma
progestagen determinations, is maintained in some of these mares
while the endometrial cups remain active but in others the corpora lutea regress within as little as 20 days after removal of
the pregnancy, despite the continued presence of high concentrations of PMSG in the blood. The reason for the difference between individual mares is unknown but whether luteal function is
maintained or not, the mares will normally only return to oestrus
and ovulate after the endometrial cups have regressed completely
and PMSG has disappeared from the blood (Allen, 1970$_b$; Mitchell,
1971). This facultative delay in cyclical activity following
loss of the pregnancy after Day 40 constitutes a considerable
practical problem in equine stud management and because they have
two sets of endometrial cups the problem is greater in mares from
which twin conceptuses have been aborted. Since the mare is unable to successfully carry twins to term it is normal equine
veterinary practice to abort mares carrying twins after positive
rectal diagnosis of their existence in the uterus at 42 - 45 days
of gestation. However, the cups have already begun to develop
by this stage and they continue to secrete large amounts of PMSG
for the next 80 - 90 days. Even if luteolysis occurs spontaneously in these animals or is artificially induced by repeated
treatment with prostaglandins, the mare's ovaries rapidly become
small, hard and completely inactive and she enters a prolonged
period of complete ovarian inactivity akin to deep winter anoestrous. Normal follicular activity will not recur until PMSG has
disappeared completely from the blood, and only then if it is
not already too late in the season to coincide with the normal
winter anoestrous period. This evidence therefore seems to indicate that in the absence of additional unknown foetal stimuli,
PMSG may not function as a luteotrophin. It is certainly not
gonadotrophic in the normal sense and very strongly depresses
rather than stimulates ovarian function. Whether this suppressant
action is exerted on the ovaries or is mediated via the hypo-
thalomic-pituitary axis remains to be shown.

One further anomaly as regards a luteotrophic role for PMSG
in the mare concerns the lack of any apparent correlation between

PMSG concentrations in the serum and luteal activity as determined by peripheral plasma progesterone measurements. In mares
carrying both intraspecific horse and interspecific mule conceptuses, and in donkeys carrying intraspecific donkey conceptuses,
despite great variations in maximum PMSG levels in the serum of
these three types of pregnancy, progestagen concentrations are
essentially similar and normally range from 5 to 35 ng/ml between
40 and 150 days of gestation (Allen, 1975$_a$). In the donkey carrying the interspecific hinny conceptus, however, where PMSG levels
are much higher than in donkeys carrying normal donkey conceptuses,
progesterone production is enormously increased with concentrations
as high as 300 to 800 ng/ml being measured in the peripheral
plasma (Allen, 1975$_b$). Experiments involving the infusion of
^{3}H-progesterone into pregnant horses and donkeys have demonstrated
that virtually all the additional progesterone present in the
donkey carrying the hinny conceptus results from increased ovarian
activity, with very little if any being secreted by the placenta
and in the absence of any increase in plasma progesterone binding
proteins (Sheldrick et al., 1977). Thus, the ovaries of these
animals, which between them may contain as many as 30 fully developed corpora lutea or large, luteinised follicles, are massively
stimulated by the PMSG produced by this type of conceptus.

It is therefore tempting to speculate that this striking
difference in ovarian response between the pregnant mare and
donkey may result from the type, and not simply the total amount,
of PMSG which the mother is confronted with during each type of
pregnancy. Thus the female donkey which is 'genetically prepared'
to receive relatively small quantities of a gonadotrophin with
a very low content of FSH-like activity when carrying a normal
donkey conceptus, responds almost uncontrollably when confronted
unexpectedly with the larger amounts of PMSG containing a much
higher FSH-like content that are produced by the hinny endometrial
cup cells.

CONCLUSIONS

Although PMSG is undoubtedly the most widely and extensively

used of all available gonadotrophin hormones, its biological
role in the pregnant mare remains an enigma. The essential
evolutionary question of why the equine foetus should go to so
much morphological bother and immunological risk to ensure the
production for such a relatively short period of time, of such
large amounts of glycoprotein hormone which appears to be of such
minimal physiological significance in the maintenance of pregnancy, has yet to be answered.

DESCRIPTION OF PLATES

PLATE 1

Fig. 1. Uterus and ovaries taken from a pregnant mare at 70 days of gestation.
The uterus has been opened and the conceptus removed to reveal the circle of
endometrial cups (arrowed) in the pregnant horn.

Fig. 2. Histological section of an endometrial cup taken from a mare at 50
days of gestation. The cup is composed of large, binucleate, decidual-like
cells clumped together within the maternal endometrial stroma. Most of the
endometrial glands in the vicinity of the cup are swollen and distended with
exocrine secretion. Maternal leucocytes are beginning to accumulate at the
periphery of the cup tissue (x 85)

Fig. 3. Section of a degenerating endometrial cup at 130 days of gestation.
The cup tissue is now completely necrotic and in the process of being sloughed
off the surface of the endometrium by a process akin to cell-mediated graft
rejection (x 66)

PLATE 2

Fig. 1. Pregnant horn of the uterus of a mare carrying a mule conceptus
opened by hysterectomy at 60 days of gestation to show the endometrial cups.
These are small and pale and already completely necrotic.

Fig 2. Low power histological section of an endometrial cup taken from the
uterus illustrated in Figure 1. The whole of the cup tissue is already
completely necrotic and the stroma beneath is densely packed with accumulated
maternal leucocytes. (x 66)

Fig. 3. Uterus and ovaries taken from a donkey carrying a hinny conceptus at
60 days of gestation. The pregnant horn is opened to show the circle of large
and healthy looking endometrial cups surrounding the embryo. Both maternal
ovaries are extremely large and contain numerous secondary corpora lutea and
heavily luteinised follicles.

Fig. 4. Low power histological section of an endometrial cup recovered from
the uterus illustrated in Figure 3. In marked contrast to the situation in
Figure 2 (mule at 60 days), the majority of the large, decidual-like cup cells
are still intact and actively secreting PMSG. Large numbers of leucocytes are
accumulated in the stroma but these are not invading the cup tissue. (x 34)

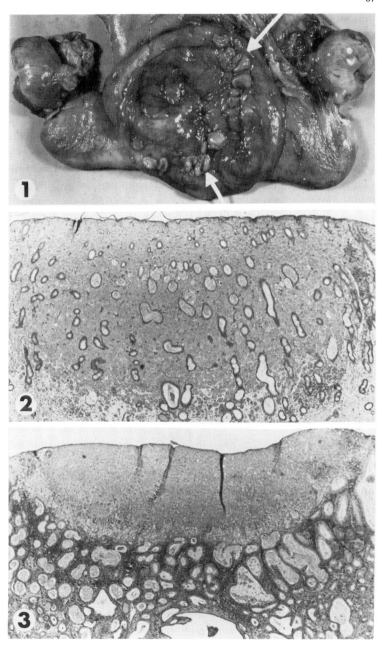

Plate 1.

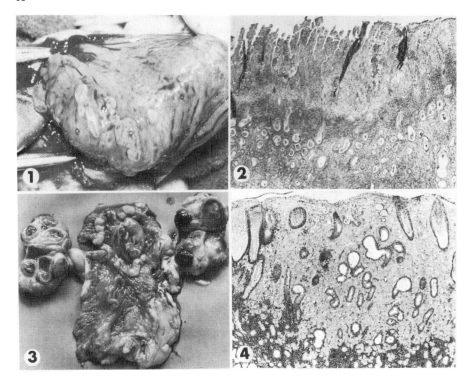

Plate 2.

REFERENCES

Allen, W.E. 1975 Ovarian changes during early pregnancy in pony mares in relation to PMSG production. J. Reprod. Fert. Suppl. 23, 425

Allen, W.R. 1969$_a$ The immunological measurement of pregnant mare serum gonadotrophin. J. Endocr. 43, 593

Allen, W.R. 1969$_b$ Factors influencing pregnant mare serum gonadotrophin production. Nature, Lond. 223, 64

Allen, W.R. 1970$_a$ Equine gonadotrophins. Ph.D. Thesis. University of Cambridge.

Allen, W.R. 1970$_b$ Endocrinology of early pregnancy in the mare. Eq. Vet. J., 2, 64.

Allen, W.R. 1975$_a$ Immunological aspects of the equine endometrial cup re-action. In: Immunobiology of Trophoblast. Ed. R.G. Edwards, C.W.S. Howe and M.H. Johnson. University Press: Cambridge.

Allen, W.R. 1975$_b$ The influence of foetal genotype upon endometrial cup development and PMSG and progestagen production in equids. J. Reprod. Fert. Suppl. 23, 405.

Allen, W.R., Hamilton, D.W. and Moor, R.M. 1973 The origin of equine endo-metrial cups. II. Invasion of the endometrium by trophoblast. Anat. Rec. 177, 485.

Allen, W.R. and Moor, R.M. 1972 The origin of the equine endometrial cups. I. Production of PMSG by foetal trophoblast cells. J. Reprod. Fert. 29, 313.

Allen, W.R. and Rossdale, P.D. 1976 Use of prostaglandins in the mare. Vet. Rec. 98, 389.

Amoroso, E.C. 1955 Endocrinology of pregnancy. Brit. Med. Bull. 11, 117

Amoroso, E.C., Hancock, J.L. and Rowlands, I.W. 1948 Ovarian activity in the pregnant mare. Nature, Lond. 161, 355.

Aylward, F. and Ottaway, C.W. 1945 The collection and examination of plasma from pregnant mares for gonadotrophic hormone. J. Comp. Path. Ther. 55, 159.

Bain, A.M. 1967 The ovaries of the mare during early pregnancy. Vet. Rec. 80, 229.

Bellows, R.A. and Short, R.E. 1972 Superovulation and multiple births in beef cattle. Tenth Biennial Symposium on Animal Reproduction. 34, (Suppl. 1) 67

Bielanski, W., Ewy, Z. and Pigoniowa, H. 1956 Differences in the level of gonadotrophin in the serum of pregnant mares. In: Third International Congress on Animal Reproduction, 1956, Cambridge, p.110.

Clegg, M.T., Boda, J.M. and Cole, H.H. 1954 The endometrial cups and allanto-chorionic pouches in the mare with emphasis on the source of equine gonadotrophin. Endocrinology, 54, 448.

Clegg, M.T., Cole, H.H., Howard, C.B. and Pigon, H. 1962 The influence of foetal genotype on equine gonadotrophin secretion. J. Endocr. 25, 245.

Cole, H.H. 1936 On the biological properties of mare gonadotrophin hormone. Am. J. Anat. 59, 299.

Cole, H.H. 1938 High gonadotrophic hormone concentration in pregnant ponies. Proc. Soc. Exp. Biol. Med. 38, 193.

Cole, H.H., Bigelow, M., Finkel, J. and Rupp, G.P. 1967 Biological half-life of endogenous PMS following hysterectomy and studies on losses in urine and milk. Endocrinol, 81, 927.

Cole, H.H. and Goss, H.T. 1943 The source of equine gonadotrophin. In: Essays in Biology in Honour of H.M. Evans, p. 107. University of California Press: California.

Cole, H.H. and Hart, G.H. 1930 The potency of blood serum of mares in progressive stages of pregnancy in effecting the sexual maturity of the immature rat. Am. J. Physiol. 93, 57.

Cole, H.H., Howell, C.E. and Hart, G.H. 1931 The changes occurring in the ovary of the mare during pregnancy. Anat. Rec. 49, 199.

Day, F.T. and Rowlands, I.W. 1940 The time and rate of appearance of gonadotrophin in the serum of pregnant mares. J. Endocr. 2, 255.

Day, F.T. and Rowlands, U.W. 1947 Serum gonadotrophin in Welsh and Shetland ponies. J. Endocr. 5, 1.

Evans, M.J. and Irvine, C.H.G. 1975 Serum concentrations of FSH, LH and progesterone during the oestrous cycle and early pregnancy in the mare. J. Reprod. Fert. Suppl. 23, 193.

Ewart, J.C. 1897 A critical period in the development of the horse. Adam and Charles Black: London.

Gay, V.L., Midgley, A.R. and Niswender, G.D. 1970 Patterns of gonadotrophin secretion associated with ovulation. Fed. Proc. Amer. Soc. Exp. Biol. 29, 1880.

Gospodarowicz, D. 1972 Purification and physicochemical properties of the pregnant mare serum gonadotrophin (PMSG). Endocr. 91, 101

Gospodarowicz, D. and Papkoff, H. 1967 A simple method for the isolation of pregnant mare serum gonadotrophin. Endocrinol. 80, 699

Hamilton, D.W., Allen, W.R. and Moor, R.M. 1973 The origin of equine endometrial cups. III. Light and electron microscopic study of fully developed equine endometrial cups. Anat. Rec. 177, 503.

Legault-Demare, J., Clauser, H. and Jutisz, M. 1961 Purification de la gonadotropine serique de jumet gravide (PMSG) par distribution a contre courant. Bull. Soc. Chim. Biol. 43, 897.

McIntosh, J.E.A., Moor, R.M. and Allen, W.R. 1975 Pregnant mare serum gonadotrophin: Rate of clearance from the circulation of sheep. J. Reprod. Fert. 44, 95.

Mitchell, D. 1971 Early foetal death and a serum gonadotrophin test for pregnancy in the mare. Can. Vet. J. 12, 41.

Newcomb, R.N. 1976 Investigation of factors affecting superovulation, egg recovery and transfer in cattle. MSc Thesis, University of Cambridge.

Papkoff, H. 1974 Chemical and biological properties of the sub-units of pregnant mare serum gonadotrophin. Biochem. Biophys. Res. Com. 58, 397.

Parlow, A.F. and Ward, D.N. 1961 Rate of disappearance of LH, PMS and HCG from plasma. In: Human Pituitary Gonadotrophins. p 204. Ed. A. Albert, Charles G. Thomas. Springfield, Illinois.

Polge, C. and Rowson, L.E.A. 1975 Recent progress in techniques for increasing reproductive potential in farm animals. In: Proceedings of the III World Conference on Animal Reproduction, p 634, Ed. R.L. Reid. University Press: Sydney.

Rowlands, I.W. 1949 Serum gonadotrophin and ovarian activity in the pregnant mare. J. Endocr. 6, 184.

Rowlands, I.1963 Levels of gonadotropin in tissues and fluids with emphasis on domestic animals. In: Gonadotropins: Their chemical and biological properties and secretory control. Ed. H.H. Cole. W. Freeman & Co: Sàn Francisco.

Schams, D., Menzer, Ch., Schallenberger, E., Hoffman, B., Hahn, J. and Hahn, R. 1977 Some studies on pregnant mare serum gonadotropin (PMSG) and on endocrine responses after application for superovulation in cattle. In: EEC Symposium on Embryo Transfer in Cattle Volume 2 (This volume).

Schauder, W. 1912 Untersuchungen über die eihaule und embryotrophe des pferdes. Arch. Anat. Physiol. 259

Sheldrick, E.L., Wright, P.J., Allen, W.R. and Heap, R.B. 1977 Metabolic clearance rate, production rate and source of progesterone in donkeys with foetuses of different genotypes. J. Reprod. Fert. (in press).

Short, R.V. 1959 Progesterone in blood. IV. Progesterone in the blood of mares. J. Endocr. 19, 207.

Squires, E.L. and Ginther, O.J. 1975 Follicular and luteal development in pregnant mares. J. Reprod. Fert. Suppl. 23, 429.

Stewart, F., Allen, W.R. and Moor, R.M. 1976 Pregnant mare serum gonadotrophin: Ratio of follicle stimulating hormone and luteinising hormone activities measured by radioreceptor assay. J. Endocr. 71, 371

Stewart, F., Allen, W.R. and Moor, R.M. 1977 Influence on foetal genotype on the follicle stimulating hormone : luteinising ratio of pregnant mare serum gonadotrophin. J. Endocr. 73, 419.

Van Niekerk, C.H. 1973 Morphological and physiological changes in the genital system of mares. DVSc Thesis, University of Pretoria.

Van Rensburg, S.J. and Van Niekerk, C.H. 1968 Ovarian function, follicular oestradiol-17 and luteal progesterone and 20 -hydroxy-pregn-4-ene-3-one in cycling and pregnant equines. Onderstepoort, J. Vet. Res. 33, 195.

RELATIONSHIP OF PMSG TO THE PITUITARY GONADOTROPHINS

Harold Papkoff

Hormone Research Laboratory and Reproductive Endocrinology Center
University of California, San Francisco, California 94143, USA

ABSTRACT

PMSG resembles pituitary LH and FSH both chemically and biologically. Isolated from the serum, it is a glycoprotein, but possesses much more carbohydrate (45%) than LH or FSH. Like the pituitary glycoprotein hormones, PMSG consists of two chemically dissimilar subunits (α and β). Individually the PMSG subunits are biologically inactive. One of the subunits (alpha) is hormonally non-specific and can be complexed with either LH-β, FSH-β, or HCG-β, regenerating the biological activity expected of the beta subunit. The amino acid composition of PMSG-α is very similar to either LH, FSH, or TSH alpha; the composition of PMSG-β is closer to LH-β. The preponderance of carbohydrate is associated with PMSG-β. Highly purified PMSG is biologically active as both an LH and an FSH in standard, specific, in vivo assays, newly developed in vitro assays, and in radioreceptor tests for LH and FSH. Antisera raised in rabbits against PMSG show a strong cross-reaction with LH but less so with FSH.

Highly purified material can also be isolated from endometrial cup tissue (PMEG) instead of serum. Comparative biochemical and biological properties of PMEG with PMSG will be presented.

INTRODUCTION

The presence of a gonadotrophin substance in the serum of pregnant mares was first described by Cole and Hart (1930). A few years later, this gonadotrophin (PMSG), was shown to possess an unusually long circulating half-life (6 days) in the mare (Catchpole, Cole, and Pearson, 1935) and other experimental animals. In the years which followed, PMSG was determined to behave biologically in various tests like both pituitary LH and FSH. Whether these actions were a function of the same or separable molecules could not be established with certainty, however, until highly purified preparations of PMSG were obtained many years later. Meanwhile, much progress has been made in the isolation, characterisation, and study of the pituitary gonadotrophins. Although studies on PMSG have not progressed to the same extent, it seems worthy at this time to review the biochemical properties of PMSG in relationship to what is known of pituitary LH and FSH.

PURIFICATION

Although there were many procedures described for the purification of PMSG in the period 1940 - 1960 (reviewed by Legault-Demare and Clauser, 1961), these were by and large tedious, resultant in very low yields, and produced materials of uncertain purity. With the advent of newer methods of protein purification and characterisation (ion-exchangers, gel-filtration, etc.) we (Gospodarowicz and Papkoff, 1967) were able to develop a new method for the purification of PMSG which proved to be simple, rapid, and yielded 50 - 80% of the hormone initially present in the serum. It was later shown (Schams and Papkoff, 1972; Gospodarowicz, 1972) that partially purified commercial preparations of PMSG (1000 - 2000 iu/mg) could be readily purified with the same methodology.

In brief, the procedure is as follows: serum (or plasma) is adjusted to pH 3.0 by the addition of metaphosphoric acid

which precipitates copious quantities of inert proteins. The
supernatant fluid is adjusted to pH 4.5 and alcohol added to a
concentration of 50%, resulting in the further precipitation
of inert or low activity proteins. Crude PMSG is precipitated
by increasing the alcohol concentration to 75%. This material
is further purified by chromatography on sulfoethyl-Sephadex,
C-50, followed by gel filtration on Sephadex G-100, to yield
the final, purified PMSG. With minor modification, the basic
procedure can be applied to the purification of the hormone
from its tissue of origin, the endometrial cups (PMEG) (Papkoff,
Farmer, and Cole, unpublished).

PROPERTIES

Biological

PMSG can be biologically evaluated by assays measuring
total gonadotrophins (ie both LH and FSH activity) or by
specific tests for LH and FSH. Highly purified preparations
have specific activities which range from 10 500 - 18 000 iu/mg
(Gospodarowicz and Papkoff, 1967; Schams and Papkoff, 1972;
Gospodarowicz, 1972). When compared to pituitary gonadotrophins
(ovine) in specific bioassays, we have found PMSG to be 75 - 135
x NIH-FSH-S1 and 1.5-2 x NIH-LH-S1. These activities are
comparable to those obtained with various species of highly
purified pituitary LH and FSH (Sairam and Papkoff, 1974). PMSG
is also found to be very potent in in vitro assays such as the
stimulation of testosterone from dispersed rat interstitial
cells (Dufau, et al., 1976; Farmer, et al., 1977), in
stimulating cyclic AMP production from isolated rat Sertoli
cells (Farmer and Papkoff, unpublished), and in receptor
binding assays for LH and FSH (Stewart, et al., 1976). Thus,
there is little doubt that PMSG has the activities of both LH
and FSH in the above mentioned systems.

Biochemical characterisation of PMSG

All known pituitary gonadotrophins as well as HCG are
glycoproteins. PMSG is no exception to this observation
although it stands by itself in terms of its very high total

carbohydrate content (41 - 45%). Table 1 summarises the sugar
content of PMSG and compares the data to values obtained
(Landefeld. et al., 1972; Landefeld and McShan, 1974$_1$) for
equine LH and FSH. PMSG contains neutral sugars (mannose,
galactose, and fucose) to the extent of 13.0%, hexosamines
(glucosame and galactosamine) which total 17.6%, and a very
high content (10.8%) of sialic acid. The high sialic acid
content is undoubtedly responsible for the very acidic nature
of PMSG. Both equine LH and FSH (Table 1) have about 25%
carbohydrate, and each category of sugar is present to a lesser
degree than in PMSG. It is also of interest that equine LH
possesses such a high content of sialic acid (7.9%) insofar
as many other species of LH contain little or none, whereas all
species of FSH contain significant quantities of sialic acid
(Sairam and Papkoff, 1974). Sialic acid, it may be mentioned
here, appears to be related to the circulating half-life of a
glycoprotein hormone and the very high content present in PMSG
is consistent with its known long half-life in the circulation.
The high content of sialic acid present in equine LH and FSH
suggests that they, too, have long survival times.

TABLE 1

CARBOHYDRATE CONTENT* OF PMSG AND ITS SUBUNITS COMPARED TO EQUINE** LH,
FSH, AND THEIR SUBUNITS

	Neutral sugars	Hexosamine	Sialic acid	Total
PMSG	13.0	17.6	10.8	41.4
LH	8.9	8.6	7.9	25.4
FSH	9.7	7.7	6.8	24.2
PMSG-α	8.4	5.7	4.5	18.6
LH-α	7.5	7.3	5.0	19.8
FSH-α	7.4	6.7	5.6	19.7
PMSG-β	15.8	18.2	21.3	55.3
LH-β	6.9	6.5	9.5	22.9
FSH-β	9.8	7.8	7.4	25.0

* g/100 g glycoprotein

** Equine data of Landefeld et al., 1972; Landefeld and McShan, 1974.

In Table 2 the amino acid composition of PMSG is shown and compared with data for equine LH and FSH (Landefeld and McShan, 1974 [1, 2]). PMSG has an amino acid composition which is typical of the glycoprotein hormones in that the content of cystine and proline is high and the content of histidine, methionine, and tyrosine is relatively low. When comparisons are made between PMSG and the LH and FSH compositions shown (Table 2), there is a greater similarity with LH than with FSH. This is particularly true with respect to the contents of lysine, aspartic acid, threonine, proline, and tyrosine.

TABLE 2

AMINO ACID COMPOSITION* OF PMSG COMPARED WITH EQUINE LH AND FSH

Amino Acid	PMSG	LH**	FSH**
Lysine	5.4	5.2	7.0
Histidine	2.5	1.9	2.5
Arginine	6.1	4.5	3.8
Aspartic	5.5	5.3	7.3
Threonine	8.6	8.4	9.7
Serine	8.0	9.6	6.8
Glutamic	8.4	7.8	7.9
Proline	12.5	10.4	7.8
Glycine	5.1	6.7	6.7
Alanine	7.5	8.4	6.7
Half-cystine	7.5	7.9	9.3
Valine	4.6	5.6	5.8
Methionine	1.7	1.6	1.1
Isoleucine	4.8	4.8	4.5
Leucine	6.3	6.3	5.3
Tyrosine	2.4	1.8	4.2
Phenylalanine	4.0	4.2	4.1

* Calculated as residues/100 residues analysed.
** Data of Landefeld and McShan (1974) recalculated.

Terminal group analysis of PMSG by the dansyl technique
reveal two amino acids, phenylalanine and serine, to be the
major N-terminal residues; traces of others are often found,
suggesting a degree of microheterogeneity. Landefeld et al.
(1972) also found phenylalanine and serine to be N-terminal
residues of equine LH. Unfortunately, no data is available
for equine FSH. Experiments to reveal the carboxyl terminal
residues of PMSG have been more difficult. The hormone (both
native and oxidised is resistant to the action of carboxypeptidase, and only by hydrazinolysis is isoleucine shown to be
present. This residue as well as serine was shown by Landefeld
et al. (1972) to be C-terminal in equine LH.

The carbohydrate, amino acid, and terminal group data on
PMSG show that it is a typical glycoprotein hormone and in
many respects bears a closer resemblance to LH than FSH. The
presence of two N-terminal residues is indicative of the
presence of two polypeptide chains (ie subunits) as is the
case with all of the glycoprotein hormones.

Subunit nature of PMSG

Although it was expected that PMSG consisted of subunits,
it is of interest that we cannot readily detect dissociation
of PMSG by sedimentation studies in the ultracentrifuge as was
the case with pituitary LH and FSH. Thus, at pH 8.2, a
sedimentation coefficient of 3.55 S is obtained, yet under
conditions where dissociation would be expected, as at pH 1.3,
PMSG has an $S_{20,w}$ of 3.05. Nevertheless, we have shown
(Papkoff, 1974) that by incubating PMSG in 10 M urea followed
by chromatography on DEAE-Sephadex A 25, effective separation
of the subunits is achieved. Further purification is accomplished
by gel filtration on columns of Sephadex G-100 in 0.05 M NH_4HCO_3.
Intact PMSG elutes with a Ve/Vo of 1.28 while PMSG-β is somewhat more retarded ($Ve/Vo=1.35$) and PMSG-α is considerably
retarded ($Ve/Vo=2.1$). It is clear from this behavior that
there is a great difference in the hydrodynamic volumes of the
PMSG subunits, much more so than is observed with pituitary
gonadotrophin subunits where the two are of nearly equal size.

Comparison of the carbohydrate content (Table 1) of the
PMSG subunits with each other and with the corresponding LH
and FSH subunits reveal two interesting aspects. First, there
is clearly an enormous difference in carbohydrate content
between PMSG-α and PMSG-β (18.6% vs 55.3%). Secondly, it is
apparent that PMSG-α has a sugar content that is very similar
to LH-α and FSH-α . PMSG-β , however, has 2.2 - 3.0 times as
much sugar as any of the alpha or beta subunits. Of its total
of 55.3% carbohydrate, sialic acid is present in the greatest
amount (21.3%).

In Tables 3 and 4, the amino acid composition of PMSG-α
and PMSG-β is shown and compared with the counterpart LH and
FSH subunits. It is apparent that all three alpha subunits
have a very similar composition (Table 3) which is to be expected
in view of the studies on other species of gonadotrophins
previously developed in many laboratories over the past decade.
Examination of the beta subunit composition (Table 4) shows
beyond doubt that PMSG-β is much more like LH-β than FSH-β ,
although absolute identity does not appear to be likely.

Terminal group analysis shows PMSG-α to begin with the
sequence NH_2-PHe-Pro-(Gly or Pro).. and to terminate with
isoleucine at the carboxyl terminus. PMSG-β has the sequence
NH_2-Ser-Pro-Gly... and as yet no C-terminal residue has been
detected with either carboxypeptidase or hydrazinolysis (Papkoff,
et al., unpublished).

Physical studies on PMSG and its subunits have involved
analyses by the technique of circular dichroism (CD). Examination
of the CD spectra obtained shows that neither PMSG nor its
subunits have any appreciable secondary structure (α -helix).
Analysis of the side chain region (250 - 350 nm) of the spectra
show clearly that the subunits when present in the intact PMSG
have a different conformation than when analysed individually.
Further, evidence was obtained to suggest that one or more
tyrosine residues are inaccessible in the intact PMSG as

compared to the individual subunits. Very similar data by this
technique has previously been reported for ovine LH (Bewley,
et al., 1972).

TABLE 3

AMINO ACID COMPOSITION* OF THE ALPHA SUBUNITS OF PMSG, EQUINE LH, AND
EQUINE FSH

Amino acid	PMSG-α	LH-α**	FSH-α**
Lysine	9.4	6.3	8.4
Histidine	3.0	2.7	3.2
Arginine	4.9	4.0	4.2
Aspartic	6.5	6.2	6.8
Threonine	9.9	9.4	9.3
Serine	5.0	6.9	4.7
Glutamic	8.7	9.1	9.2
Proline	6.4	7.3	7.3
Glycine	4.5	5.4	5.6
Alanine	4.4	6.5	5.1
Half-cystine†	10.1	8.7	9.0
Valine	5.0	6.0	5.8
Methionine†	2.0	1.1	1.1
Isoleucine	4.6	5.6	5.3
Leucine	4.3	5.3	4.9
Tyrosine	5.1	4.2	4.4
Phenylalanine	6.0	5.6	5.7

*   Calculated as residues/100 residues analysed.

**   Data of Landefeld and McShan (1974) recalculated.

†   Determined as cysteic acid and methionine sulfone in performic acid
    oxidised samples of PMSG-α .

TABLE 4

AMINO ACID COMPOSITION* OF THE BETA SUBUNITS OF PMSG, EQUINE LH, AND
EQUINE FSH

Amino acid	PMSG-β	LH-β **	FSH-β **
Lysine	3.1	2.7	5.9
Histidine	1.6	1.3	2.7
Arginine	6.4	4.8	2.5
Aspartic	4.8	4.2	9.6
Threonine	7.7	8.1	11.1
Serine	10.0	9.7	7.2
Glutamic	7.2	6.6	7.2
Proline	15.9	14.9	5.8
Glycine	6.0	5.5	7.9
Alanine	9.1	10.4	6.8
Half-cystine†	6.9	8.9	8.5
Valine	4.7	5.3	6.5
Methionine†	1.9	1.3	0.9
Isoleucine	4.7	5.1	4.4
Leucine	7.3	6.9	5.0
Tyrosine	1.6	1.5	5.7
Phenylalanine	3.0	2.9	2.4

* Calculated as residues/100 residues analysed.

** Data of Landefeld and McShan (1974) recalculated.

† Determined as cysteic acid and methionine sulfone in performic acid
oxidised samples of PMSG-β

We have previously (Schams and Papkoff, 1972) described an antiserum prepared in rabbits against PMSG. Employing the agar double diffusion technique (Ouchterlony) we are able to show a line of precipitation with PMSG and also with equine LH which forms a line of identity with PMSG. Equine FSH is either weakly or not reactive with this antiserum as are other species of gonadotrophins (human, ovine, porcine). It is of interest that when the PMSG subunits are tested, only PMSG-β cross-reacts, forming a line of identity with PMSG (Papkoff, unpublished).

Biological activity of PMSG subunits and subunit recombinants

Although this paper is intended to emphasise biochemical aspects of PMSG, it is of interest and importance to briefly describe the structure-function relationships of the PMSG subunits in terms of both LH and FSH activity. These data are presented in Table 5; summarised from Papkoff (1974). It is seen that the individual subunits are of very low activity compared to PMSG when tested in both LH and FSH bioassays. The residual activity (2.5 - 6%) probably represents cross-contaminati with each other subunit. As in the case of the ovine LH sub-units (Papkoff et al., 1973) further purification would be expected to diminish this activity. Upon recombination of PMSG-α and PMSG-β in a 1 : 1 ratio (not optimal) it is seen that significant amounts of both LH activity (27.4%) and FSH activity (33.6%) are regenerated. Interestingly the restoration of each activity is comparable. In addition, Table 5 shows that PMSG-α can be combined with either ovine LH-β , ovine FSH-β , or HCG-β with regeneration of the activity expected of the beta subunit employed. Thus, as in the case of the other glycoprotein hormones, one of the subunits (PMSG-β) is hormonally specific, and one (PMSG-α) is exchangeable and probably very similar to other gonadotrophin alphas. For reasons not yet understood, PMSG-β , while able to complex with PMSG-α , does not readily recombine with the alpha subunits of other gonadotrophins. Perhaps the most important observation in these studies is the fact that both LH and FSH activity are lost and regained in a comparable manner which is in keeping with both of these activities being intrinsic to the same molecule.

TABLE 5

SUMMARY OF LH AND FSH ACTIVITY* OF PMSG SUBUNITS AND SUBUNIT
RECOMBINATIONS**

Preparation	LH Potency (%)	FSH Potency (%)
PMSG	100	100
PMSG-α	3.6	2.5
PMSG-β	6.2	5.0
PMSG-α + PMSG-β	27.4	33.6
PMSG-α + oLH-β	20.6	---
PMSG-α + oFSH-β	---	16.0
PMSG-α + HCG-β	33.2	---

*   Assayed by the OAAD test for LH and the HCG augmentation test for
     FSH; potencies expressed in terms of PMSG (15 000 iu/mg) taken as 100%.

** Recombinations of PMSG-β with oLH-α, oFSH-α, and HCG-α were not
     significantly active.

Endometrial cup gonadotrophin (PMEG)

It has been firmly established (Allen and Moor, 1972) that
PMSG is produced by foetal trophoblast cells which give rise
to the endometrial cups in the uterus of the pregnant mare. It
was mentioned earlier that the purification scheme we have
described for serum can be employed with minor modification to
obtain highly purified preparations of the gonadotrophin present
in the endometrial cups. A series of endometrial cup preparations
have been fractionated and some of the properties of the PMEG
preparations obtained have been determined. These materials
appear to be somewhat less homogeneous than PMSG preparations
and there seems to be a greater degree of microheterogeneity as
evidenced by end group analysis and disc electrophoresis. In
all cases (four different preparations), the PMEGs had both LH
and FSH activity, but significantly lower than that observed
with PMSG. LH activity ranged from 23 - 54% of PMSG and FSH
activity from 11 - 41%. Of further interest is the indication

that the ratio of LH : FSH differed from preparation to
preparation and from PMSG, varying from 1.3 - 2.2. The amino
acid composition of the PMEGs are similar one to another and
to PMSG, although differences are not ruled out. Also, they
fully cross-react with antiserum against PMSG. The notable
difference between the PMEGs and PMSG is in the content of
carbohydrate. This varied for each preparation and ranged
from 12.7% to 30% compared to 41% for PMSG. These observations
suggest that the carbohydrate is very important for the full
manifestation of biological potency. In addition, the suggestion
is made that the ratio of LH : FSH activity can vary dependent
on carbohydrate composition. Stewart et al. (1976), employing
specific receptor assays for LH and FSH, have also suggested
that the ratio of activities of PMEG may vary from PMSG.

ACKNOWLEDGEMENTS

Much of the work reported in this paper is a summarisation
of studies performed in recent years in my laboratory and
entailed the collaboration of a number of colleagues. In
particular, I would like to thank and note the encouragement
of Prof. C.H. Li, and the participation of Drs. D. Schams,
T.A. Bewley, J. Ramachandran, and S.W. Farmer. The work on
PMSG has been supported in part by a grant from the National
Institute of Child Health and Human Development, National
Institutes of Health (HD-05722) and a Rockefeller Center Grant.

REFERENCES

Allen, W.R. and Moor, R.M. 1972. The origin of equine and endometrial cups. I. Production of PMSG by foetal trophoblast cells. J. of Reprod. and Fertl. 29, 313-316.

Bewley, T.A., Sairam, M.R. and Li, C.H. 1972. Circular dichroism of ovine interstitial cell stimulating hormone and its subunits. Biochemistry 11, 932-936.

Catchpole, H.R., Cole, H.H., and Pearson, P.G. 1935. Studies on the rate of disappearance and fate of mare gonadotropic hormone following intravenous injection. Am. J. Physiol. 112, 21-26.

Cole, H.H. and Hart, G.H. 1930. The potency of blood serum of mares in progressive stages of pregnancy in effecting the sexual maturity of the immature rat. Am. J. Physiol. 93, 57-68.

Dufau, M.I., Pock, R., Neubauer, A. and Catt, K.J. 1976. In vitro bioassay of LH in human serum: The rat interstitial cell testosterone (RICT) assay. J. Clin. Endocrinol. Metab. 42, 958-969.

Farmer, S.W., Suyama, A. and Papkoff, H. 1977. Effect of diverse mammalian and non-mammalian gonadotrophins on isolated Leydig cells. Gen. Comp. Endocrinol. In press.

Gospodarowicz, D. 1972. Purification and physicochemical properties of the pregnant mare serum gonadotrophin (PMSG). Endocrinology 91, 101-106.

Gospodarowicz, D. and Papkoff, H. 1967. A simple method for the isolation of pregnant mare serum gonadotrophin. Endocrinology 80, 699-702.

Landefeld, T.D., Grimek, H.J. and McShan, W.H. 1972. End group and carbohydrate analysis of equine LH. Biochem. Biophys. Res. Commun. 46, 463-469.

Landefeld, T.D. and McShan, W.H. 1974[1] Isolation and characterisation of subunits from equine pituitary follicle-stimulating hormone. J. Biol. Chem. 249, 3527-3531.

Landefeld, T.D. and McShan, W.H. 1974[2] Equine luteinising hormone and its subunits. Isolation and physicochemical properties. Biochemistry 13, 1389-1393.

Legault-Demare, J. and Clauser, H. 1961. Gonadotrophine serique de jument gravide (PMSG). in 'Etudes d'endocrinologie', pp. 69-91, Ed. R. Courrier. Paris: Hermann.

Papkoff, H. 1974. Chemical and biological properties of the subunits of pregnant mare serum gonadotrophin. Biochem. Biophys. Res. Commun. 58, 397-404.

Papkoff, H., Sairam, M.R., Farmer, S.W. and Li, C.H. 1973. Studies on
the structure and function of interstitial cell-stimulating hormone.
Rec. Prog. Horm. Res. 29, 563-590.

Sairam, M.R. and Papkoff, H. 1974. Chemistry of pituitary gonadotrophins.
in "Handbook of Physiology", pp. 111-131, Ed. E. Knobil and W.H.
Sawyer, Washington, D.C. : Am. Physiol. Soc.

Schams, D. and Papkoff, H. 1972. Chemical and immunological studies on
pregnant mare serum gonadotrophin. Biochem. Biophys. Acta 263,
139-148.

Stewart, F., Allen, W.R., and Moor, R.M. 1976. Pregnant mare serum
gonadotrophin: Ratio of follicle-stimulating hormone and luteinising
hormone activities measured by radioreceptor assay. J. Endocrinology
71, 371-382.

THE PRODUCTION AND STANDARDISATION OF PMSG

E.J. Passeron

ELEA, Saladillo 2452.   1440 Buenos Aires, Argentina.

ABSTRACT

Pregnancy is detected immunologically by routine blood sampling of breeding mares. Individual PMSG levels and packed cell volumes are monitored while bleeding is in progress. Pooled plasma is processed by acetone fractional precipitation giving a final sterile product of 2 000 to 3 000 IU/mg potency as determined by bioassay. The finished product is stable even when subjected to temperatures much higher than those usually encountered during shelf life. Shelf life at room temperature is well over five years. Small batches of PMSG further purified by liquid chromatography to a potency of 10 000 to 12 000 IU/mg are used as antigen in antiserum production. When purified PMSG and endometrial cup gonadotrophin are assayed by a complement fixation test consecutively, using antisera raised against one and the other of the hormones, the response curves obtained are parallel.

INTRODUCTION

ELEA laboratories has been producing PMSG commercially since 1939, the year the Company began its operations. During these years we have relied heavily on the work of many scientists, some of whom we are fortunate enough to have as speakers at this meeting. In what follows I shall try to outline the method we use for obtaining pregnant mares' serum, its processing to commercial and higher purity, the controls to which it is subjected and our facilities for processing and testing. I will also describe a serological comparison between PMSG and the hormone from endometrial cups.

BREEDING THE MARES

Our production process starts in the early spring, sometimes in the last weeks of winter. According to production schedules from 3 000 to 12 000 mares are mated. Stallions performing satisfactorily are left on the breeding fields. Otherwise they are replaced. The criterion is established by following the number of pregnancies due to each stallion. Our average mare is an animal of four to twelve years of age. Each mare is examined clinically and gynaecologically twice every year and each receives an annual vaccination against strangles, equine abortion and encephalomyelitis* as well as an annual treatment for intestinal parasites. They feed on natural grasslands and artificial pastures in the spring and autumn. In winter and summer they are complementarily fed oats, corn grain and sorghum as necessary. They are kept in lots of about 100 mares to every 300 acres. Stallions are approximately matched to the average size of the mares in the lot. Stallions failing to form their own herd under these conditions are replaced.

SCREENING OF PREGNANCIES

Two months after the beginning of the mating each lot of mares is run through the chute. A sample of blood is collected

with a sterilised needle. The labelled samples are analysed
at our serological laboratory by a complement fixation test
detecting 5 IU/ml of blood. Positive samples are recorded, the
data incorporated into an individual donor mare file card and
the identity numbers of the mares radiotransmitted to the breeding fields. The donor mares are screened out.

BLEEDING

Blood is collected with a sterile trochar from the jugular
vein after careful shaving of the neck. This is preceded when
necessary by local anaesthesia. A labelled sample of each
bleeding is retained. Packed cell volume and PMSG level are
used to monitor the bleeding. The measurements are written
into the individual donor mare file card.

PLASMA PRODUCTION CONTROL

The progress in the collection of plasma is carefully
followed. The decision to continue or discontinue bleeding
is taken on the basis of PMSG level evolution (Cole and Hart,
1942; Allen, 1969) and red cell packed volume. The efficiency
of the plasma collection process as a whole is monitored by
comparing among other variables, the amount of PMSG actually
obtained against the amount expected.

The data are combined in such a way as to obtain efficiency
ratios which signal the changes necessary to meet production
demands.

Careful observation of the stallions on the field comple-
ments the statistics kept on each animal.

Pooled plasma is stored in a cold room at the site of
collection, centrifuged and then sent to the processing facilities.

On arrival each pool of plasma is stored, sampled and

weighed; PMSG concentration is assayed serologically and then biologically and the total amount of units received for processing at the plant is recorded. The plasma is again centrifuged and its processing begins.

PLASMA PROCESSING

The method of Cartland and Nelson (1937) as modified by Goss and Cole (1940) is used to obtain a crude assaying 50 to 100 IU/mg. This method as is well known, is based on the acetone precipitation of inactive protein at a pH of approximately 9 by adding acetone to the serum so that a final concentration of 50% is obtained.

The precipitate is allowed to settle overnight and the slurry is passed through a filter press.

An interesting observation was made by Rimington and Rowlands (1944) on the correlation between PMSG serum concentration and PMSG recovery when precipitating inert protein with metaphosphoric acid. According to these authors richer serum yields less recovery of PMSG, so they recommend dilution of the serum with water in equal volumes. We have verified this observation during our own production operations.

The cake is dried, pelleted and used as animal feed. The clear liquid is acidified and pumped to conical bottomed tanks where it is allowed to stand overnight. A small quantity of inactive precipitate separates at this stage and is centrifuged off after having decanted the greater part of the liquid to a contiguous battery of conical bottomed tanks where the acetone concentration is taken up to 70%.

After coagulation and settling, the clear supernatant is syphoned off and passed through a centrifuge which collects the smaller active particles. The greater part of the activity now in the bottom cone is collected in a smaller centrifuge. All acetone solutions are fractionally distilled and the recovered acetone recycled.

FURTHER PURIFICATION OF THE CRUDE

The crude is repeatedly washed and centrifuged. The remaining active precipitate is extracted with aqueous ethanol by turbine mixing and left in the cold. Depyrogenisation by calcium phosphate co-precipitation if necessary and dialysis follow. From this point on, the following operations are performed under a laminar flow hood; sterile filtration, precipitation and washings. The active residue is washed with acetone, and peroxide free ether. It is dried under a vacuum over calcium chloride and phosphorus pentoxide or freeze-dried according to batch size. A sample is taken for sterility, pyrogen testing and biological assay. The purity ranges from 1 500 to 3000 IU/mg.

TESTING THE PRODUCT

The potency of the powder is established by bioassay. We use the ovarian weight method. For 2 x 2 assays ten litters, each consisting of 4 immature female rats from our animal house are selected. Doses are assigned at random to each animal in the litter. Rats with the same assigned dose of each preparation being tested are put into the same cage. Each animal is assigned a number using a colour and earmark code. After all the rats have been sorted into their cages doses are injected. Each dose is subdivided into three equal injections given at 24 h intervals. After the third injection, that is after completion of the dose, the animals are allowed to rest with food and water ad libitum for 24 h. They are then killed with ether in the same order as they were injected.

Five technicians operate simultaneously usually on 100 to 300 rats. One technician is in charge of the ether and the abdominal incision. Three or four carry out ovariectomy controlling the number of the operated rat against the number on the slide glass on which the ovaries are deposited. Another technician ensures that the ovaries have been adequately separated from fat or other adhering tissue and covers the glands with saline moistened cotton while they await weighing.

Weighing of the ovaries is performed by pairs to the nearest 1/10 mg on a torsion balance. The heaviest pairs weigh around 300 mg. The weighings are recorded by another operator. The whole group is under professional surveillance.

When more than one sample is assayed, which is the usual situation, none of the operators know the correspondence between sample numbers and sample identity until the end of the assay. Usually for definitive potency estimates three 3 x 3 point assays are carried out. Assays are subjected to the usual variance analysis (Finney, 1952) with linearity and parallelism test of regression lines and homogeneity test of assay results for pooling. Precision indices of 0.1 and confidence intervals of approximately $\pm$ 10% to 20% are consistently attained. The standard used is either the Second International Standard for Serum Gonadotrophin or the National PMSG Standard. The latter has been compared against the Second International Standard by the National Institute for Food and Drug Control on a collaborative basis with ELEA and authoritative foreign laboratories experienced in bioassay. It contains 4 000 IU per ampoule (Toth, Rosenkrantz and Glanczspigel, 1971).

ANIMAL HOUSE

Due to the importance of bioassay in the control of PMSG production and final quality standardisation, a brief description of our animal house seems appropriate.

The total rat population is around 10 000. The animal house is separated from our main building and it is subjected to a 12 h light circadian cycle. The rat strain has been inbred for eight years. There are five breeding rooms which contain

approximately 150 cages each. Each cage houses a male and a
female rat which have been specifically selected for breeding
from second-on litters of the same breeding room.

Some characteristics of our animals are shown in Table 1.

TABLE 1

RAT CHARACTERISTICS

Litter	11	Vaginal opening	35 days
Retained	8	Testicular descent	35 days
Weaned at	21 days	Sexual maturity	55 days
Weight at weaning	35 g	Mating	120 days

The average production of the animal house is around 1 200
rats per week. Most of these are used as weanlings for PMSG,
HCG or HMG bioassays but a certain number of males are kept
until they reach a weight of 150 - 200 g for hypophysectomy in
ACTH bioassay.

The animal house also includes facilities for pyrogen and
safety tests. Guinea pigs are also raised, mainly for the
production of complement.

FURTHER CONTROL ON THE FINISHED PRODUCT

The stability of freeze-dried PMSG preparations has been
studied by us together with the National Institute for Food and
Drug Control during the preparation of the National PMSG Standard. The product is very stable even at temperatures well above
normal room temperature. The shelf life is therefore very long.
Our experience on some of our commercial batches is that potency
is maintained for at least five years. This is illustrated by
the data in Table 2 (Toth, 1977).

TABLE 2

PMSG STABILITY DATA POTENCY AND LIMITS IN IU/MG

Lot No	Assay date	Stored at	Potency	Limits p = 0.05
A70p18B	Mar 30 1972	Room temp	2 293	2 179-2 412
Same	Apr 14 1977	Room temp	2 184	1 993-2 408

HIGHER PURITY PMSG

This is produced from a 200 - 600 IU/mg starting material
using the methods of Gospodarowicz and Papkoff (1967) and Schams
and Papkoff (1972). A final potency of 10 000 to 12 000 IU/mg
is consistently obtained. This material is freeze-dried in
ampoules and used by us to obtain immune rabbit antiserum for
our mare pregnancy test.

PMEG

From the endometrial cup secretion we extract an active
crude assaying around 400 IU/mg. We follow this with DEAE
Cellulose and SP-Sephadex C-50 chromatography. During this
purification it is interesting to observe that if the process
is followed using an immunoassay with a PMSG (serum hormone)
antiserum and a bioassay (ovarian weight) using a PMSG standard
in both cases, the ratio of serological to biological activity
decreases as the purification proceeds. At first the ration
has a value of about 4; when final purity is reached, the value
has fallen to about 1.2. The greater decrease is observed in
the DEAE Cellulose chromatography step. (Isler and Passeron,
1975). This serological to biological activity ratio decrease
is not observed when monitoring PMSG purification in the same
way.

Antiserum against purified endometrial cup gonadotrophin
has been used by us in conjunction with anti-PMSG serum to study

cross-reactivity by immunoelectrophoresis, gel immunoprecipit-
ation and complement fixation techniques on the actual material
that we could make commercially available. This matter has
already been thoroughly studied in the paper by Papkoff and
Schams (1971) proving the similarity of both hormones. In our
hands both preparations have shown qualitatively similar properties.

TABLE 3

DILUTION OF ANTIGEN

Antigen	Antiserum	at 50% haemolysis	Slopes
PMSG	PMSG	1/33	4.54
	PMEG	1/34	4.50
PMEG	PMSG	1/134	4.67
	PMEG	1/151	4.23

These are well illustrated by the results of a complement
fixation assay (Costantini and Passeron, 1975) which show that
the same estimate of the concentration of a 100 IU/ml solution
of PMSG is obtained whether the antiserum used be anti-PMSG or
anti-endometrial cup gonadotrophin; furthermore both response
curves are parallel. The same is observed when a solution of
100 IU/ml of endometrial cup gonadotrophin is assayed against
the same two antisera. Again the two response regression lines
are parallel, and they are also parallel to the regression
lines obtained in the PMSG assay, as may be observed in Table
3.

REFERENCES

Allen, W.R. 1969. 'Factors influencing pregnant mare serum production'.
    Nature, Vol. 223, July 5, 1969.

Cartland, G.F. and Nelson, J.W. 1937. 'Preparation and purification of
    extracts containing PMSG'. J. Biol. Chem. 119, 59.

Cole, H.H. and Hart, G.H. 1942. J. Am. Vet. Med. Ass. 101, 124.

Costantini, H. and Passeron, E.J. 1975. ELEA unpublished data.

Finney, D.J. 1952. 'Statistical method in biological assay' Hafner
    Publishing Co. New York.

Gospodarowicz, D. and Papkoff, H. 1967. 'A simple method for the isolation
    of PMSG'. Endocrinology 80, 699.

Isler, G. and Passeron, E.J. 1975. ELEA unpublished work.

Papkoff, H. and Schams, D. 1971. 'Comparison of the properties of pregnant
    mare serum (PMSG) and endometrial cup (PMEG) gonadotrophins' 53rd
    Meeting, The Endocrinol. Soc., S. Fco. June 24-26.

Rimington, C. and Rowlands, I.W. 1941, 1944. 'Serum gonadotrophin' Bioch.
    J. 35, 736. 1941. Bioch. J. 38, 54, 1944.

Schams, D. and Papkoff, H. 1972. 'Chemical and immunochemical studies on
    pregnant mare serum gonadotrophin' Bioch. Bioph. Acta 236, 139.

Toth, A. 1977. ELEA unpublished results.

Toth, A., Rosenkranz, A. and Glanczpigel, R. 1971. 'PMSG-Preliminary Biolog-
    ical assays for the establishment of a national PMSG standard' 3rd
    Argentine Congress for Endocrinology and Metabolism Dec. 9. 1971.

DISCUSSION

S. Willadsen *(UK)*

Dr. Ryan, you mentioned that when you are superovulating animals, you are probably rescuing follicles and that you do not recruit anything from the population of follicles which is coming up. You also said that this was so because one only treated the animals during a very short part of just one cycle. What would you propose then to increase the recruitment in earlier cycles, for superovulation later on?

R.J. Ryan *(USA)*

I would suggest trying to do the following: perhaps use oestrogen, FSH and not PMSG for cycle one, cycle two, cycle three, and then PMSG in cycle four. It seems to me very possible that one of the effects of the mid-cycle surge of FSH is that it starts recruiting, or maintaining, some of those smaller follicles. Unfortunately, we have very little information about the hormone responses or metabolic characteristics of the preantral follicle. We have even less information, I think, about what happens in atresia. We desperately need information on both of these. The reason I have suggested not using PMSG is that it has a lot of HCG or LH-like activity. I think the latter can be detrimental to follicles.

J.P. Gosling *(Ireland)*

The granulosa cells increase in their responsiveness to LH and HCG, and you show an increase in the responsiveness of the cyclase. Is there any increase in steroid secretion?

R.J. Ryan

Oh yes, progesterone in particular, and in whole follicles, oestrogen as well. In the isolated granulosa cell you do get an increase in secretion of progesterone, but you get very little progesterone from granulosa cells in small follicles.

J.C. Mariana *(France)*

To which precise stage of follicle growth do you limit the action of the gonadotrophin hormone?

R.J. Ryan

I think there are two stages. Firstly, I think follicle stimulating hormone has its action primarily on the preantral granulosa cell and certainly the induction of the formation of the antrum and the secretion of follicular fluid are clearly FSH mediated responses. In the large granulosa cell and luteal cell, LH is predominantly the determinant action. Now there are data that suggest that between the small antral follicle granulosa cell and the large antral follicle granulosa cell, the FSH also has an effect in terms of induction of an aromatising enzyme. This is the cell hypothesis for oestrogen production. But the theca cell outside the follicle is responsive to LH for production of androgen.

D. Schams (West Germany)

Thank you for those questions; may we now turn to Dr. Baker's paper.

S. Willadsen

I think the last experiment you mentioned is very interesting, but I think one has to be even more careful than you were. The reason I say this is that we are doing some experiments with sheep follicles which were harvested at various times after induced LH surge. The oocytes were dissected out, transferred into an oestrous ewe, which was then inseminated. The oocytes, or the embryos, were then recovered on Day 7 and it was quite clear that you had to leave the oocytes within the follicle for at least 12 hours after LH, otherwise you did not get any normal development. After some 12 hours we got 50% normal development, and by 18 hours, which is, of course, about 6 - 8 hours before ovulation would have occurred, completely normal development.

Chris Polge then suggested delaying insemination with these animals. We looked at sperm penetration in these oocytes and they were all polyspermic - all the ones that didn't develop; we postponed insemination in the recipient animal and have found that the animals which have had oocytes transplanted from follicles 6 hours after HCG, and only inseminated 8 hours after transfer of the oocyte, would then have a proportion of normal embryos. What I am getting at is - did you ever try to delay insemination in these animals? Presumably you inseminated at about the time of ovulation.

The second thing I am interested in is, did these oocytes, once they were released in the oviduct, go on, and, if they did not, why do you think they did not?

T.G. Baker *UK)*

Certainly, we inseminated the animals straight away but we were just under the impression that this process released eggs which were immature.

S. Willadsen

Did you characterise their incompetence in any other way?

T.G. Baker

No, we were interested in the follicle.

S. Willadsen

But wouldn't you think it strange that those oocytes would not be able to complete meiosis if you had kept them in an oviduct?

T.G. Baker

All I can say is that we haven't done that. I don't know what would happen to them. It's a good point! I can perhaps tell you in a roundabout fashion that is, if you remove the eggs from mouse follicles of different size, then you find that oocytes from preantral follicles, or very early antral follicles, will block at the GD stage for a very long time - up to 60 or 70 hours. We never, in fact, found an oocyte go beyond that. However, if you take them from medium size antral follicles they reach meta-phase 1 and then they block. If you let the follicle go to full size, then they can reach metaphase 2. It is as if, in fact, the oocyte is reading in some way what is happening in the follicle, that there are maturation events that are affecting other components.

S. Willadsen

But you think your oocytes were really derived from follicles which wouldn't have ovulated normally, that they were
immature follicles?

T.G. Baker

They were large follicles, but they were immature along
the lines which I have demonstrated.

A.J. Breeuwsma *(Netherlands)*

I would like to comment in support of your findings. We
have injected sows at Day 17 of the cycle, with a slightly higher
dose; simultaneous injection of HCG and PMSG and achieved up to 70
to 80 ovulations. We only inseminated the sows in oestrus and
then we achieved full blastocyst development. When we opened
up the uterus at Day 19 after insemination we found 60 or 70
fully developed blastocysts.

T.G. Baker

What doses were you using?

A.J. Breeuwsma

Up to 1 200 iu PMSG and up to 800 iu HCG, simultaneously.

T.G. Baker

It may well be that you pushed the oestrogen levels up
quite a bit; I think that is probably the way it works.

P. Mauleon *(France)*

Do you think that the reserve of primordial follicles is
an homogeneous pool?

T.G. Baker

We have no evidence to the contrary at the moment. Are
you thinking of Henderson and Edwards' original ideas, or what?
The idea of Henderson and Edwards was that there was a production
line of primordial follicles; those that were formed first would
be the first ones ovulated, those that were formed last would be

ovulated later. This may well be a reasonable hypothesis for
species like the cow or the human with very long gestation periods, but I would like anyone to explain to me how it could work
with the rat where you have 98% synchrony of germ cells and a
period of oogenesis lasting six days. They have got to be very
sharp germ cells to show any programming at all, except in short
periods. So, in my view, there is no evidence and one must
assume that there are other factors at work. We just don't know
how a primordial follicle is selected anyway to proceed through
follicular development. So I don't see a selection mechanism.

I. Wilmut *(UK)*

Could I follow on from that by asking you about this data
where I think a constant percentage of oocytes begins to develop
and move along. I think you referred to it in your paper. Could
you comment on that? You seem to me to be implying that there
might be some positive input to this system coming from the primordial follicles.

T.G. Baker

Frankly, I was side-stepping the issue. We don't know.
Again, whether we can extrapolate from that mouse data to higher
species, would be a worrying factor to me. The observations
showed that the number of follicles in each class appeared to be
relatively constant. However, that is in a species where follicles don't really start to grow until a week or two after birth.
One has to remember that in the human ovary, long before birth
you can find follicles, and the process seems to be relatively
jumbled up thus I do not know whether there is a controlling
mechanism in that way.

I. Wilmut

You mentioned the data in the human, where there are still
primordial follicles after the menopause. One thought that
occurred to me when you were talking about that was whether you
have a certain population of primordial follicles to support the
development up to the next stage?

T.G. Baker

This is what is speculated. A lot of people have said
this - again, with very little evidence. The story for the human
is that probably, when the number of primordial follicles reaches
a certain low number, it cannot sustain further follicular growth
so you don't get graafian follicles and therefore your feedback
system is broken and you get high gonadotrophin to try to com-
pensate. You then can go through a period where you might get
an ovulation occasionally. In other words, sometimes the gonadotrophin is effective in giving an ovulation but normally the
process stops very quickly. Whether this is really related to
the number of primordial follicles, or whether it is related to
associated interstitial tissue, or something like this, we just
don't know. It is all speculation. There is a lot of that in
the literature but there is very little hard data.

D. Schams

We must now pass on to the next subject: the paper by
Dr. Allen.

J.C. Mariana

Do you think that there is an androgen regulation of PMS
secretion during the pregnancy?

W.R. Allen *(UK)*

No, I don't think there is although we have no evidence
for or against any form of androgen control but the fact that
the girdle cells will grow so well and be so productive in culture
when you are adding no exogenous hormones to them indicates,
to me at least, that there is no particular hormonal control
with PMSG.

H. Papkoff *(USA)*

Perhaps this is a naive question but has anybody ever
tried to superovulate mares using PMSG?

W.R. Allen

Yes, I have, and many other have tried too, with a total

lack of success. I have personally given injections of 100 000
units of PMSG on three occasions, which was becoming an expen-
sive pastime, and I got one follicle - a big follicle - but only
one! There seems to be some block in the ovary. We have tried
many gonadotrophic preparations, purified and unpurified, from
all sorts of species - sheep, cows, pigs - you name it - but
with a total lack of success. There is a group in Wisconsin now
which, by giving daily injections of a crude extract of mare
pituitaries for no less than 14 days, can get ovulations approach-
ing two, three and sometimes four but still nothing like the
other species and one must be left with the feeling that the
horse's ovary is very resistant at least to its own gonadotrophins,
and I think to all gonadotrophins.

L.E. Rowson *(UK)*

It isn't really very surprising that you don't get responses to PMS when you think of the level of something like 100
units per ml in the circulating blood; you would have to give
an absolutely massive dose to get a response.

W.R. Allen

Yes, in terms of the fact that we only get one, two or
three secondary ovulations from these levels. I think from
the case of the donkey carrying the hinny there is obviously a
very necessary block that the mare must have because if she were
to respond to her own PMS, like the donkey apparently responds
to the hinny's PMS, she would rupture the ovary.

L.E. Rowson

In your donkey foetuses where you get an extremely low
FSH level, you talk about the function of the secondary corpora
lutea to maintain pregnancy. In these animals where you get an
extremely low FSH level, do you, in fact, get more abortions?

W.R. Allen

No. I think the general view is that mares will give a
higher foaling rate carrying mules than they will carrying normal horses, which goes against all forms of immunological

argument, but I think it is a fact. As I showed in my slide, no matter what the PMS level as such, the progesterone level is about the same in these animals - it doesn't correlate in any way with the number of secondary corpora lutea. Again, donkeys, carrying donkey foetuses, have a very low FSH or PMS, but they have about the same amount of secondary corpora lutea. I have it in my mind that it is the pituitary FSH which governs how many follicles will develop.

H. Hardie *(Belgium)*

Have you any theories about the actual function of PMS?

W.R. Allen

I have lots of ideas but nothing to support them at all; I tend to like to think of it as some form of protective agent in the mare. There is now a lot of work being directed towards trying to show that perhaps HCG in the human, and perhaps the other gonadotrophins which are now being discovered in cows and sheep and other species, that the principal function of these is somehow (and we have no idea how, I don't think) to act in the immunology of pregnancy.

H. Hardie

But the pregnancy went on quite happily in the case of the donkey foetuses?

W.R. Allen

Well, under very low levels, certainly it does, and, as I mentioned, in these mares we have a second concurrent mule pregnancy and we could not detect PMS and they carried on to term with no trouble.

S. Willadsen

Do you think there is a direct relationship between the secretion of PMSG by the cells and the actual concentration you get in the blood? In other words, could you have a local effect of PMSG in the donkey situation which you didn't pick up on serum levels?

W.R. Allen

I suppose it is not beyond the bounds of possibility but I can't envisage how it would work, it's secreted directly into the blood stream. It is collected from the endometrial cup.

S. Willadsen

And there is no indication of an immune response by the mare to the PMSG?

W.R. Allen

Well, we see this localised immune response; we are now getting some results that indicate that that is a peripheral response, that she is certainly producing antibodies very soon after 40 days. Whether that is to the PMS as a molecule, or whether it is to antigens on the cell surface of the cells that are producing the PMS, we don't know.

D. Schams

Perhaps we could have a question or two to Professor Papkoff.

R.J. Ryan

Is there any indication that the mobility of the FSH portion of the molecule is different from the mobility of the LH portion of the molecule? Can you oxidise, or reduce, or modify in one way or another, to destroy FSH but not LH?

H. Papkoff *(USA)*

We haven't studied this extensively although it is something we want to do. When you do something drastic to the molecule such as oxidise with performic acid, you will cleave all the disulphide bonds. We certainly destroyed both activities. If we treat with neuraminidase, as I mentioned earlier, we certainly diminish the activity using in vivo assays; we don't do anything to the LH activity in vitro but there seems to be some sort of change in the type of response we obtained in what we think was an FSH activity. So, the possibility exists, I think, that we may be able to alter the two activities chemically if we

know the right reagents to use and the right troops to get
involved.

R.J. Ryan

The reason I ask is that we have very preliminary data
where we used Professor Papkoff's PMSG. If you label it with
125_I and use the standard radiolabelling technique, what you
find is that you don't get any binding with small granulosa cells
but the same batch of label binds magnificently to granulosa
cells from a large follicle, which suggests to us that either
there are no FSH receptors in the small follicle or that there
is some differential effect on the PMSG molecule by the iodination.

H. Papkoff

That's really quite interesting. I must admit that in
the studies we have been doing on the chemical properties, I am
more and more impressed by how similar PMSG is to an LH rather
than an FSH. This is primarily with respect to the peptide portions and so on. It may come to a point where the carbohydrate
is perhaps the influence which determines whether a given structure predominates in FSH or not.

J. Saumande

Do you know the activity of alpha and beta sub-units for
PMS in the in vitro bioassay for FSH and LH?

H. Papkoff

That's on a slide here; it was around 3 - 5% but I would
say that this probably represents cross-contamination. I think
our experience now with the pituitary glycoprotein hormones is
that the individual sub-units do not seem to have any significant activity themselves. The biological activity is really a
function of both of the sub-units.

J. Saumande

But what about the in vitro bioassay?

H. Papkoff

Well, the limited number of in vitro assays that we have done suggest that it is also very, very low. I don't think there is any intrinsic activity in any of the sub-units at least in the systems we have studied.

W.R. Allen

You have already mentioned adding neuraminidase. Is it possible to do a form of control desialation of the molecule, and would this reflect on FSH and LH activity? Why I ask is particularly because of your final comment on this very nice difference in ratio, or activity of endometrial cup secretions and when you consider what endometrial cup secretion is, this PMS is incubating in a lovely system and, I should think, presumably being degraded to a certain extent; I would have thought that PMS recovered at, say, 70 days from endometrial cup secretion would be very different from that recovered at up to 30 days where it has had about 50 days to sit around and be metabolised or acted upon.

H. Papkoff

I am very interested in doing the experiment which you suggest, to try to get a graded series of preparations with varying amounts of sialic acid. We tried this unsuccessfully, because it is extremely difficult to inactivate neuraminidase. What we are now trying to do is to prepare an antisera against the neuraminidase which we can throw in at appropriate times and stop the reaction.

D. Schams

I guess we have to stop here and move on to Dr. Passeron's paper.

K. Betteridge *(Canada)*

What is the youngest age of mare that you breed?

E.J. Passeron *(Argentina)*

Our mares are aged between 4 and 12 years.

W.R. Allen

    I was fascinated by the problems you must have in managing this number of mares. I was particularly interested in your finding about the lower PMSG production the later in the year the mare conceives. How accurate is your diagnosis of the stage of pregnancy when the blood is taken?

E.J. Passeron

    Well, our complement fixation test is detecting between 2 and 5 iu per ml of blood.

W.R. Allen

    What I was getting at is that initially you put your stallions in and take mares out two months later. How frequently then are you testing the remaining mares? Could your low PMSG production be explained by leaving it too late in pregnancy in those mares?

E.J. Passeron

    It is confounded inextricably with the fact that time is passing and those mares may have mated in the previous month. I am not contending that that particular graph you are referring to may be the reflection of an actual physiological process; I am just showing it as a way of conveying to you the idea of the things you have to take into account.

D. Wishart *(UK)*

    I have used considerable quantities of PMS in sheep and cattle studies and, particularly as far as sheep are concerned, it is inconvenient to have small glass ampoules. I have been told that if PMS is put up in rubber-stoppered multi-dose vials, then the potency of FSH in particular, is likely to decline quite quickly. Is that true?

E.J. Passeron

    We haven't found that to be true. I must say that of the sheep trials I have shown, the smaller bottles, the ampoules you saw in the slide, are actually used 100% in Argentina for medicinal purposes. All of the material used for veterinary purposes is exported and that is exported in the dry powdered form. So, even if the potency were to decline, which I have not found, you would not have the problem because that is not the form in which the material is exported. Most of the material you are using would have been exported in the dry powdered form.

S. Willadsen

A purely practical question: when you have made up a dose of PMSG for injection, and there is something left over in your bottle, if you have been using saline, how long can you store PMSG which has already been dissolved and under what conditions should you store it?

E.J. Passeron

We have found that it is not convenient to store dissolved PMSG. I would say you shouldn't keep it after dissolving unless you keep it very well cooled.

S. Willadsen

At what temperature?

E.J. Passeron

You should put it in an ice box immediately. I wouldn't use it at the nominal potency for more than a week.

S. Willadsen

Could you store it for one or two days?

E.J. Passeron

Yes, I think you could, but I think it would depend very much on the final concentration that you had built it up to - the actual concentration in your injectable fluid.

S. Willadsen

Yes. If you were using a concentration of 250 iu per ml?

E.J. Passeron

I would not go over one week in the ice box with that solution.

S. Willadsen

But you wouldn't expect it to be damaged by storage for one day?

E.J. Passeron

No, I don't think it would.

D. Schams

Perhaps I could ask a short question. Do you have any data or is it more a feeling than experimental results that PMSG in solution is only stable for so short a time, because it is in contrast to other hormones? You can store very diluted LH in a freezer for a long time and it is still active.

E.J. Passeron

I think we do have data on that. I was referring to what happens when we are testing and we are going through a bioassay. We can, of course, keep it for a few days in the ice box with no problems. The question raised here was whether it can be stored and keep all its nominal potency - I personally would not guarantee that.

M.T. Kane *(Ireland)*

This is maybe a somewhat awkward question but is it true, in fact, that most of the PMSG that you export is sold to drug companies for resale under their own names? Secondly, is there much of a difference in price, do you know? Thirdly, would you be prepared to sell it direct to people who required very large quantities?

E.J. Passeron

This is indeed an awkward question! I think it is one on which I should not go into details just now. It is no secret that we do export most of our hormones and that this is bought

by the drug companies. As to the future form of commercial
organisation, well, fortunately, I am way out of that area - I
am sorry I can't answer your question.

W.R. Allen

I think perhaps I can add something to the question of
stability in solution; it is only by immunoassay, of course, not
by biological assay, but we have found no loss on immunoassay
in storing PMS at 100 units per ml and 500 units per ml at 4°,
20° and 37° over two days. At 37° we started to get a loss,
particularly noticeable in the solution of 100 units per ml, at
about four days.

Another point I would like to comment on is your very
interesting observation on the increased loss of recovery the
higher the potency was initially. Would you have any ideas why
this should be, biochemically or physiologically?

E.J. Passeron

No, personally, I don't. I don't know whether Professor
Papkoff has anything to say on the subject.

H. Papkoff

I really can't add to that. I probably have not had as
much experience with different factors. I don't understand it.

W.R. Allen

Didn't Cole in the 1930s find that if he extracted PMS in
any form to get powdered PMS and then made that up in saline for
injecting rats, as compared to dissolving it in non-pregnant
mare's serum, he got an enhanced response if it was dissolved in
the non-pregnant mare's serum?

E.J. Passeron

How can that be related to its stability?

W.R. Allen

If you took your purified (or unpurified) material and

dissolved it with water or saline to inject the rat you would
get a particular response with a certain weight of material. If
you dissolve that same material in serum from a non-pregnant mare
and inject it into the rat, you increase the ovarian response
in the rat, indicating, of course, that serum proteins, or something, have an enhancing effect which perhaps may explain the
apparent loss of activity, or loss of recovery in saline.

E.J. Passeron

Would that be compatible with the fact that you do actually
get high recovery if you add water to the serum? If you add an
equal volume of water so you dilute the serum to one half of its
initial concentration, the protection which other serum proteins
would give to PMSG would be apparently less under these conditions
I don't see the correlation with your observation and this fact.

W.R. Allen

I'm not sure I see it myself! What the question really is
that I am posing, is, when one measures a level of, let's say,
100 units per ml in serum, you finish up with what seems to be a
low percentage recovery. Is one's initial assay, at 100 units
per ml, not, in fact, overestimated? It never had, in fact, a
high concentration of PMSG molecules as such.

E.J. Passeron

I think there is very strong coincidence between our data
and yours ....

D. Schams

May I interrupt please. We are running short of time and
I think we had better move on to the general discussion. First
of all we will discuss the two papers that deal with the basic
principles of ovulation. The second topic is PMSG, the physiology, biochemistry and production.

On the first subject, I think there are two points which
are not very clear: one is that Dr. Ryan has shown a lot of data
for the rat and also some hormonal profiles but we have to realise

that while the basic principles may be the same, that at least
from the profiles, there seem to be some species differences,
we must realise that the rat is not a cow without horns; it is
a little different! Would you like to comment, Dr. Ryan?

R.J. Ryan

I would like to comment on one thing that Dr. Baker has
said. It may illustrate the difference.

There is a very interesting association between the follicle
and the corpus luteum in the human. He suggested that it would
be of interest to look for that in the cow. My hunch is that you
won't find that association in the cow and the reason is that in
the human the corpus luteum makes oestrogen; in the cow, it does
not. So, I think the things we talked about are general patterns
but for specificities I think you are going to have to look at
the cow. You are going to have to do those same experiments
again to define each one of these sets of circumstances to decide
what is applicable to your system.

D. Schams

There is the big difference that in the human there is only
one follicle growing during the cycle whilst in the cow there
are many follicles coming and going; the FSH shows a wave-like
pattern, not just one single peak. However, the second point
which I think is the key to our problem with superovulation is,
what is the selection basis of the primordial follicle to react
to stimulation or not?

T.G. Baker

There is no evidence that the primordial follicle responds
to any hormone except some recent comment that in the woman there
is some evidence that all follicular stages can respond, (they
don't necessarily respond but they can respond), to gonadotrophins.
I don't know what this is based on - it is certainly not true of
the species I have looked at.

R.J. Ryan

I think, as a generalisation, you can say that those follicles that are not surrounded by an organised theca will not respond to a hormone. If you have got an organised theca it will, and, by and large, you show up with LH receptors on those theca cells. The FSH receptors tend to be confined to the granulosa cells. The LH receptor doesn't show up on the granulosa cell until very late in this process, until it is a preovulatory follicle.

B. Hoffman *(West Germany)*

I would like to ask Dr. Ryan a question. You mentioned the role of oestrogens to induce receptor sensitivity, especially for FSH. Would you speculate that there is a more cyclic pattern of oestrogen release necessary or do you think that a basal level will do the job? If it is a basal level, where does it come from, especially when ovarian activity starts during puberty?

Another question is, could you see a role for the excessive amount of oestrogen produced during the latter half of gestation on ovarian activity in the foetus, especially in the cow where there is a lot of oestrogen produced?

R.J. Ryan

The point is well taken. Most of the experiments that have been done on showing oestrogen to produce FSH recpetors, have required pharmacologic doses of oestrogen. Now, the argument goes something like this. The reasons for using pharmacologic doses is that you have to achieve a particular concentration in the ovary because there is evidence that this effect of oestrogen is not mediated through the pituitary since it occurs in the hypophysectomised animal. There is further evidence that it is a local effect because if you increase the oestrogen in one ovary, that ovary develops more response to FSH than the contralateral ovary. Now, the idea is that you have to mimic what an intra-ovarian concentration is. Where it comes from is anybody's guess. I don't think there is any evidence that this particular growing follicle with its developing granulosa and its theca, is necessarily the source of the oestrogen for the induction of the FSH receptors. It could be another vesicular follicle from a previous generation. I don't know that just because you can demonstrate FSH and LH receptors in a particular growing follicle, it is necessarily proof that oestrogen is being produced in that follicle. So I am not very sure about the source of it. If somebody else knows I'd be delighted to hear it.

S. Willadsen

On the same subject, I noticed that Dr. Baker said that he thought oestrogen was of little significance in the in vitro culture system that is currently being used by Dr. Moore to mature oocytes. May I ask why you think this?

T.G. Baker

No, I wasn't suggesting that at all. What I was saying was that if you want to mature follicles in vitro, both from the pig and the sheep, you have to give FSH, LH and oestradiol, in the ratio of 2 : 1 : 1. Whether that oestradiol is actually required or whether it would have been made by the follicle, had the follicle been in vivo, is an unanswered question. You have got a very large follicle in a culture system; we don't know whether that follicle can make oestradiol in the amount that it would have done in vivo, and keep it within the follicular fluid. I was merely questioning where the oestrogen comes from and whether it is required.

J.P. Gosling

I am very interested in the finding that Dr. Ryan reported on the role of prolactin in the maintenance or, as he said, development of LH receptors in the corpus luteum: I take it that all of this work has been done in the rat. Has he any evidence on this role of prolactin in any other species?

R.J. Ryan

No, I haven't. The work is Joanne Richards' work from the University of Michigan. As far as I know it has only been

done with the rat. I would like to make a speculation though,
I think that prolactin might have a very subtle role in a variety
of species. We have recently obtained some data from examining
pig follicular fluid which indicates that the only lipoprotein
that is present in follicular fluid is high density lipoprotein.
This is of interest because it has been shown in other cell lines
that low density lipoprotein is the mechanism for cholesterol
transport. So we thought that since the follicular fluid lacks
this, the high density lipoprotein has to play this role for
cholesterol transport into the granulosa cell. It turns out that
there are highly specific, high activity, high density, lipoprotein receptors on granulosa cells, also on adrenal cells and
also on corpora luteal cells. Now, one of the known actions of
prolactin on most steroid secreting cells, is the increase in
cholesterol content, whether you can show an effect on progesterone or not is another matter. You can generally show an increase of cholesterol content of the organ and this does not
seem to be quite so species specific as it is in terms of a
leuteotrophic action. So I would like to suggest the possibility
that one of the things that prolactin does is to regulate not
only LH receptors in the rat but perhaps high density lipoprotein
receptors in a variety of other species. But that's pure speculation.

K. Betteridge

Moving on to Dr. Allen's paper, I would like to ask about
progesterone production by girdle cells. I believe you find this
in vitro and I wonder whether there is any significant progesterone production by the various cells in vivo. You said that when
you remove the embryo, progesterone levels came right down although the PMS levels stayed up. How significant is the production
of progesterone in vivo?

W.R. Allen

I can't really answer that I'm afraid. We do know that
the girdle cells secrete progesterone, as you suggested, for a
few days in culture, for eight days or so, but then that dies
off very quickly. However, we are not giving them any special

things to make them secrete progesterone. I imagine there is a
shortage of metabolites of some description, but I don't know. In
vivo, we do know that in the hinny endometrial cup, the ones that
produce the very high levels of progesterone, their girdle cells
secrete more progesterone. But then, of course, that is masked
by the fact that the blood contains high progesterone. In collaboration with Brian Heap we have done some injecting of labelled
progesterone, looking at half-lives, and we find that in the hinny
situation the great bulk of all progesterone is coming from the
anlagen. So, I would have liked to have ascribed a good physiological role for progesterone production by the cups but I can't.
It's minimal and we have never been able to measure it specifically from cups in vivo.

R.J. Ryan

I would just like to make a comment concerning the potential
immunosuppressibility with PMSG. Several years ago a man named
Papkoff described the phenomenon with HCG. It has subsequently
been shown, however, that the immunosuppressant of HCG is not due
to the HCG per se but is something contained in the crude HCG
preparation that we use. Highly purified HCG or its alpha/beta
sub-units are not immunosuppressant. There has, however, been a
report quite recently, from Japan, on a microbial protease inhibitor that blocks blast transformation of lymphocytes. Blast
transformation of lymphocytes is the immunosuppression assay
that is used to show immunosuppressant activity of HCG. Following this up, quite recently, in our laboratory, we have looked
at commercial HCG preparations and what we find is that protease
inhibitor is present in HCG and we are now in the process of
seeing if that accounts for the immunosuppression activity. If
that is the case, my hunch is that PMSG doesn't have immunosuppressant activity but maybe something in the preparation does.

W.R. Allen

Well, perhaps I can add to that. Certainly crude PMS does
have an immunosuppressing activity on blast transformation. We
haven't yet done the definitive thing of putting the purified
material in - I will have to talk to Professor Papkoff about that.

But I think that is a very interesting finding. Could you say again what it was?

R.J. Ryan

Well, it's something smaller than HCG and it appears to have protease inhibitory activity. It doesn't seem to be the usual kind of protease inhibitor that you find in human serum.

W.R. Allen

This is something that you think is being produced specifically in pregnancy and hasn't come out ....

R.J. Ryan

I can't say that for certain. It is something that you find in urinary extracts containing HCG.

D. Schams

I would like to stimulate another point of discussion. How about the ratios of PMSG, LH, FSH; if you look at the literature and you take the data presented today, earlier data report a higher LH activity and lower FSH activity and today it seems that it is about the same. Are there any special comments?

R. Newcomb (UK)

Could I make a comment in relation to the effects of different ratios on the superovulatory response in cattle? Dr. Allen did mention that we have looked at five batches of PMSG with FSH:LH ratios varying from 0.9 to 1.3 and we found that there was no difference in response in terms of ovulation rate, in cattle. Now, recently, with the support of Intervet, we took three batches of PMSG with FSH ratios varying from .9 to 1.2. We have done a very controlled study and we have found absolutely no difference whatsoever, attributable to batch, in numbers of ovulations or numbers of follicles.

J. Sreenan (Ireland)

Could I just add to what Ray Newcomb has said. We have also tested five discrete batches of PMSG. These were selected as different mare collections and not to be confused with packing batches. These five batches were tested in heifers and total ovarian response (ovulations plus large follicles), was the same for all five. One batch only had a lower ovulation rate.

D. Wishart

I wonder if I could ask the two previous contributors to the discussion whether they have noticed differences in the size of the ovaries in superovulated cattle - not so much the ovulation rate but the size of the ovaries, varying with the different batches of PMSG?

R. Newcomb

It is not something which we have specifically looked for I must admit and therefore I have no comment at all on it.

J. Sreenan

I cannot comment because we have only counted and measured follicles and corpora lutea. We do not measure ovarian size.

A.J. Breeuwsma

To come back on this FSH : LH ratio, I think we have looked at the same five batches which have been examined now for various reasons. We are using a bioassay and there we find a ratio of approximately 1 : 12, which is pretty constant between the various batches which, in fact, are mare batches and not packing batches. So I think the FSH : LH ratio test applied by Dr. Allen and the bioassay test are different although they are both constant.

D. Schams

We also have ratios of about 1 : 12 and 1 : 1.2. What is the real relationship? Is it due to the different assays used? Are there any special comments?

H. Papkoff

I really think that this is almost a problem of comparing apples and oranges. You can't! The closest approach I think if you want to express PMSG in terms of an FSH or an LH, should

be using the ascorbic acid depletion test and the Steelman -
Pohley test using the ovine standards. In fact, quite frequently
we observed lack of parallelism when we used these standards.
The data I presented here is from a few assays where we did get
parallelism. I am not sure how valid it really is: I can only
say that, as I see it, purified PMSG is comparable in activity
to a pituitary LH or a pituitary FSH. I think it is quite an
old story that different bioassays give different measures of
LH and FSH activity.

W.R. Allen

I agree, we find also that PMS gives a much better degree
of parallelism with human standards than with ovine standards.
I wonder, with regard to the difference between assay methods
in vivo and in vitro, whether Dr. Baker or Dr. Ryan could comment on the very long half-life of PMS and what effect this may
be having on follicular development, ovulation and ovarian weight,
when you are going to inject into a rat a hormone that is going
to hang around for thirty hours plus, as compared with a pituitary
gonadotrophin which presumably has a much shorter half-life?

R.J. Ryan

I think with a molecule that has such potent activity as
both LH and FSH, that what kind of response you are going to get
initially is going to depend on what kind of receptor this particular tissue has. If it happens to be an FSH responsive tissue,
you are going to get an FSH response initially. My hunch is,
from the rat data, that you will then induce LH receptors and if
it sticks around longer you are going to start getting an LH
response later. However, if you injected into a rat that has
predominantly LH receptors and no FSH receptors, you are going
to desensitise the LH system and get no response from the FSH.

T.G. Baker

There was a paper some years ago in which the authors documented what happened when you injected PMSG into mice. There
was an almost immediate ovulation in a lot of the animals. These
were follicles that were sitting around and which responded

effectively to an LH-like activity. Later on they got another response - many of the animals grew follicles; they were responding to an FSH-like activity. Some of those follicles ovulated prematurely, some of them ovulated at the right time. It looked as if the molecule can behave in different ways at different times in the same animal.

W.R. Allen

Maybe that would account for the disparity between in vitro and in vivo bioassays and perhaps we can't even try to compare the results of the two.

D. Schams

I am sorry but our time is over. I would like to thank all the speakers and close this session. Thank you.

SOME STUDIES ON PREGNANT MARE SERUM GONADOTROPHIN (PMSG) AND ON ENDOCRINE RESPONSES AFTER APPLICATION FOR SUPEROVULATION IN CATTLE

D. Schams, Ch, Menzer, E. Schallenberger[2], B. Hoffmann[1]
J. Hahn, and R. Hahn[3].

1. Institut für Physiologie, Südd. Versuchs- und
Forschungsanstalt für Milchwirtschaft Weihenstephan,
TU München, D-8050 Freising

2. Klinik für Geburtshilfe und Gynäkologie des Rindes der
Tierärztlichen Hochschule, D-3000 Hannover

3. Besamungsstation, D-8530 Neustadt a.d.Aisch. BRD.

ABSTRACT

In order to elucidate characteristics of the action of PMSG in superovulation in cattle we attempted four different approaches:

1. One highly purified and 4 relatively crude commonly used PMSG batches were characterised for their sialic acid content and biological LH and FSH activity measured by radioreceptor-assays in vitro. There was no significant difference between the batches.

2. A radioimmunoassay for determination of PMSG in plasma was developed. The disappearance rate of one crude batch of PMSG as measured in two heifers after infusion showed two components with different 'slopes': a faster one with a half-life of 40.0 or 51.2 h and afterwards a slower one with a half-life of 118.4 or 123.2 h respectively.

3. No antibody titre against PMSG in peripheral blood was measurable in 6 animals after repeated injections (im) of 3 000 or 12 000 iu PMSG respectively over a 5 or 7 week interval.

4. Hormonal profiles for LH, FSH, PMSG, progesterone and oestrogens in peripheral blood (sampling interval 6 h) during superovulation experiments were evaluated radioimmunologically. Concentrations of PMSG were highest about 12 h after the injection of PMSG and were still measurable 10 days later. There was no difference in the heights of LH and FSH peak values compared to the pattern in normal cycling animals. Animals with a good

response (16.0 $\pm$ 7.2 corpora lutea) showed approximately parallel pre-ovulatory LH and FSH peaks coinciding with the beginning of oestrus instead of a few hours later as in poorly responding animals (3.6 $\pm$ 2.8 corpora lutea) or during normal oestrous cycles. The increase of progesterone concentrations over 1 ng/ml plasma occurred in animals with a good response 48 h after the preovulatory LH surge, after 72 h in poorly responding animals and after 132 h during the normal oestrous cycle.

INTRODUCTION

The success of egg transfer techniques in cattle depends extensively on reliable methods for the induction of super-ovulation at predetermined times. The most common method is the stimulation of follicular growth and ovulation in the ovaries by means of PMSG, a hormone which is available in sufficient quantities at reasonable costs. One of its advantages is its slower inactivation rate in peripheral blood compared to pituitary FSH and LH preparations. However, the range of response to PMSG has proven to be a major limitation in the practice of superovulation in cattle. Until now the reasons for the extreme variability in the number of ovulations after administration of PMSG were unknown. Possible factors include the influence of different endocrine statuses of the individual animals or of the different batches of PMSG.

In an effort to elucidate this topic further we tried to make four different approaches:

In vitro studies of some physico-chemical properties of PMSG: (Experiment I)

In vivo studies of the half-life of PMSG in cattle (Experiment II)

Examination of whether repeated injections provoke antibody formation (Experiment III)

Establishing some hormone profiles after PMSG-administration (Experiment IV).

The information from these pilot studies carried out so for with a limited number of experiments and animals is to some extent preliminary, nevertheless we think it is of interest to present some new points for discussion.

ANALYTICAL PROCEDURES

Characterisation of the biological LH and FSH activity in vitro

LH and FSH activities were evaluated by means of radio-receptor assays as described for LH by Schams and Menzer (1977)

and developed for FSH by Kruip and Schams (unpublished data).
The receptor source for FSH was bull testis homogenate (1 500 g
precipitate). Furthermore a highly purified sheep FSH
preparation from our laboratory was used for the labelling
procedure by means of the lactoperoxidase method. LH activity
is expressed in ng of a bovine LH pituitary preparation (bLH-DSA)
having a biological activity of 1.0 times NIH-LH-S_1. FSH
activity is expressed in ng of an ovine FSH preparation (NIH-FSH-
S_{12}) having a biological activity of 1.24 units/mg NIH-FSH-S_1.

Determination of sialic acid content

Sialic acid content (N-acetyl-neuramic acid) was determined
according to the method of Warren (1959).

Radioimmunoassay for serum PMSG in cattle

The antiserum was raised in rabbits after immunisation with
a crude PMSG preparation (Anteron[R], Schering); it could be used
at a final dilution of 1 : 40 000. With regard to specificity
this antiserum showed a weak cross reaction with highly
purified bovine pituitary LH and FSH reaching a plateau at a
concentration of 5 - 10 ng/ml. This cross reaction could be
eliminated by the addition of a constant amount of bovine LH
or FSH to each test tube within the assay. No cross reaction
exists with other anterior pituitary hormones (bovine prolactin,
TSH, growth hormone), human chorionic gonadotrophin (HCG) and
the α - and β- subunits of sheep LH. For labelling with [125]Iodine
a highly purified PMSG preparation with a biological activity
of 10 000 iu/mg as described by Schams and Papkoff (1972) was
used. Two different labelling procedures were carried out which
proved to be of identical value:

a) A modification of the chloramine-T iodination-method
according to Greenwood et al. (1963).

b) An enzymatic iodination method with lactoperoxidase as
described for FSH by Schams and Schallenberger (1976).

Within the assay bovine serum proteins showed unspecific
influences which could be eliminated by the addition of a constant

amount of PMSG-free bovine serum to each tube of the standard
curve. Recovery experiments (addition of PMSG in a range
between 1 - 20 miu to bovine serum) in the assay resulted in a
mean recovery in all experiments of 90.5 $\pm$ 3.2%. The
reproducibility (interassay variability) was controlled by
determining PMSG during the whole experimental series in serum
samples from pregnant mares, yielding a coefficient of variation
(CV) of 11.1%. Intraassay coefficient of variation was
calculated as 10.7%. Inhibition curves obtained with serum
from pregnant mares and PMSG treated cattle were parallel to
the standard. The average dilution curves of the used PMSG-
standard obtained from 20 different assays is shown in Figure 1.

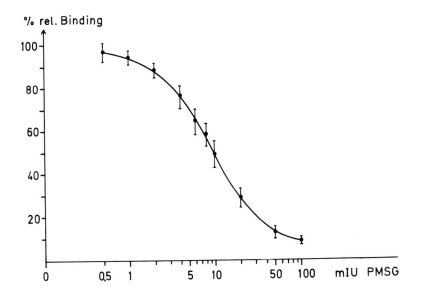

Fig. 1.   Radioimmunoassay for PMSG-inhibition curve (mean $\pm$ SD, n = 20
assays).

Determination of antibody titre against PMSG in blood serum of cattle

For this experiment two methods were used:

1) Blood serum was incubated with ^{125}-Iodine labelled PMSG and after 4 days incubation the gammaglobulin fraction was precipitated by means of an antiserum against bovine gammaglobulin (raised in rabbit) as second antibody. The ability of this second antibody to precipitate the total gammaglobulin fraction in bovine serum was proven by means of ^{125}-Iodine labelled bovine gammaglobulin: labelled bovine gammaglobulin, added to bovine serum could be totally precipitated after a serum dilution of 1 : 1 000.

2) Serum samples obtained from the PMSG treated cattle were incubated with graded concentrations of cold PMSG (standard curve) for 24 h; afterwards a constant amount of PMSG antibodies obtained from rabbits was added and incubated for a further 24 h; then labelled PMSG was added to the tubes and incubated for a further two days. The complex of iodinated and cold PMSG bound to the rabbit antibodies was precipitated by means of a second antibody raised against rabbit gammaglobulin. The same procedure was also followed with sera of untreated cattle. If there had been any antibodies against PMSG in the sera of treated animals the recovery of the added cold PMSG would have been less, compared to the sera of untreated animals.

Determination of LH, FSH, progesterone and oestrogens in blood

These hormones were anlaysed radioimmunologically as described in the following papers: LH (Schams and Karg, 1969), FSH (Schams and Schallenberger, 1976) progesterone (Hoffmann et al., 1973), oestrogens (Hoffmann, 1972). The potency of LH was furthermore determined by means of a radioreceptor assay (Schams and Menzer, 1977).

EXPERIMENTS AND RESULTS

Experiment I:

Characterisation of sialic acid content and hormonal activity
of PMSG in vitro

This experiment was carried out in order to get some
characteristics for 4 different batches of PMSG which are used
in superovulation studies within this EEC-project in comparison
with a highly purified preparation. These batches of PMSG were:
GE243 = 1817 (biological activity 1750 iu/mg protein) GE 251 =
1853 (biological activity 1980 iu/mg protein), GE 254 = 1858
(biological activity 1900 iu/mg protein), one batch widely used
for superovulation (Can-D, biological activity 1900 iu/mg
protein), and a highly purified preparation, PMSG-Org.-B
(biological activity 10 000 iu/mg protein).

The results are summarised in Table 1. There was about
the same content of sialic acid in the 4 crude batches assayed.
Only the highly purified preparation showed a higher content.
Furthermore no difference between the batches could be detected
with respect to in vitro biological LH and FSH activity. Only
the PMSG preparation Can-D showed some lower absolute content
of LH and FSH activity.

Experiment II:
Determination of the half-life of PMSG in peripheral blood
serum of cattle

Two heifers of the local Brown Swiss breed received an
infusion of 12 000 iu PMSG (batch Can-D), dissolved in 3 litres
of saline, during a period of three hours to provide
information about the half-life after PMSG application. Blood
sampling was performed by means of an indwelling catheter before
and during infusion at 30 min intervals, for 12 h after the end
of infusion at 2 h intervals, for a further 5 days at 6 h
intervals and continuing for 12 days at 12 h intervals. Figure
2 shows the half-life of PMSG, measured by RIA, in the blood of
the two heifers after the infusion of 12 000 iu PMSG. In both
animals two components of half-life are obvious. A shorter
one with 40.0 or 51.2 h and then a longer one with 118.4 or 123.2
h duration respectively.

TABLE 1

CHARACTERISATION OF FIVE DIFFERENT BATCHES OF PMSG FOR SIALIC ACID CONTENT AND IN VITRO BIOLOGICAL ACTIVITY

PMSG-batch	Sialic acid (g/100 g protein)		In vitro bioassay activity (receptorassay, n = 4) of one IU PMSG			
			L H equiv. ng bLH-DSA		F S H equiv. ng NIH-FSH-S$_{12}$	
Ge 243 (1750 iu/mg)	5.3	(5.05–5.8)	58	(56–63)	337	(271–401)
Ge 251 (1980 iu/mg)	6.05	(5.35–6.75)	53	(53–57)	301	(259–339)
Ge 254 (1900 iu/mg)	4.8	(4.5 – 5.0)	62	(59–70)	319	(290–347)
Can-D (1900 iu/mg)	5.1	(4.55–5.1)	47	(45–50)	269	(250–303)
Org. B (10 000 iu/mg)	10.2		55	(53–58)	331	(255–496)
CV %	10 (crude preparations)		10.2		8.8	

Experiment III:

Antibody production after repeated injections of PMSG

The experiment was performed with 6 animals (4 heifers
of the local Brown Swiss Breed and 2 Jersey cows). Three
animals were injected im five times with 3 000 iu PMSG (batch
Can-D) at 35 day intervals and a further three animals received
3 times 3 000 iu PMSG at 55 day intervals and an additional
6 times a provocative dose of 12 000 iu PMSG (batch Can-D) at
35 day intervals. Two days after the PMSG administration
luteolysis was induced with 0.5 mg of a prostaglandin analogue
(Estrumate[R], ICI 80996). Peripheral blood for antibody titre
determination was collected at two week intervals. Further
blood samples for progesterone determination as a parameter
for successful stimulation of the ovaries by means of PMSG
injection were taken at about 5 day intervals. Sera obtained
after each PMSG injection were assayed for their binding
capacity of ^{125}J-PMSG. No binding of labelled PMSG could be
detected. Samples also assayed with the second method showed
no different pattern to sera of untreated cattle. From these
data we conclude that no antibodies against PMSG had been
raised even after repeated provocative high PMSG injections.

Experiment IV:
Hormonal profiles after induction of superovulation with PMSG

Experimental animals and blood collection:
Eight heifers and two cows of the 'Fleckvieh' and 'Gelbvieh'
breeds were used for these experiments performed at the AI
Station, Neustadt/Aisch. The animals received a single
intramuscular injection of 1 500 - 3 100 iu PMSG (Anteron[R],
Schering) between Days 6 and 14 of the oestrous cycle. Forty-
eight h later each animal received an intramuscular injection
of 500 µg prostaglandin $F_2\alpha$ - analogue (Estrumate[R], ICI 80996).
About 8 - 9 days after the prostaglandin injection the number
of corpora lutea was recorded in the course of surgical egg
removal.

Blood was collected from the jugular vein by means of

indwelling catheters in kaolinised or heparinised plastic
centrifuge tubes and the serum or plasma obtained was stored
at -18°C until further use. Sampling was routinely carried
out at 6 h intervals during a period of two days before until
7 days after PMSG application and during the following three
days at 12 h intervals. In Table 2 detailed information about
timing of treatments, dose of PMSG used, basal and peak values
of gonadotrophins and response to stimulation are summarised.
The range of response to the PMSG injection is quite variable.
In most cases the animals responding well had also got cystic
appearing follicles with a diameter $\geqslant$ 20 mm.

All animals (except No. 219) exhibited a functional corpus
luteum at the time of treatment. After PMSG injection in 7
animals a further increase of progesterone values could be
seen. Luteolysis was induced in all animals by means of the
prostaglandin $F_2\alpha$ -analogue (PgA). About 2 - 3 days after PgA
injection oestrous symptoms could be observed in all animals
in different correlation with distinct preovulatory LH and FSH
peaks. For example, the profiles in peripheral blood for LH
(radioimmunoassay RIA and radioreceptorassay, RRA), FSH,
progesterone and oestrogens (free oestradiol-17β and oestrone)
of three animals (two with good and one with bad response) are
shown in Figures 3, 4 and 5. In animals with good responses
(average 16.0 $\pm$ 7.2 CL) preovulatory LH and FSH peaks coincide
with the beginning of oestrus whereas in animals with weak
responses (average 3.6 $\pm$ 2.8 CL) peaks occur about 8 hours
later. Preovulatory gonadotrophin peaks are preceded in animals
with good responses by clear oestrogen peaks in contrast to ones
with weak responses (see for example Figures 3 and 4). As shown
in Table 2 there was no correlation between absolute values of
LH and FSH peaks and ovarian response. The LH values measured
by RIA or RRA show a very close physiological correlation
(Figures 3, 4 and 5). Progesterone concentrations of cattle
with good responses rise about 1 ng/ml 48 h after the preovulatory LH surge compared to 72 h in animals with weak
responses and 132 h in normal cycling animals (Figure 6). In
general a good ovarian response (recorded by the number of

TABLE 2

PARAMETERS CHARACTERISING RESPONSE AFTER PMSG-ADMINISTRATION

Parameters		H176	H219	H195	C224	H200	H185	H194	H216	C225	H187
PMSG injection	day of cycle	12	12?	14	12	12	6	11	10	14	11
	dose (iu)	2300	2000	1800	3000	2000	1900	1700	1500	3100	2100
Pg-F$_2$a-analogue injection	day of cycle	14	14?	16	14	14	8	13	12	16	13
Corpora lutea	n	1	1	3	6	6	8	14	16	18	26
Cystic-appearing follicles ($\geq$20 mm $\emptyset$)	n	-	-	-	1	2	-	1	-	3	2
Progesterone (ng/ml)	before PMSG inj.	6.6	0.6	5.3	12.5	7.4	4.7	7.1	13.5	7.5	6.5
	around Pg-F$_2$a-analogue inj.	16.4	0.7	11.1	10.8	14.7	6.4	12.8	18.1	14.7	12.9
	during oestrus	0.3-0.5	0.3-0.4	0.4-0.6	1.4-2.3	0.6-0.7	0.3-0.4	0.4-0.6	0.9-1.4	0.6-0.9	0.4-1.0
	8-9 days after Pg-F$_2$a-analogue	4.1	5.3	8.2	23.0	24.6	16.3	76.2	23.7	30.0	5.0
LH (ng/ml)	base value	0.6	1.4	0.6	0.8	0.5	0.7	0.6	0.9	0.7	1.5
	preovulatory peak value	22.0	16.3	22.3	15.4	19.3	7.7	15.6	7.9	16.7	25.9
FSH (ng/ml)	base value	81	97	120	113	97	94	113	145	142	84
	preovulatory peak value	256	197	327	258	223	186	235	217	217	197
LH and FSH peak	time (h) after Pg-F$_2$a-analogue inj.	63	75	63	45	45	45	39	39	33	39
Oestrus	begin after Pg-F$_2$a-analogue inj. (h)	45	no detection	39	33	27	39	39	33	27	39
	duration (h)	24	30	30	22	24	30	24	18	24	30

H = heifer; C = cow

corpora lutea) was also characterised by significantly higher
progesterone values compared to a bad one, however in animals
with good responses there seems to be no clear correlation
between number of visible corpora lutea and progesterone
values.

In Figure 7 for example the radioimmunologically measured
PMSG values of 4 animals are presented. Concentrations of PMSG
are highest about 12 h after the injection and remain still
measurable about 10 days afterwards. There is no correlation
between the absolute values of PMSG, the dose of PMSG given im
and the response of the ovaries.

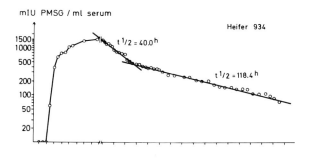

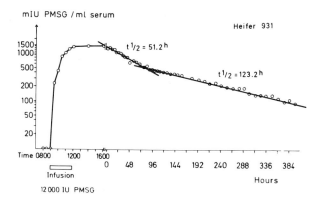

Fig. 2. Disappearance time for PMSG in peripheral blood of two heifers
after infusion of 12 000 iu PMSG during 3 hours.

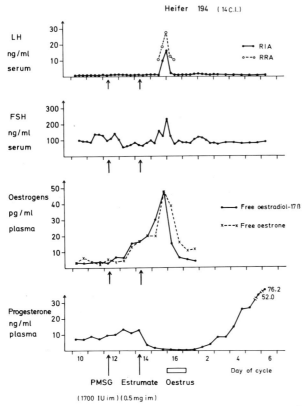

Fig. 3.   Hormonal profiles for LH, FSH, oestrogens and progesterone in
heifer number 194 before and after stimulation with 1 700 iu PMSG

DISCUSSION

Some workers postulate that the superovulatory effect of
PMSG can vary with the particular batch employed.   In our
experiments (see Table 1) there was no difference recognisable
between the 5 batches of PMSG assayed except the higher sialic
acid content of the highest purified batch which may be due to
the calculation model based on the content of sialic acid/weight
unit of protein.   The results concerning the in vitro biological
LH and FSH activity of PMSG confirm those of Stewart et al. (1976)
who measured LH and FSH concentrations in five different
commercially available batches of PMSG.   Thus variation of half-

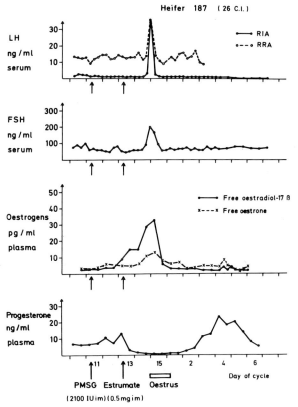

Fig. 4. Hormonal profiles for LH, FSH, oestrogens and progesterone in
heifers number 187 before and after stimulation with 2 100 iu PMSG.

life (as a result of sialic acid content) and a different LH
and FSH activity may be excluded as explanation for stimulation
differences. Consequently the variation in ovarian response
to PMSG seems to be dependent rather on the individual endocrine
responsiveness of the animal. There is a lack of proper
information about the duration of action of PMSG in cattle.
Since an exact half life of this hormone could not be calculated
after intramuscular injection of PMSG due to an unknown resorption time, the disappearance time had to be checked after
intravenous infusion. Two components could be calculated which
are in good agreement for the two experimental animals. The
component with the longer half-life agrees very well with the

136

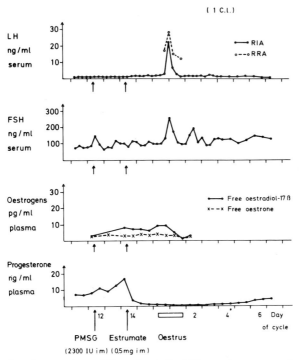

Fig. 5. Hormonal profiles for LH, FSH, oestrogens and progesterone in
         heifer number 176 before and after stimulation with 2 300 iu PMSG.

one obtained in the mare by means of bioassay (Cole et al.,
1967) and is much longer than the one observed in ewes where
$t_{1/2}$ = 21 h (McIntosh et al., 1975). This reduced half-life is
probably related to the high sialic acid content of PMSG
(Schams and Papkoff, 1972; Gospodarowicz, 1972). The content
of sialic acid is important in regulating the clearance of a
glycoprotein because only after removal of the sialic acid
residues can the otherwise intact molecule be taken up by
specific (calcium dependent) receptors in the parenchymal cells
of the liver (Hudgin et al., 1974). It could be shown clearly
by Morell et al. (1971) that even partial desialylation of FSH,
HCG and most normal plasma glycoproteins results in their
prompt removal from the circulation.

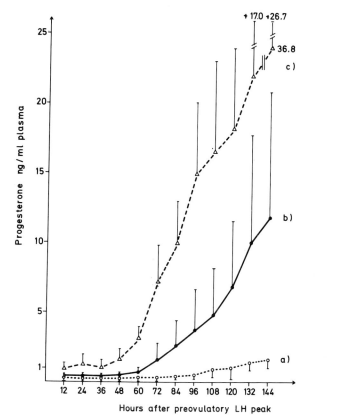

Fig. 6. Increase of progesterone concentration (mean + SD) after the preovulatory LH peak a) of 12 animals during $\bar{1}$5 normal oestrous cycles b) of 5 weak responding animals (3.6 + 2.8 C.l.) after PMSG application (no cyst-like follicles) c) of 5 good responding animals (16.0 + 7.2 C.l.) after PMSG application (included 4 animals with cyst-like follicles $\geq$ 20 mm $\emptyset$).

The very long half-life for PMSG raises the question of a possible antibody production in treated cattle especially after repeated administration. In our experiments about the effect of repeated injections of PMSG on antibody production we have not been able to demonstrate any binding of labelled PMSG. Nevertheless we can not exclude completely the possibility that there are no antibodies produced against PMSG as we can not record antibodies at a very low titre in peripheral blood with our method. We conclude that the production of antibodies against PMSG in cattle appears to be extremely difficult even after repeated injections with doses usually given in super-ovulation studies. Our results seem to be contrary to data

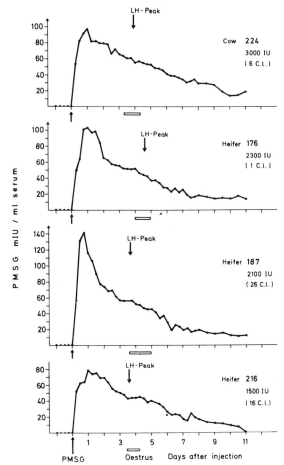

Fig. 7.   Concentrations of PMSG in blood serum of 4 heifers after a single
intramuscular injection of PMSG.

reported by Jainudeen et al. (1966) who demonstrated by means
of a rat bioassay an antigonadotrophic activity after repeated
injections of PMSG.   Blood serum from cows treated with PMSG
and injected in rats inhibited the follicular stimulating
properties of PMSG, but had no adverse effect on follicular
development and ovulation resulting from endogenously secreted
gonadotrophins.   These authors could further demonstrate an
unresponsiveness of the ovaries after the second injection of
PMSG in cattle.   In earlier experiments Willett et al. (1953)
could also find a refractoriness to gonadotrophins (including
PMSG) after repeated injections.   Since this effect could not
be overcome completely by' long periods of rest they suggest
that antibodies against gonadotrophins are not the cause of the

refractoriness. Saumande and Chupin (1977) measured the
ovarian response in heifers during 8 injections of PMSG at 7 - 9
week intervals. In each of these eight heifers, the number of
ovulations decreased after the second treatment, remained low
until the sixth one but then response was in some way reversible.
Heifers which showed a poor response to the first treatment
continued to show a poor response throughout the experiment.
In our studies we could not find a general tendency for reduced
stimulation after 5 - 7 times repeated treatment with PMSG;
for example progesterone peak concentrations after treatment
ranged between lower than 5 up to 50 ng/ml plasma. As after
the 5th or 7th injection the response was comparable with the
first treatments we conclude that there had not been antibodies raised against PMSG in bovine blood. This supports our
findings of the binding studies with labelled PMSG. We think
that antibodies against gonadotrophins are not so much involved
in any possible decrease in ovarian responsiveness as other
unknown factors in the ovary itself (for example exhaustion of
available follicles which are capable of being stimulated by
PMSG or interactions in the gonadotrophin feed-back mechanisms).

The administration of PMSG to heifers and cows has marked
effects on the ovarian morphology (number of follicles and
ovulation rate). The response seems not to be dose dependent.
There is not a big difference between hormonal profiles around
oestrus and ovulation after PMSG stimulation and during the
normal cycle (Schams et al., 1977). Only after a good response
(as indicated by a high number of corpora lutea) do concentrations
of free oestradiol-17β and oestrone rise higher just before
oestrus perhaps due to the higher number of growing follicles.
The pronounced oestrogen production might also be one reason
for. the earlier LH and FSH surge which coincides with the
beginning of oestrus contrary to normal cycling animals where
the LH peak occurs about 8 h later (Schams and Butz, 1972).
The data obtained concerning LH, progesterone and oestrogen
responses agree well with data from Henricks et al. (1973)
after PMSG injection in beef cows.

There was no consistent relationship between the absolute concentrations of LH and FSH during oestrus and ovulation rate which indicates that the number of ovulations is not dependent on higher amounts than those measurable during the normal cycle. These data do not agree with results reported by Saumande and Pelletier (1975) who found a correlation of both LH and oestrogens with the number of ovulations in PMSG treated cattle. LH values obtained by radioimmuno- and radioreceptor-assay showed a very close physiological correlation. The difference in absolute basal levels can also be found during the normal oestrous cycle and seems to be more a methodological problem than a physiological one (Schams and Menzer, 1977).

To our knowledge no data exist at the present time about PMSG levels in cattle blood after stimulation. It was surprising that PMSG values (which have an equivalent in activity (RRA) of about 1.0 - 1.5 ng bovine LH) can still be measured 10 days after application. These PMSG levels in blood serum may have some influence on the ovaries beside stimulation of follicular growth. For instance the function of an existing corpus luteum after PMSG injection and before prostaglandin application can be stimulated. In 7 animals in our experiments progesterone values increased after PMSG application, most likely due to the LH component of PMSG. Also the rapid increase of progesterone values after the LH peak in animals with good responses (see Figure 6) could at least partly be due to the effect of PMSG in addition to the number of growing corpora lutea. This would also be an explanation why cyst-like follicles can be found mainly in animals with good responses. The sharp increase of progesterone values after the LH peak may block ovulation of follicles which are still growing due to further stimulation by the FSH-activity of PMSG.

ACKNOWLEDGEMENT:

This project was supported by the Commission of the European Communities. We thank the following for generous gifts: National Institute of Health, Bethesda, USA (NIH-FSH-S$_{12}$, NIH-FSB-B$_1$); Intervet Int. Ltd., Oss, The Netherlands (PMSG GE 243, 251, 254); Imperial Chemical Industries, UK (Estrumate, ICI 80996).

Cole, H.H., Bigelow, M., Finkel, J. and Rupp, G.R. 1967. Biological half-life of endogenous PMSG following hysterectomy and studies on losses in urine and milk. Endocrinology 81, 927-930.

Gospodarowicz, D. 1972. Purification and physicochemical properties of the pregnant mare serum gonadotrophin (PMSG). Endocrinology 91, 101-106.

Greenwood, F.C., Hunter, W.M. and Glover, J.S. 1963. The preparation of ^{131}-J labelled human growth hormone of high specific radioactivity. Biochem. J. 89, 114-123.

Henricks, D.M., Hill, J.R., Dickey, J.F. and Lamond, D.R. 1973. Plasma hormone levels in beef cows with induced multiple ovulations. J. Reprod. Fert. 35, 225-233.

Hoffmann, B. 1972. Die Bestimmung von Öestrogenen. In: Kaiser E. (Ed). Fortschritte der klinischen Chemie: Enzyme und Hormone, 491-500 Verlag der Weiner Medizinischen Akademie.

Hoffmann, B., Kyrein, H.J. and Ender, M.L. 1973. An efficient procedure for the determination of progesterone by radioimmunoassay applied to bovine peripheral plasma. Hormone Res. 4, 302-310.

Hudgin, R.L., Pricer, W.E., Ashwell, G., Stockert, R.J. and Morell, A.G. 1974. The isolation and properties of a rabbit liver binding protein specific for asialoglycoproteins. J. Biol. Chem. 249, 5536-5543.

Jainudeen, M.R., Hafez, E.S.E., Gollinck, P.D. and Moustafa, L.A. 1966. Antigonadotrophins in the serum of cows following repeated therapeutic pregnant mare serum injections. Am. J. Vet. Res. 27, 669-675.

McIntosh, J.E.A., Moor, R.M. and Allen, W.R. 1975. Pregnant mare serum gonadotrophin: Rate of clearance from the circulation of sheep. J. Reprod. Fert. 44, 95-100.

Morell, A.G., Gregoriadis, G., Scheinberg, I.H., Heitzman, J. and Ashwell, G. 1971. The role of sialic acid in determining the survival of glycoproteins in the circulation. J. Biol. Chem. 246, 1461-1467.

Saumande, J. and Pelletier, J. 1975. Relationship of plasma levels of oestradiol-17β and luteinising hormone with ovulation rate in super-ovulated cattle. J. Endocr. 64, 189-190.

Saumande, J. and Chupin, D. 1977. Superovulation: A limit to egg transfer in cattle. Theriogenology 7, 141-149.

Schams, D. and Karg, H. 1969. Radioimmunologische LH-Bestimmung im Blutserum vom Rind unter besonderer Berücksichtigung des Brunstzyklus Acta endocr. (Kbh.) 61, 96-103.

Schams, D. and Papkoff, H. 1972. Chemical and immunological studies on
    pregnant mare serum gonadotrophin. Biochim. Biophys. Acta 263, 139-148.

Schams, D. and Butz, E.D. 1972. Zeitliche Beziehungen zwischen Brunstsymptomen,
    elektrischen Widerstandsmessungen des Vaginalschleims, präovulatorischer
    Ausschüttung des Luteinisierungshormons und Ovulation beim Rind.
    Zuchthyg. 7, 49-56.

Schams, D. and Schallenberger, E. 1976. Heterologous radioimmunoassay
    for bovine follicle - stimulating hormone and its application
    during the oestrous cycle in cattle Acta endocr. (Kbh.) 81, 461-473.

Schams, D. and Menzer, Ch. 1977. Determination of luteinising hormone
    (LH) in bovine blood by radioligand receptor assay and comparison
    with radioimmunological evaluation. Acta endocr. (Kbh.) in press.

Schams, D., Schallenberger, E., Hoffmann, B. and Karg, H. 1977. The
    oestrous cycle of the cow: Hormonal parameters and time relation-
    ships concerning oestrus, ovulation and electrical resistance of
    the vaginal mucus. Acta endocr. (Kbh.) 86,180-192.

Stewart, F., Allen, W.R. and Moor, R.M. 1976. Pregnant mare serum
    gonadotrophin: Ratio of follicle-stimulating hormone and luteinising
    hormone activities measured by radioreceptorassay. J. Endocr. 71,
    371-382.

Warren, L., 1959. The thiobarbituric acid assay of sialic acids. J.
    Biol. Chem. 234, 1971-1975.

Willett, E.L., Buckner, P.J. and McShan, W.H. 1953. Refractoriness of cows
    repeatedly superovulated with gonadotrophins. J. Dairy Sci. 36,
    1083-1088.

OVARIAN RESPONSES IN RELATION TO ENDOCRINE STATUS FOLLOWING PMSG STIMULATION IN THE COW

J.M. Sreenan, D. Beehan & * J.P. Gosling.
The Agricultural Institute, Belclare, Tuam, Galway, Ireland.
* The Biochemistry Dept., University College, Galway.

ABSTRACT

   A total of 325 heifers were used in a series of experiments to determine some of the factors affecting ovarian responses to superovulation (PMSG) treatments in the cow.

   Superovulation was carried out in conjunction with oestrus synchronisation treatments. Administration of PMSG at 48 h prior to the end of progesterone or cronolone treatment or 48 h prior to prostaglandin analogue administration resulted in similar ovarian responses. The interval elapsing between PMSG administration and oestrus onset significantly affected follicular and total ovarian responses. The day of cycle at PMSG administration affected the ovarian response with higher ($P<0.01$) mean ovulation rates on Days 9, 10, 11 and 14 than on Days 7, 8 or 12.

   Plasma progesterone level at the time of PMSG administration or for three days afterwards was not correlated with ovarian response. Total oestrogen level at the time of PMSG administration was not correlated with ovarian response but on the third day following PMSG a significant positive correlation was established ($P<0.01$). The peak level of preovulatory LH was not correlated with subsequent ovulation rate but there was a significant negative correlation between ovulation rate and the interval between oestrus onset and LH peak ($P<0.05$). Maximal LH output (peak level + the level at -2 and +2 h) was negatively correlated with the interval between oestrus onset and LH peak ($P<0.001$), and positively correlated with ovulation rate.

INTRODUCTION

The dependence of the future development of practical egg transfer programmes in cattle breeding, on consistent methods of fertilised egg supply is well recognised. Induction of multiple ovulations following the use of gonadotrophins (FSH preparations or PMSG) is the normal method of egg supply, but most authors report wide variation in follicular development and ovulation responses following superovulation treatments.

Some of the factors involved in ovarian response to super-ovulation treatments in the cow have been recently reviewed (Sreenan and Beehan, 1976; Betteridge, 1977). These reviews show that response is affected by breed, gonadotrophin preparation (mainly PMSG) and dose level. There would seem to be little evidence to indicate a significant difference in batches of PMSG preparations from the same manufacturing source. Stewart et al. (1976) found no difference in FSH : LH ratio in six different batches of PMSG. Beehan and Sreenan (1977) examined five discrete batches of PMSG from the same manufacturing source for their effect on ovarian response in heifers and reported no batch effect on either follicular development or total ovarian response (follicles + ovulations). One of the five batches resulted in a lower ovulation rate (P <0.01) with the remaining four effecting similar ovulation responses. Recent studies have indicated that season does not seem to affect ovarian response to superovulatory doses of PMSG in the cow (Beehan and Sreenan, 1977; Betteridge, 1977 and others).

It would seem therefore that most of the variation in ovarian response to superovulation treatments is due to individual animal variation. The interval elapsing between PMSG administration and subsequent oestrus onset in some studies has been reported to result in increased ovulation rates (Scanlon et al., 1968) while Moore (1975$_1$) has reported no effect of this interval on ovulation rate. Recent studies where PMSG has been administered during the luteal phase followed by prostaglandin to induce luteolysis have indicated

that the day of administration may be critical, with lower
ovulation responses being obtained in the earlier cycle phase
(Days 3 - 8) compared with later stages (Days 9 - 12) (Phillipo
and Rowson, 1975; Newcomb and Rowson, 1976; Sreenan and Gosling,
1977). However, the effect of the endocrine status of the animal on the day of PMSG administration and during this interval
from PMSG to oestrus onset on ovarian response is not clearly
understood and little data is available on this aspect.

The experiments reported on in this communication were
carried out to determine the effects of the PMSG to oestrus
interval, day of PMSG administration, endogenous progesterone,
oestrogen and LH levels on ovarian response to superovulation
treatments in the cow.

MATERIALS AND METHODS

Beef heifers of Hereford cross breed types were used in
all experiments. All heifers were sexually mature within a
weight range of 305 - 386 kg and within an age range of 18 -
28 months. PMSG (Folligon : Intervet) was administered intramuscularly in 5 ml sterile distilled water. Ovarian measurements were carried out at laparotomy between 3 and 7 days from
the day or expected day of oestrus (Day 0). Eggs were collected by in vitro recovery techniques and were assessed for
fertilisation rate and normality. These were then used in a
series of transfer studies. In this communication the results
are confined to ovarian responses.

Experiment 1
A total of 285 donor heifers were used in this experiment
to examine the relationship between the PMSG - oestrus interval
and day of PMSG administration on ovarian response. Superovulation treatments were carried out in conjunction with oestrous synchronisation techniques to control the PMSG to oestrus
onset interval. Three superovulation-synchronisation regimes
were used as follows :

1. PMSG administered 48 h before removal of a 9-day progestin (progesterone or fluorogestone acetate) intra-vaginal pessary.

2. PMSG administered 48 h before removal of a 9-day progestin (Norgestomet; G.D. Searle & Co.) subcutaneous implant.

3. PMSG administered on any day between Days 7 and 14 inclusive followed by administration of 500 µg prostaglandin analogue (Cloprostenol; ICI Ltd.) at 48 h and again at 72 h. PMSG doses used were 1500, 2000 and 2500 iu but the data presented here has been analysed to correct to a standard dose level.

Oestrous and ovarian responses were determined for each animal. Preliminary data from this experiment has been presented (EEC Seminar, 'Egg Transfer in Cattle', Cambridge, 1975).

Experiment 2

A total of 40 heifers were used in this study to examine the relationship between endogenous progesterone and oestrogen (total oestrogens) levels on the day of PMSG administration and during the interval from PMSG - oestrus on ovarian response; also to determine the relationship between the pre-ovulatory LH surge and the ovarian response to superovulation with PMSG. All heifers received 500 µg Cloprostenol 48 h after PMSG (2000 iu) administration and again at 72 h. Oestrous and ovarian responses were determined for each animal.

Blood sampling for progesterone and oestrogen was carried out twice daily at 0.800 and 16.00 h from the day before PMSG administration until 10 days after oestrus or the expected date of oestrus. Blood sampling for LH determination was carried out at two hourly intervals from 12 hours before the expected onset of oestrus until six hours after the end of standing oestrus. The onset of oestrus was predictable from the time of prostaglandin analogue administration.

Following centrifugation all plasma was stored at $-20^{\circ}C$ until required for assay.

Statistical Analysis

The data were analysed by least squares procedures (Harvey, 1960) with logarithmic transformation where appropriate. Treatment means were compared using Duncan's new multiple range test (Kramer, 1951).

RESULTS

Experiment 1

Oestrous and ovarian responses following the three super-ovulation methods used were not different (Table 1).

TABLE 1

SUPEROVULATION METHOD AND OESTROUS AND OVARIAN RESPONSES

	PMSG Progestin pessary	PMSG Progestin implant	PMSG Prosta-glandin
No. heifers	98	64	123
Oestrous reponse (%)	90	88	98
Follicles ($\geqslant$10 mm)	2.2	2.4	2.9
Ovulations	9.5	9.2	9.1
Total ovarian response	12.8	13.1	14.0

A total of 264 (93%) of these donors came in oestrus after the superovulation treatment. The interval elapsing between PMSG administration and oestrus onset varied from two to five days and the relationship between this interval and ovarian response is shown in Table 2.

The number of unovulated follicles ($\geqslant$ 10 mm) was significantly negatively correlated with the PMSG to oestrus interval ($P<0.01$). This interval also negatively affected the total response ($P<0.05$) but did not affect ovulation rate.

TABLE 2

CORRELATION BETWEEN THE PMSG TO OESTRUS INTERVAL AND OVARIAN RESPONSE

	r
Follicles ($\geqslant$ 10 mm)	- 0.4051**
Ovulations	+ 0.0829
Total ovarian response	- 0.1234*

* $P < 0.05$; ** $P < 0.01$

The effect of the day of the cycle on which PMSG was
administered for 109 of the PMSG-prostaglandin treated heifers
is shown in Table 3.

TABLE 3

EFFECT OF DAY OF PMSG ADMINISTRATION ON OVARIAN RESPONSE (LEAST SQUARE
MEANS)

	Day of cycle at PMSG						
	7	8	9	10	11	12	14
No. of heifers	5	10	18	45	21	6	4
Follicles ($\geqslant$ 10 mm)	5.52	2.25	1.79	1.86	3.47	2.26	2.88
Ovulations	4.99^a	5.16^a	13.35^b	13.89^b	16.71^b	7.64^a	16.23^b
Total ovarian response	12.4	9.2	14.8	15.9	21.6	11.2	16.0

a vs b, $P < 0.01$

Day of PMSG administration did not affect the proportion
of unovulated follicles ($P > 0.05$) but it did have a significant effect on ovulation rate ($P < 0.01$). Mean ovulation rates
were higher on Days 9, 10, 11 and 14 than on Days 7, 8 or 12.

Experiment 2

To examine the effect of cycle phase further, the relationship between progesterone and oestrogen levels at and near the
time of PMSG administration and LH levels at the treatment
oestrus were examined in relation to ovarian response.

Progesterone

Progesterone levels were similar to those previously reported (Sreenan and Gosling, 1976) and showed a slight increase from the time of PMSG administration until the prostaglandin analogue was administered. The effect of endogenous progesterone level on ovulation rate is presented in Table 4 for a number of days.

TABLE 4

CORRELATION BETWEEN PLASMA PROGESTERONE LEVEL OVULATION RATE

	r
Day of PMSG	− .0839
+ 1 day	.1478
+ 2 day	.1631
+ 3 day	.1080
All days	.0674

There was no evidence for a significant correlation between plasma progesterone level and ovulation rate on any of the days examined here, up to the day of oestrus. However, on Days 1 and 2 following oestrus a significant correlation ($P < 0.05$) existed between plasma progesterone level and ovulation rate.

Total oestrogens

Mean total oestrogen levels varied from a level of 35.6 pg/ml the day before PMSG to a level of 50 pg/ml 72 h after PMSG. The relationship between total oestrogens and ovulation rate at and near the time of PMSG administration is shown in Table 5.

TABLE 5

CORRELATION BETWEEN PLASMA OESTROGEN LEVEL AND OVULATION RATE

	r
- 1 day	.0824
Day of PMSG	.2705
+ 2 days	.0389
+ 3 days	.5582**

** $P < 0.01$

From the day prior to PMSG administration until two days after, there was no evidence of a significant correlation between oestrogen level and ovulation rate. However, on the third day following PMSG there was a significant correlation (P <0.01) between oestrogen level on that day and subsequent ovulation rate.

LH

LH levels varied from base-line levels of 0 - 2 ng/ml to peak levels in a range of 14 - 150 ng/ml. The peak level of the preovulatory LH surge was not significantly correlated with ovulation rate. However, a negative correlation did exist ($P < 0.05$) between ovulation rate and the interval between oestrus onset and LH peak (Figure 1).

Shorter intervals between oestrus onset and LH peak were associated with higher ovulation rates.

When the relationship between maximal LH output (represented by the peak level plus the levels at -2 h and + 2 h) and the interval between oestrus onset - LH peak was examined, a significant negative correlation existed ($P < 0.001$), as shown in Figure 2. High maximal LH values were associated with shorter intervals from oestrus onset to LH peak.

When the relationship between ovulation rate and maximal

LH output was then examined, a positive correlation (P≤0.05) was found to exist as is shown in Figure 3.

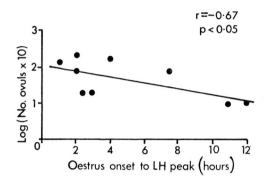

Fig. 1. Relationship between ovulation rate and interval between oestrus onset and LH peak.

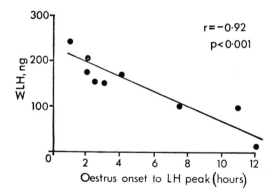

Fig. 2. Relationship between maximal LH output (ie peak value + values at -2 and + 2h) and interval between oestrus and LH peak.

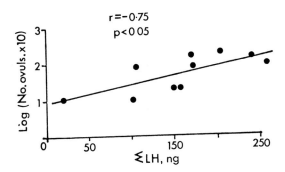

Fig. 3. Relationship between ovulation rate and maximal
LH output (ie peak value + values at -2 and
+2 h).

DISCUSSION

The data presented in Experiment 1 show that the interval
that elapses between PMSG administration and onset of oestrus
significantly affects the proportion of stimulated but unovulated follicles ($\geqslant$ 10 mm). As this interval increases, a higher
proportion of stimulated follicles ovulate. Because, however,
the total ovarian response decreases with interval, the ovulation rate is not affected. The data of Moore (1975$_1$) also indicates that in their studies using either PMSG or HAP (Horse anterior pituitary extract) this interval did not affect ovulation
rate. The highest total ovarian responses were recorded in
animals with the shortest PMSG - oestrus intervals. The initial
higher follicular stimulation in these animals may have resulted
in the production of higher oestrogen levels at an early stage
causing early onset of oestrus and LH discharge. Consequently
not all follicles may have reached the capacity to respond
fully to LH and ovulate. Failure of these stimulated follicles
to bind sufficient quantity of LH during the preovulatory surge
may be the reason for their failure to ovulate. The longer

154

intervals may allow the time for a higher proportion of follicles to reach the stage of being able to respond ie to bind
sufficient quantities of LH and therefore ovulate.

The day of the cycle (in the luteal phase) on which PMSG
was administered had a significant effect on subsequent ovulation rate (P< 0.01) with lowest mean ovulation rates occurring
on Days 7 and 8. Similar responses have been recorded by
Phillipo and Rowson (1975) and Newcomb and Rowson (1976). However, in the data presented here there is also evidence of a
lower mean ovulation response following PMSG administration on
Day 12 rising again on Day 14. It is probable that this reflects distinct waves of follicular growth.

The data presented from Experiment 2, clearly indicates
that the increase in ovulation rate during the mid luteal phase
of the cycle moving from Days 7 and 8 to 9, 10 and 11 is not
simply a reflection of the increased progesterone levels evident at this stage of the cycle. No correlation exists between
progesterone level and ovulation rate. The addition of exogenous progesterone at different cycle stages (Days 3 and 10) has
not influenced ovarian response either (Sreenan and Gosling,
1977). Progesterone levels were only correlated with ovulation
rates after ovulation had occurred. On the first and second
days after oestrus, a positive correlation (P< 0.05) existed
between plasma progesterone and ovulation rate.

The data presented on total oestrogen levels show that
plasma oestrogen is not an indicator of the potential responding
follicle population at the time of PMSG administration. No
evidence was found for a significant correlation between total
oestrogen level and subsequent ovulation rate, from a 24 h
period prior to PMSG administration until 72 h afterwards. At
72 h after PMSG a positive correlation (P <0.01) was evident.
This relationship reflects the follicular growth that had
occurred as a result of stimulation with PMSG.

High oestrogen levels before oestrus in superovulated cows have been reported previously (Lemon and Saumande, 1972; Saumande and Pelletier, 1975; Halford et al., 1975 and others). However, the significant correlation that does exist 72 h after PMSG stimulation is not an indicator of the potential follicle population to respond to stimulation, it is rather a measure of the response.

The data presented in Experiment 2 on pre-ovulatory LH levels and their relationships with ovulation rate, indicate a highly significant correlation ($P < 0.001$) between maximal LH output (ie the peak level, plus the levels at -2 h and +2 h) and the interval between oestrus onset and LH peak. Animals with shorter oestrus onset - LH peak intervals had higher maximal LH levels. Animals with similarly short oestrus onset - LH peak intervals had higher ovulation responses ($P < 0.05$) which indicated a positive correlation between maximal LH output and ovulation rate which was shown in Figure 3. ($P < 0.05$). It would seem that while the peak level of LH was not significantly correlated, in this study, with ovulation rate and this is similar to the data of Henricks et al. (1973) that LH output over a 6 h period (including the peak level) affected ovulation rate, with greater LH output resulting in higher ovulation rates.

From the data presented it is clear that optimum superovulation response would be achieved by administering the gonadotrophin during the mid luteal phase in particular on Days 9, 10 and 11. Also a longer interval from PMSG to oestrus onset (4 - 5 days) will ensure a lower proportion of large unovulated follicles. This can be achieved using synchronisation techniques. While no positive relationship was evident between plasma progesterone or oestrogen level at the time of PMSG administration and subsequent ovulation rate, further study is required on the endocrine status of the animal at this time in order to determine the endocrine status for optimum ovarian. The variation in response to superovulation treatments depends both on the level of the total ovarian response and the prop-

ortion of this follicular response that eventually ovulates.
The positive relationships between maximal LH output and ovulation rate and also the relationships between LH output and
interval from oestrus onset - LH peak indicate the need for
further study of the role of the LH pre-ovulatory surge in
superovulated animals.

ACKNOWLEDGEMENTS

The authors wish to thank Mr. A. McDonagh, Mr. G. Morris,
Mr. D. Morris, and Mr. P. Creaven for excellent technical
assistance in the oestrus and ovarian studies; Mrs. A. Glynn
for progesterone determinations; Ms. Greta Tisdall for the
oestrogen determinations and Mr. G. McLoughlin for the LH
determinations.

Thanks are due to Dr. J.P. Hanrahan for advice on statistical procedures.

These studies have been supported by the Commission of the
European Communities.

REFERENCES

Beehan, D. and Sreenan, J.M., 1977. Factors affecting superovulation
    response to PMSG in the cow. Proc. Soc. Study Fert., Dublin, July
    11-15, 1977.

Betteridge, K.J. 1977. 'Superovulation' - In 'Embryo Transfer in Farm
    Animals' Canada. Dept. Agric. Monograph No. 16, pp 1-9.

Hallford, D.M., Turman, E.J., Wetteman, R.P. and Pope, C.E., 1975. Plasma
    LH and oestradiol in the bovine after PMSG J. Anim. Sc. 41, 356 Abstr.

Harvey, W.R. 1960. Least squares analysis of data. USDA Publ. No. ARS-
    20-8.

Henricks, D.M., Hill, J.R. (Jr), Dickey, J.F. and Lamond, D.R., 1973.
    Plasma hormone levels in beef cows with induced multiple ovulations.
    J. Reprod. Fert. 35, 225-233.

Kramer, C.Y. 1951. Extension of multiple range tests to group correlated
    adjusted means. Biometrics Vol. 13, 13-14.

Lemon, M. and Saumande, J. 1972. Oestradiol - 17 β and progesterone after
    induction of superovulation by PMSG in cattle. J. Reprod. Fert. 31,
    501 - 502. Abstr.

Moore, N.W. 1975_1. The control of time of oestrus and ovulation and the
    induction of superovulation in cattle Aust. J. Agric. Res. 26, 295 -
    304.

Newcomb, R. and Rowson, L.E.A. 1976. Multiple ovulation, egg transplantat-
    ion: towards twinning. In principles of cattle production ed. H.
    Swan and W.H. Broster, Butterworths, London. pp 59 - 83.

Phillipo, M. and Rowson, L.E.A. 1975. Prostaglandins and superovulation
    in the bovine. Ann. Biol. Anim. Biochim. Biophys. 15; 233-240.

Saumande, J. and Pelletier, J. 1975. Relationship of plasma levels of
    oestradiol 17-B, and luteinising hormone with ovulation rate in
    superovulated cattle. J. Endocrinol., 64, 189 - 190.

Scanlon, P.F., Sreenan, J.M. and Gordon, I. 1968. Hormonal induction of
    superovulation in cattle. J. Agric. Sci. Camb., 70, 179-182.

Sreenan, J.M. and Beehan, D. 1976. Methods of induction of superovulation
    in the cow and transfer results. Proc. EEC Conference 'Egg Transfer
    in Cattle'. CEC, Eur. 5491. pp 19-34.

Sreenan, J.M. and Gosling, J.P. 1976. Peripheral plasma progesterone levels
    in cycling and pregnant beef heifers. Ir. Vet. J. 29, 105-108.

Sreenan, J.M. and Gosling, J.P. 1977. The effect of cycle stage and plasma
progesterone level on the induction of multiple ovulations in heifers.
J. Reprod. Fert. 50, 367-369.

Stewart, F., Allen, W.R. and Moor, R.M. 1976. Pregnant mare serum gonadotrophin: ratio of follicle stimulating hormone and luteinising
hormone activities measured by radioreceptor assays. J. Endocrinol.
71 : 371-382.

SUPEROVULATION OF CATTLE WITH PREGNANT MARE'S SERUM GONADOTROPHIN AND FOLLICLE STIMULATING HORMONE

G.E. Seidel, Jr., R.P. Elsden, L.D. Nelson and R.A. Bowen.
Animal Reproduction Laboratory, Colorado State University,
Fort Collins, Colorado 80523 USA.

ABSTRACT

A review of published literature provides little evidence of differences in efficacy between PMSG and extracts of the anterior pituitary for superovulating cattle, but there are very few published experiments comparing such treatments. New experiments are presented comparing regimens using PMSG and FSH-LH. In some experiments, no differences were found in the number of normal embryos produced, but in other experiments, the FSH-LH regimen was superior to PMSG. It may be useful to alternate gonadotrophin treatments when valuable cattle are superovulated repeatedly.

INTRODUCTION

Excellent reviews on superovulating cattle are available,
eg Foote and Onuma (1970), Gordon (1975), Sreenan and Beehan
(1976), and Betteridge (1977). From these reviews, it is evident that, although much of the early work was done with extracts
of the anterior pituitary, most of the recent work has been
with extracts from the blood of pregnant mares. In this paper
we will briefly review this work with emphasis on the methodology of using pituitary extracts. We will also present data
from recent experiments comparing pituitary extracts with
extracts from pregnant mare's serum for superovulating cattle.

REVIEW

One problem in designing superovulation treatments is that
our knowledge of factors controlling normal follicular growth
in cattle (and other species) is fragmentary. Follicle stimulating hormone (FSH) is generally assumed to be involved;
luteinising hormone (LH) is also probably involved; prolactin
may play a role; and numerous steroids and prostaglandins are
also implicated in the process (Richards and Midgley, 1976).

In some respects, literature on superovulation is difficult
to interpret because the pituitary preparations used to superovulate cattle are far from chemically pure. Most are relatively crude extracts. Therefore, one is always using mixtures
of hormones, often in different ratios from preparation to
preparation. In general, standards have not been used that
allow comparisons from one publication to another, especially
concerning 'contaminating' hormones. To complicate interpret-
ations further the extracts have usually been made from pituitaries of domestic species other than cattle and, in some
cases, the species was not specified. Extracts from pregnant
mare's serum (PMSG) also vary in degree of purity and characteristics of response among batches (Church and Shea, 1976).

Superovulation with PMSG is usually accomplished with a

single injection whereas three to ten injections of pituitary
hormones are usually used because of their short half lives
(see reviews mentioned above). However, Wildt et al. (1975)
were able to produce limited, but reliable superovulation with
a single injection of pituitary extract followed by HCG. In
our opinion, considerable work remains to be done with pituitary extracts concerning dose, frequency of administration, and
timing of administration of pituitary extracts relative to
luteal regression.

The first report of superovulating cattle dealt with experiments utilising pituitary extracts (Casida et al., 1943).
From the reviews cited above, it is clear that superovulation
with pituitary gonadotrophins, can be successful. Probably
the main reason that PMSG has evolved as the superovulatory
gonadotrophin of choice is its availability and convenience of
administration (because a single injection suffices) and not
its greater efficacy. Betteridge (1977) rightly points out
the lack of critical data comparing PMSG to pituitary gonadotrophins. Moore (1975) carried out some of the best comparative
experiments and found no differences between PMSG and pituitary
extracts. However, the timing and frequency of injections,
ratio of FSH to LH, and dose rates may not have been optimal.

RECENT EXPERIMENTS

Over the past few years we have been conducting studies
(unpublished) in which cattle were superovulated with PMSG or
pituitary extracts. We have used FSH and LH preparations (Armour) that are commercially available. Each hormone is considerably contaminated with the other (Laster, 1972) and with
other substances. Nevertheless, we established dose rates based
on the labelled potencies which are set according to the
manufacturer's standards. The FSH is extracted from porcine
and the LH from equine pituitaries. We have been using a 5 :
1 ratio of FSH to LH preparations because Onuma et al. (1969)
found that while the FSH preparation produced good follicular
growth in calves, the follicles did not ovulate as well as

TABLE 1

SUPEROVULATION TREATMENTS FOR CATTLE[a]

Experiment	No.	Superovulation Treatment[b]	No. Ovulations[c]	No. Transferrable embryos	Total ova	% Recovery	% Normal embryos
I	50	1800 iu PMSG day 15 (oestrus = day O)	14.4	4.7	7.8	54	60
	24	4,3,2,1,.5 mg FSH + 20% LH 2X daily days 15-19	6.7	4.4	5.3	79	83
	10	1800 iu PMSG day 10 + PG	10.8	5.5	6.3	56	87
	13	4,3,2,1,.5 mg FSH + 20% LH 2X daily days 10-14 + PG	12.8	4.1	8.8	69	47
II	6	1800 iu PMSG day 15	15.3	4.8	8.5	43	51
	7	1000 iu PMSG days 14 and 15	14.4	.2	2.0	29	52
	7	5,4,3,2 mg FSH + 20% LH 2X daily days 15-19	15.5	2.6	4.1	37	50
	8	1800 iu PMSG day 10 + PG	11.8	4.9	6.2	58	70
	7	1000 iu PMSG days 9 and 10 + PG	18.5	4.2	10.0	49	49
	6	5,4,3,2 mg FSH + 20% LH 2X daily days 10-14 + PG	22.5	10.7	12.2	54	81
III	24	1800 iu PMSG 10 + PG	11.1	5.7	6.9	60	74
	24	5,4,3,2, 2 mg FSH + 20% LH 2X daily days 10-14 + PG	11.0	5.8	7.9	72	76
	28	5,4,3,2,2 mg FSH 2X daily days 10-14 + PG	12.4	5.3	7.2	55	67
IV	20	1800 to 2700 iu PMSG day 10 + PG	6.2[d]	2.3[e]	3.4	55	69
	28	5,4,3,2,2 mg FSH + 20% LH 2X daily days 10-14 + PG or 1.5X gonadotrophin	10.5	7.7[f]	8.8	78	87

cont./.

TABLE 1 continued.

[a] Seidel et al., unpublished experiments.

[b] When treatments began on day 10, $PGF_2\alpha$ was injected on day 12 (15–30 mg) and again 4 to 9 h later (7.5 to 15 mg).

[c] Least-square means are presented for experiments II, III and IV.

[d] Estimated by rectal palpation.

[e,f] These resulted in 1.1 and 4.5 pregnancies per donor respectively.

those stimulated by PMSG. Seidel et al. (1971) were able to
overcome this by using a mixture of FSH and LH preparations.
There is also other evidence that both FSH and LH are required
for optimal follicular growth: cow (Laster, 1972), human (Berger et al., 1972), and laboratory rodents (Labhsetwar, 1970 and
Richards and Midgley, 1976).

In preliminary experiments, ten injections of 2 mg FSH
with .5 mg LH at half-day intervals beginning on day 15 of the
oestrous cycle and without any exogenous ovulatory hormone resulted in fewer ovulations and fertilised ova than a single
injection of 1800 iu of PMSG given on day 15, also without any
exogenous ovulatory treatment. The experiments described in
Table 1 were done subsequently.

The cattle in Experiment I were primarily Simmental-Here-
ford cross-bred heifers, but included a few three-quarter-blood
Maine Anjou and pureblood Simmental heifers. Most were in a
commercial embryo transfer programme; the unequal numbers are
due both to a limited availability of hormones and to chance.
Embryos were recovered surgically. There were no differences
among hormonal treatments in the number of normal embryos
recovered, indicating that the FSH-LH regimen is a suitable
alternative to PMSG. It may be wise to alternate these regimens when donors are to be superovulated repeatedly in order to
minimise antibody formation. The FSH-LH treatment without
prostaglandin resulted in recovery of a high percentage of
normal embryos relative to the ovulation rate.

The data of Experiment I include 17 donors from which no
attempt was made to recover ova. These animals either did not
display behavioral oestrus or had very poor follicular development and, therefore, were not inseminated. They were distributed evenly among treatments. For the purposes of analysis
these heifers were assigned one ovulation and zero embryos.

In Experiments II and III Hereford-Angus cross-bred heif-

ers were treated and surgical recovery attempted in all cases.
If a heifer failed to show behavioral oestrus in Experiment II,
she was given HCG 6 days after the initiation of treatment and
inseminated as if oestrus began at that time.

In Experiment III heifers not showing behavioral oestrus
by 3 days after prostaglandin $F_2\alpha$ ($PGF_2\alpha$) injection were inseminated as if they had shown oestrus 2 1/2 days after $PGF_2\alpha$

In Experiment II, two treatments stand out: 1000 iu PMSG
on each of Days 14 and 15 produced very few normal embryos and
the FSH-LH-$PGF_2\alpha$ treatment produced 10.7 normal embryos per
donor. However, the advantage of FSH-LH over PMSG was not
significant in Experiment III. Further, there was no significant difference when the LH was omitted from the FSH treatment.
From these experiments, it can be concluded that the FSH and
FSH-LH treatments are as efficacious as the PMSG treatment.
However, there may be other FSH-LH or PMSG regimens that are
superior to those tested.

It should be noted that with the exception of FSH-LH-$PGF_2\alpha$
treatments of Experiments II and III, similar treatments in
different experiments resulted in remarkably similar responses.
For example, the numbers of transferable embryos per donor
treated with PMSG-$PGF_2\alpha$ were 5.5, 4.9, and 5.7 for Experiments
I, II and III.

Experiment IV is an analysis of survey data from donors
in a commercial embryo transfer programme. The donors consisted of cows and heifers of numerous breeds. They were superovulated up to three times each at approximately 2-month intervals. Doses were determined by size, age, and previous responses. Donors were randomly assigned to one regimen or the
other for the first superovulation. PMSG and FSH-LH regimens
were alternated within the same donor. There was no significant decline in the number of normal embryos recovered with the
second and third superovulation, although only a few donors

were superovulated three times. All donors treated are included in the calculations. Recoveries were done non-surgically (Elsden et al., 1977) and, therefore, the numbers of ovulations were estimated by rectal palpation.

Although there are limitations to survey experiments of this kind, the FSH-LH regimen was clearly superior to the PMSG regimen. Possibly higher doses of PMSG would have led to better responses. We have found the FSH-LH regimen to be amazingly consistent. In this series more than four pregnancies resulted per donor treated (range 0 - 11) with non-surgical recovery and surgical transfer. Additional experiments are required to determine optimal doses and intervals for repeated superovulation of valuable donors.

ACKNOWLEDGEMENTS

This work was supported in part by grants from Select Sires, Inc., Plains City, Ohio and the Colorado State University Experiment Station through Regional Project W-112. The THAM salt of prostaglandin $F_2\alpha$ was donated by the Upjohn Co., Kalamazoo, Michigan through Dr. James Lauderdale. Excellent technical assistance was provided by J. Bowen, L. Case, R. Carter, J. Hasler, N. Homan and S. Seidel.

REFERENCES

Berger, M.J., Taymor, M.L. Karam, K. and Nudemberg, F. 1972. The relative
    roles of exogenous and endogenous follicle stimulating hormone (FSH)
    and luteinising hormone (LH) in human follicular maturation and
    ovulation induction. Fertil. Steril. 23: 783-790.

Betteridge, K.J. 1977. Techniques and results in cattle: Superovulation.
    In: Embryo Transfer in Farm Animals, ed. K.J. Betteridge. Canada
    Department of Agriculture Monograph 16. pp. 1-9.

Casida, L.E., Meyer, R.K., McShan, W.H. and Wisnicky, W. 1943. Effects of
    pituitary gonadotrophins on the ovaries and induction of superfecun-
    dity in cattle. Am. J. Vet. Res. 4: 76-94.

Church, R.B. and Shea, B. 1976. Some aspects of bovine embryo transfer.
    In: Egg Transfer in Cattle, ed. L.E.A. Rowson. Commission of the
    European Communities. Luxembourg. EUR 5491. pp. 73-86.

Elsden, R.P., Hasler, J.F. and Seidel, Jr. G.E. 1976. Non-surgical recov-
    ery of bovine eggs. Theriogenology. 6: 523-532.

Foote, R.H. and Onuma. H. 1970. Superovulation, ovum collection, culture,
    and transfer. A review. J. Dairy Sci. 53: 1681-1692.

Gordon, I. 1975. Problems and prospects in cattle egg transfer. Irish
    Vet. J. 29: 21-30 and 39-62.

Labhsetwar, A.P. 1970. Synergism between LH and FSH in the induction of
    ovulation. J. Reprod. Fertil. 23: 517-519.

Laster, D.B. 1972. Follicular development in heifers infused with follicle
    stimulating hormone. J. Reprod. Fertil. 28: 285-289.

Moore, N.W. 1975. The control of time of oestrus and ovulation in the
    induction of superovulation in cattle. Aust. J. Agric. Res. 26: 295-
    304.

Onuma, H., Hahn, J., Maurer, R.R. and Foote, R.H. 1969. Repeated superovul-
    ation in calves. J. Anim. Sci. 28 : 634-637.

Richards, J.S. and Midgley, Jr. A.R. 1976. Protein hormone action: A key
    to understanding ovarian follicular and luteal cell development.
    Biol. Reprod. 14: 82-94.

Seidel, G.E., Jr., Larson, L.L. and Foote, R.H. 1971. Effects of age and
    gonadotrophin treatment on superovulation in the calf. J. Anim. Sci.
    33: 617-622.

Sreenan, J.M. and Beehan, F. 1976. Methods of induction of superovulation in the cow and transfer results. In: Egg Transfer in Cattle. Ed. L.E.A. Rowson. Commission of the European Communities, Luxembourg. EUR 5491. pp 19-34.

Wildt, D.E. Woody, H.D. and Dukelow, W.R. 1975. Induction of multiple ovulation in the cow with single injections of FSH and HCG. J. Reprod. Fertil. 44: 583-586.

RELATIONSHIPS BETWEEN OVARIAN STIMULATION BY PMSG AND STEROID SECRETION

J. Saumande

INRA - Station de Physiologie de la Reproduction
37380 Nouzilly France.

INTRODUCTION

For 40 years, research on superovulation in the cow has
not made any noticeable progress. The variability in response
is still very wide although many factors which affect this
variability have been shown. To solve this problem, it is not
enough to treat the animals, to observe how they come into
oestrus and to count the number of ovulations - we have to improve the methods of analysis of the effects of treatments. A
first step has been made when the consequences of PMSG treatments
on the follicular population were studied (Mariana and Machado,
1976; Mauléon and Mariana, 1977; Turnbull et al., 1977). In
the human, hormonal analysis gives very good results of the
primary reason for amenorrhea and the efficiency of its treatment. We thought that the methodology could be applied to the
problem of superovulation in the cow.

Most of the time, eggs are obtained from adult cycling
animals; however due to the interest in accelerating the
selection, experiments have been also carried out on prepubertal
animals.

1. - PREPUBERTAL ANIMALS

To obtain eggs from prepubertal animals, two types of
treatment are used:

- Since Casida (1943) injection of PMSG followed 120 h later
by an injection of HCG which is necessary to obtain ovulation.

- Testart (1972) proposed that, at PMSG injection, a vaginal
FGA pessary be inserted and removed 4 days later. Compared with

170

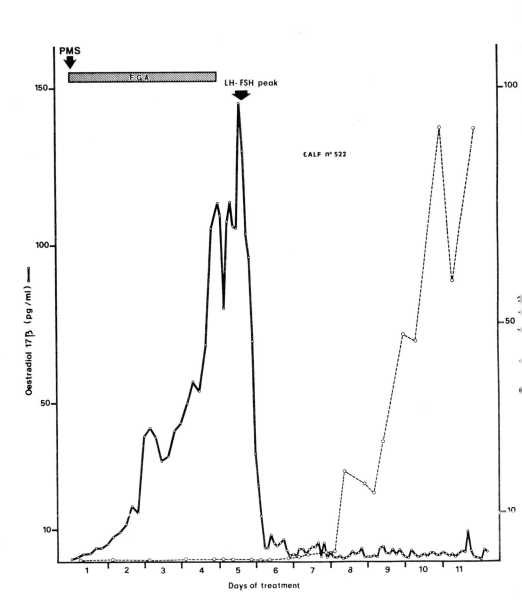

Fig. 1.

the first method, without any HCG, the percentage of non-superovulated animals decreases and the ovulations are synchronised;
the animals seem to be more sensitive to the gonadotrophin.

We have analysed hormonal levels for these two types of
treatments (Saumande and Testart, 1974; Testart et al., 1977).

1. Description of hormonal patterns (Figure 1)

Gonadotrophin hormones
In the animals which only receive PMSG, LH always stays at
the same basal level which explains the need to use HCG. In
FGA - PMSG treated heifers, 12 to 22 h after the pessary removal,
preovulatory LH - FSH releases are observed resembling those
of normal cyclic animals.

Oestradiol-17β
In all animals, 24 h after PMSG injection, an increase in
concentration is evident and is the same for all animals whether
or not treated with FGA. The maximum levels are reached 5
days after PMSG which is between 10 and 18 h after sponge
removal in the groups which received the FGA treatment. The
drop off in secretion depends on the LH discharge or injected
HCG. After ovulation, high levels of oestradiol are recorded
sporadically.

Progesterone
During the first 6 days, progesterone concentrations are
always less than 0.5 ng/ml. After ovulation, the increase in
levels is shown earlier as the number of corpora lutea increases.
At 72 h after the LH - FSH peak, progesterone levels vary from
3 ng/ml to more than 40 ng/ml compared with 1 ng/ml in cycling
treated animals; 2 days later, in many heifers, it goes up to
80 - 100 ng/ml. Part of this secretion comes from luteinised
follicles.

2. Oestradiol and ovulation rate

The magnitude of oestradiol secretion can be appreciated by the measurement of the area under the curve. This value during the preovulatory period (from PMSG injection to 48 h after sponge removal, ie after the mean time of ovulation) is highly correlated with the number of ovulations ($r = 0.92$).

The peak concentration of oestradiol recorded after sponge removal is also highly correlated with the ovulation rate ($r = 0.98$).

3. Oestradiol and follicular growth

The association between oestrogen secretion and the presence on the ovary of large follicles is well known. In our experiments, we have analysed this association more precisely.

- After PMSG injection, the profiles of follicular growth with time (as measured by the mean diameter of the 3 greatest follicles) and oestradiol levels are similar (Figure 2).

- The comparison between oestradiol levels and follicles greater than 1 mm, 3 days after PMSG injection, leads us to the following conclusions:

. in the peripheral blood, oestradiol concentrations are mainly due to the largest follicles (Table 1).
The number of 8 - 12 mm follicles and the secretion of oestradiol during the last 24h, estimated by the area under the curve, are highly correlated ($r = 0.98$); on the contrary, there is no correlation between this parameter of oestradiol secretion and either the number of 4 - 8 mm follicles ($r = 0.13$) or the total number of follicles larger than 4 mm ($r = 0.46$). The total number of follicles larger than 1 mm did not influence the oestradiol concentration observed.

. as far as oestradiol concentrations are concerned, the size of the follicle is more important than its degree of maturation* (Figure 3): three days after gonadotrophin treatment,
* Determined from the organisation of granulosa cells around the oocyte

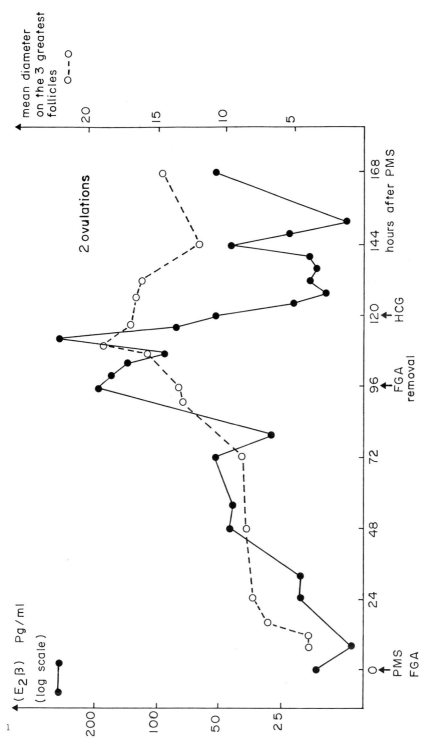

Fig. 2.

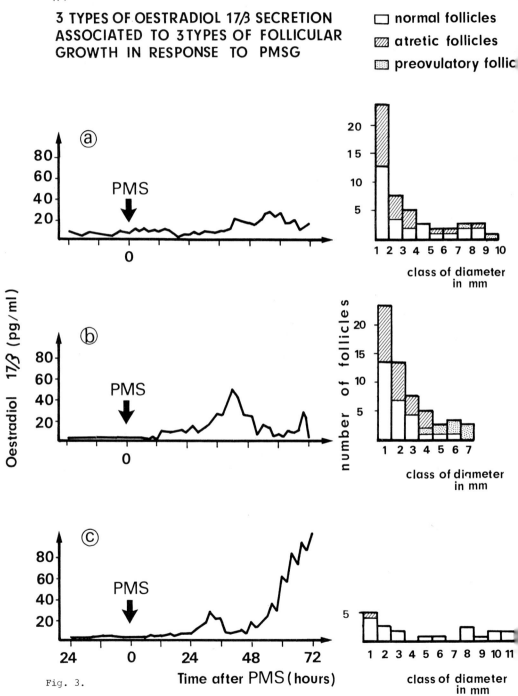

3 TYPES OF OESTRADIOL 17β SECRETION ASSOCIATED TO 3 TYPES OF FOLLICULAR GROWTH IN RESPONSE TO PMSG

☐ normal follicles
▨ atretic follicles
▨ preovulatory follic

Fig. 3.

TABLE 1

OESTRADIOL AND FOLLICULAR POPULATION IN PMSG TREATED CALVES.

Calf Number	Ovary studied	Number of follicles				Atretic (%)	Quantity of oestradiol[+]
		Follicle diameter (mm)			Total number		
		1-4	4-8	8-12			
650	R.O.	116	2	0	118	50.8	135.8
632	L.O.	47	36	3	86	37.2	221.8
642	L.O.	52	43	0	95	27.4	271.6
654	L.O.	250	0	0	250	57.6	242.0
630	L.O.	42	10	4	56	41.1	369.4
648	R.O.	63	9	1	73	41.1	382.6
656	R.O.	50	16	1	67	34.3	425.0
640	R.O.	15	3	8	26	23.1	1220.0
638	L.O.	24	28	20	72	38.9	2395.2

+ Area under the oestradiol curve during the last 24 h (pg/ml x h)

the concentration was 4 times higher for 6 x 8-11 mm follicles (Case c) than for 8 x 4-8 mm preovulatory follicles (Case b).

- Oestradiol levels measured after ovulation suggest a post-ovulatory follicular growth. The large follicles observed 5 days after ovulation secrete this oestrogen but they are not of the same generation as those which have ovulated. Let us consider the following facts:

. the correlation coefficient between oestradiol secretion and the number of corpora lutea plus large follicles is lower than that with the number of corpora lutea (Table 2):

TABLE 2

THE RELATIONSHIP BETWEEN OESTRADIOL-17β AND THE NUMBER OF CORPORA LUTEA (CL) OR THE NUMBER OF CORPORA LUTEA + LARGE FOLLICLES IN PMSG - FGA TREATED CALVES

	Correlation coefficient between oestradiol levels and the number of:	
	C.L.	C.L. + large follicles
Area under the curve	$r = 0.92$ $P < 0.01$	$r = 0.80$ $P < 0.05$
Preovulatory peak value	$r = 0.98$ $P < 0.01$	$r = 0.89$ $P < 0.01$

. after the preovulatory release of gonadotrophin, all follicles larger than 3 mm degenerate (atresia, luteinisation: Arrau, 1974) and such follicles do not secrete any oestrogens (Hay and Moor, 1975); therefore they cannot explain the post-ovulatory oestradiol secretion.

. 2 days after ovulation, the ovarian weight is correlated with the number of ovulations ($r = 0.74$; $P < 0.05$) but not at 5 days; at this time, it is correlated with the number of large follicles ($r = 0.89$; $P < 0.01$).

So, if large follicles present at observation (ie Day 5)
were not at preovulatory size at the time of ovulation, it is
not justified to calculate ovulation yield:

$$(\frac{\text{No. corpora lutea}}{\text{No. corpora lutea + large follicles}}).$$

In fact we can see the long term effect of PMSG (see part 2).

4. Some practical conclusions

The interest of this work is not limited to the description
of hormonal events induced by treatments. We have shown that
it leads us towards a better analysis of results (post-ovulatory
growth). It also helps us to have a better understanding of the
non-success of treatment, mainly the variability in the number
of ovulations and the poor quality of oocyte.

a) Variability of response

As in FGA - PMSG treated calves, there is a spontaneous
LH - FSH preovulatory discharge. We think that the quality of
the stimulation is better than with PMSG only; moreover, it has
been noticed that the gonadotrophin discharge - ovulation
interval is the same as in the adults. The FGA - PMSG treatment
provides all the necessities for the ovulation to occur.

Under these conditions, the variability in the number of
ovulations induced is no longer due to the treatment (time and
dosage of HCG) but reflects the possibility of animals to be
stimulated. The origin(s) of this variability which will be
found again in cyclic animals also exists in prepubertal animals.
We do not know if the responses to a superovulatory treatment
before and after puberty can be related, if they are, we have a
very early method to detect animals which could be used as
donors for egg transfer.

b) Quality of oocyte

Eggs obtained from prepubertal treated animals are of poor
quality even though follicular growth is good as shown by

endocrinological and histological studies. Two hypotheses can
be formulated:

- deleterious effects of high levels of steroids. The high
concentrations of oestradiol and progesterone are related to the
high number of ovulations, but for the same number they are
higher in calves than in adult animals, as shown by the slope
of the regression line between the number of ovulations and the
level of steroids (Table 3).

TABLE 3

RELATIONSHIP BETWEEN OVULATION RATE AND STEROID LEVELS IN SUPEROVULATED
CALVES AND HEIFERS

Slope of the regression line between the number of ovulations and:	Calf	Cyclic animals
. the maximum preovulatory $E_2\beta$concentration	12.987	5.661
. the progesterone concentration 3 days after ovulation	0.578	0.389

Whatever the cause, these concentrations would be an
unfavourable environment   initially for the oocytes and later
for the eggs (Chang, 1966, 1967; Humphrey, 1968).

- lack of an essential maturation factor for the oocyte.
Such a factor would appear with cyclicity. The prepubertal
ovarian follicle seems to be different from that of the adult
in the composition of its follicular fluid and its behaviour
in culture (Thibault et al., 1975).

2. - PUBERTAL ANIMALS

During an experiment of superovulation with the PMSG - PG
treatment, 14 cyclic heifers of the FFPN breed were bled every
2 hours from PMSG injection to 24 h after the beginning of
oestrus, then twice a day until Day 10 when ovaries were observed
by endoscopic examination.

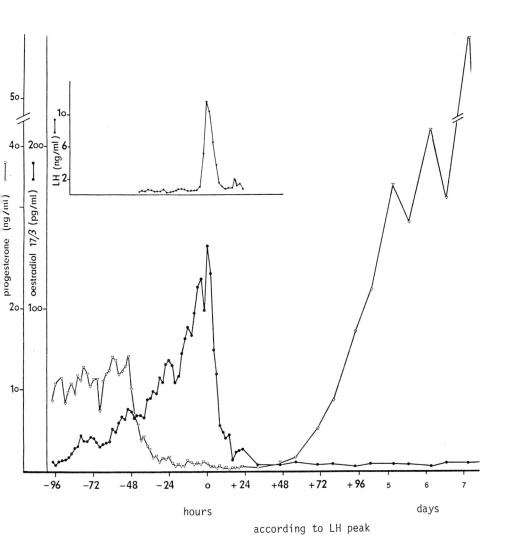

Fig. 4.

Animals received different treatments summarised in Table 4 together with results of oestrus and ovarian response. Hormonal results will be studied separately for superovulated and non-superovulated animals; superovulated animals will be considered as normal and used as reference.

1. Superovulated heifers (Figure 4)
    a) LH (Table 5)
According to the animal, basal levels are between 1.0 and 3.9 ng/ml which is the same range found in non treated cyclic animals. The preovulatory discharge appears 43.8 ± 4.9 h after PG injection; the value of the standard deviation shows that they are very well synchronised, which agrees with oestrus results.

Among animals there is a 5-fold difference in the maximum concentration and also in the area of the peak but these levels and variations do not differ from those found in non-treated animals. None of these parameters are related to the number of ovulations.

It seems that the superovulatory treatment itself or its hormonal consequences do not affect the LH pattern.

    b) Progesterone (Table 6)
    - Before ovulation: The first consequence of PMSG injection is an increase in progesterone levels. At the time of PG injection, concentrations have increased by 22 - 158%. This effect already demonstrated in PMSG - HCG treated animals (Henricks et al., 1973; Thibier and Saumande, 1975), reflects the LH activity of the hormone (Cole et al., 1932; Stewart et al., 1976).

The prostaglandin acts rapidly - 4 h after injection, the decrease in progesterone concentrations is about 50%; between 10 h and 32 h, they are 1 ng/ml. Corpus luteum regression or the loss in its activity seems incomplete: mean concentrations during the LH peak were higher than 0.5 ng/ml (in non-treated

TABLE 4

EEFECTS OF SUPEROVULATORY TREATMENT IN PMSG – PG TREATED HEIFERS

| Type of treatment | | Animal number | PG-oestrus interval (h) | Number of | |
Day of injection ƒ	Dose of PMSG (iu)			Corpora lutea	Large follicles
8	1200	41	48	14	4
		50	48	9	2
		51	72	1	2
		62	– *	2	1
8	2000	44	48	1	3
		49	– *	1	3
		55	48	27	0
16	2000	43	168	0	6
		47	48	17	2
		52	48	14	2
16	2400	45	48	20	1
		46	72	– **	– **
		48	48	12	2
		53	48	10	1

In all cases, 500 µg of PGF$_2\alpha$analogue (ICI 80996) were injected 48 h after PMSG.

* These animals did not come into oestrus during the sampling period

** Ovaries could not be observed. From the relationship between the number of corpora lutea and the progesterone and oestradiol levels we calculated respectively 19 and 15 ovulations.

ƒ Day 0 = day of oestrus.

TABLE 5

LH RESULTS IN PMSG – PG TREATED HEIFERS

Animal * number	Basal level (ng/ml)	Preovulatory peak			
		Interval from PG injection (h)	Duration (h)	Maximum value (ng/ml)	Area under the curve
43	2.7	–+	–	–	–
44	3.9	44	10	16.5	120.8
49	2.1	–+	–	–	–
51	1.7	70	14	14.4	108.6
62	1.4	–+	–	–	–
50	1.9	48	12	32.7	149.4
53	0.9	42	12	36.7	180.6
48	1.2	48	12	38.7	218.0
41	1.4	46	10	8.8	45.6
52	1.6	40	12	21.7	117.4
47	1.0	50	12	11.6	78.2
46**	0.9	46	12	6.7	33.4
45	1.1	38	12	32.0	152.6
55	2.7	36	14	21.4	125.0

* Classified according to the ovulation rate

** Ovulation rate was calculated (see Table 4)

+ No preovulatory peak recorded

TABLE 6a

PROGESTERONE RESULTS AFTER PMSG – PG TREATMENT a) EFFECT OF PMSG INJECTION

Animal* number	Concentration		Relative increase (%)
	At PMSG injection (ng/ml)	Immediately before PG injection (ng/ml)	
43	8.3	14.4	73
44	0.6	0.4	–
49	4.3	10.0	132
51	3.1	8.0	158
62	5.4	9.9	83
50	5.9	8.4	42
53	12.3	18.9	54
48	5.0	6.1	22
41	6.4	12.1	89
52	7.5	9.2	23
47	8.6	14.1	64
46**	7.5	15.6	108
45	5.8	12.3	112
55	5.7	11.9	109

* Classified according to the ovulation rate

** Ovulation rate was calculated (see Table 4)

TABLE 6b

PROGESTERONE RESULTS AFTER PMSG – PG TREATMENT b) STEROID LEVELS AFTER PG INJECTION

Animal * number	Decrease in the level 4 h after PG injection (%)	Mean concentration during LH peak (ng/ml)	Level on Day 3 (ng/ml)
43	52	+	0.4
44	–	0.24	0.9
49	62	+	++
51	66	0.05	0.7
62	47	+	1.0
50	53	0.50	2.7
53	44	0.89	6.4
48	29	1.06	3.9
41	42	1.58	5.9
52	68	0.76	6.1
47	52	0.66	5.4
46**	40	1.64	6.8
45	49	0.99	8.1
55	49	0.90	11.4

+ No LH peak recorded

++ This animal did not come into heat

* Classified according to the ovulation rate

** Ovulation rate was calculated (see Table 4)

animals, this never occurs) and for 4 heifers higher than 1
ng/ml (up to 1.6 ng/ml).

A part of the secretion could come from follicles (Thibier
and Saumande, 1975).

- after ovulation, when the number of corpora lutea increases,
the rise in progesterone concentration occurs earlier and is
quicker than normal. Steroid concentration is correlated to
the number of corpora lutea as soon as the morning of Day 1,
ie a few hours after ovulation (Figure 5).

The relationship between the number of corpora lutea and
progesterone concentration previously demonstrated on Day 10
(Lemon and Saumande, 1972) is confirmed and can be shown to
occur even earlier.

From a practical point of view, the number of ovulations
can hardly be determined from the progesterone levels on Day 1,
but on Day 3 it is possible to separate superovulated and non-
superovulated animals: by this time, maximum levels in animals
with less than 5 ovulations is 1 ng/ml; the maximum levels in
those with more than 5 ovulations is 2.7 ng/ml; such a difference
can be determined with a rapid assay used for pregnancy diagnosis.

C) Oestradiol (Table 7)

In most of the animals which are superovulated, the increase
in oestradiol concentrations can be found 24 h after PMSG
injection; it is highest in heifers which will have the greatest
ovulation rate. This increase accelerates with time and between
36 and 52 h after PG injection, the preovulatory maximum which
is correlated with the number of ovulations (r = 0.90) is
recorded.

After the preovulatory peak, the oestradiol level decreases
dramatically and 24 h - 26 h afterwards reaches the pre-treatment concentration.

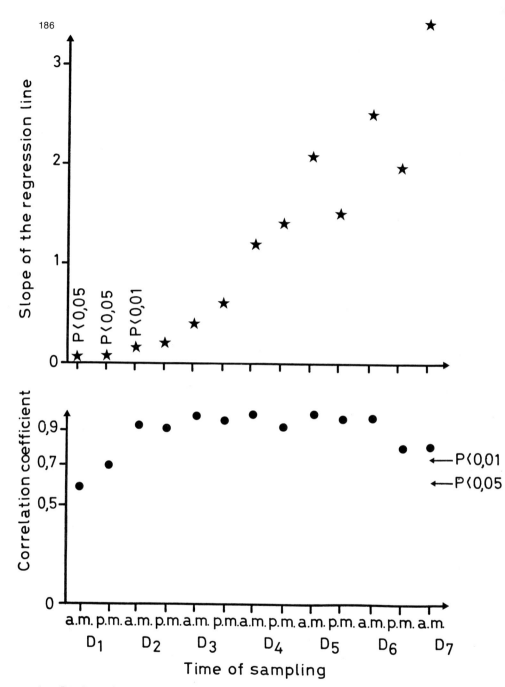

Fig. 5. The relationship between the number of corpora lutea and progesterone
levels

TABLE 7

OESTRADIOL AFTER PMS - PG TREATMENT

Animal* number	Relative increase 24 h after PMSG +	Preovulatory peak	
		Time after PG injection (h)	Maximum value (pg/ml)
43	1.82	-++	-
44	0.89	44	23.2
49	0.72	-++	-
51	0.80	72	14.8
62	0.79	-++	-
50	4.90	48	80.5
53	3.01	38	95.4
48	1.25	46	62.2
41	2.05	50	92.0
52	0.80	44	50.9
47	4.19	52	138.6
46**	0.71	48	101.2
45	3.62	38	146.9
55	3.89	36	158.8

* Classified according to the ovulation rate

** Ovulation rate was calculated

\+ Mean value of the 3 first samples
mean value of the samples taken in the following 24 h

\++ No peak recorded during the sampling period.

In two animals (46 and 53) treated on Day 16 with the highest dose of PMSG, we observed a second rise of oestradiol with a maximum 3 days after oestrus (respectively 226 pg/ml and 75.2 pg/ml). After this peak, the decrease was slower: it lasted more than 48 h, therefore probably according to a mechanism different from that induced by the ovulatory discharge of gonadotrophin.

This postovulatory increase was previously considered as the rule (Booth et al., 1975) whereas we found it only in two animals. If it is due to the largest follicles present on the ovary 10 days after oestrus, we do not think that these follicles are the follicles which have not ovulated with the others for two main reasons:

- firstly, the high correlation coefficient between the maximum oestradiol level and the number of ovulations. Such a relationship can be found only if all oestradiol secreting follicles ovulate or if a constant percentage of them do not ovulate; however, neither of these 2 conditions are found experimentally, as the number of large follicles present on the ovary when corpora lutea are counted is highly variable and not related to the number of ovulations.

- secondly, the atretic processes which affect all large follicles after the LH discharge making them unable to secrete oestrogen.

So, if these large follicles are not of the same generation as those which ovulate, we are wrong in calculating an ovulation yield from observations made 6 to 10 days after ovulation.

In fact, we think that the postovulatory increase in oestradiol-17β and the presence of large follicles which it reflects are the consequence of the long term effect of PMSG either directly or indirectly:

. directly: this postovulatory growth could be due to the proportion of PMSG still active. We know that the half-life of PMSG is very long (Saumande, 1977) and that as few as 50 iu can induce superovulation (Smith et al., 1977).

. indirectly: it has been demonstrated that PMSG injection acts on all the follicular population. The growth of follicles too small to ovulate can be stimulated sufficiently to make them large enough to secrete large amounts of oestradiol.

2 Non-Superovulated heifers

Usually following administration of PMSG animals that are non-superovulated are classified as insensitive to the dose of PMSG. The study of hormonal levels (Figure 6) enables us to be able to distinguish 3 different situations:

a) Insensitive animals

No qualitative hormonal difference from superovulated heifers can be demonstrated (No. 51). This animal is truly insensitive to the dose of PMSG used.

b) Inappropriate treatment for the physiological situation

In animal No. 44, steroid levels at time of PMSG injection are those of preovulatory period, not those of Day 8; there could have been an error in oestrus detection, or, these steroid levels could reflect a pathological situation.

We cannot conclude on the sensitivity to PMSG as the treatment used was inappropriate for the physiological state of the animal.

c) Incomplete response to the treatment

In such animals, one of many events caused a failure in ovulation:

- failure of luteolysis (heifer No. 49): this animal came into oestrus 11 days after PMSG, ie 19 days after the preceding oestrus. This case, plus the high levels of progesterone

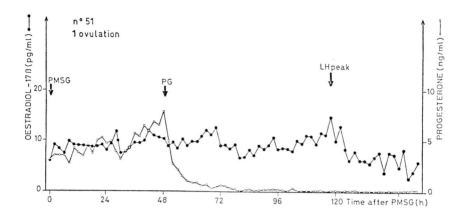

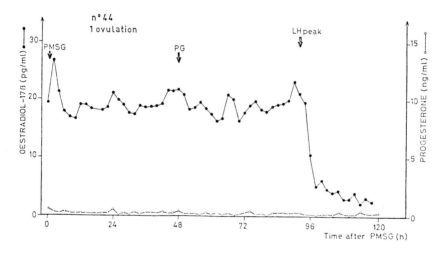

Fig. 6.

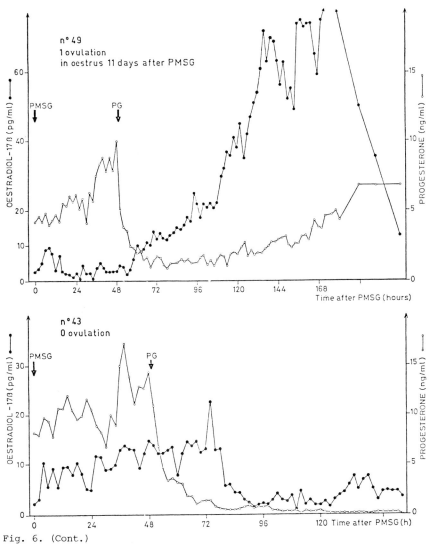

Fig. 6. (Cont.)

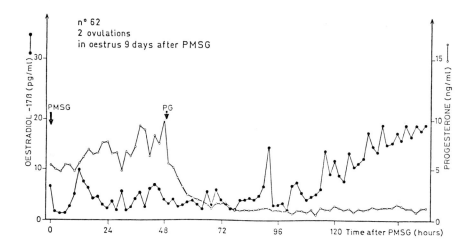

Fig. 6. (Cont.)

recorded during the LH peak of superovulated animals show that the conditions to induce luteolysis in PMSG treated animals have to be reconsidered. It is to be noticed that from oestradiol levels this animal cannot be classified as non responding to PMSG.

- lack of preovulatory LH discharge (heifer No. 43): despite normal corpus luteum regression and increase in oestradiol concentration, no LH peak was recorded.

- non synchrony of events leading to ovulation (heifer No. 62): in contrast to superovulated animals, oestradiol levels increased only when progesterone levels were low.

In a previous experiment (unpublished results) where animals were treated to induce twinning (1600 iu PMSG on Day 16, 1500 iu HCG on Day oestrus), we drew the same conclusions: few non-superovulated animals are really insensitive to PMSG. In most cases, if they fail to respond to treatment, it is because the treatment was inappropriate or because some of the mechanisms leading to ovulation did not work.

In conclusion, it can be seen that hormonal analyses have allowed us to have a better understanding of the non-super-ovulated animals and at the same time, it gives us a means to decrease the incidence of such animals.

ACKNOWLEDGMENT

The author wishes to thank L. Cahill for assistance with the preparation of this manuscript.

194

REFERENCES

Arrau, J., 1974. Ann. Biol. anim. Bioch. Biophys. 14, 633.

Booth, W.D., Newcomb, R., Strange, H., Rowson, L.E.A. and Sacher, H.B., 1975. Vet. Rec. 97, 366.

Casida, L.E., Meyer, R.K., Mc Shan, W.H. and Wisnicky, W., 1943. Am. J. Vet. Res. 4, 76.

Chang, M.C., 1966. Endocrinology 79, 939.

Chang, M.C., 1967. Endocrinology 81, 1251.

Cole, H.H., Guilbert, H.R. and Goss, H.T., 1932. Am. J. Physiol. 102, 237.

Hay, M.F. and Moor, R.M., 1975. J. Reprod, Fert. 43, 313.

Henricks, D.M., Hill, J.R., Dickey, J.F., and Lamond, D.R., 1973. J. Reprod. Fert. 35, 225.

Humphrey, K.W., 1968. J. Endocr. 42, 17.

Lemon, M. and Saumande, J., 1972. J. Reprod. Fert. 31, 501 (abstr.)

Mariana, J.C., 1970. Ann. Biol. anim. Bioch. Biophys. 10, 575.

Mariana, J.C. and Machado, J., 1976. Ann. Biol. anim. Bioch. Biophys 16, 545.

Mauléon, P. and Mariana, J.C., 1977. In Cole and Cupps, 3rd edition, p. 175. Academic Press New York, San Francisco, London.

Saumande, J., 1977. Ann. Med. Vet. 121, 449.

Saumande, J. and Testart, J., 1974. Theriogenology 2, 121.

Smith, M.F., Burrell, W.C., Shipp, L.D., Sprott, L.R., Songster, B. and Wiltbank, J.N., 1977. Animal Repoduction Science: accepted for publication.

Stewart, F., Allen, W.R. and Moor, R.M. 1976. J. Endocr. 71, 371.

Testart, J., 1972. VIIth Cong. Intern. Anim. Reprod. Artif. Insem. 1, 493.

Testart, J., Kann, G., Saumande, J. and Thibier, M., 1977. J. Reprod. Fert. 51, 329.

Thibault, C., Gerard, M. and Menezo, Y., 1975. Ann. Biol. anim. Bioch. Biophys. 15, 705.

Thibier, M. and Saumande, J., 1975. J. Steroid. Biochem. 6, 1433.

Turnbull, K.E., Braden, A.W.H. and Mattiner, P.E., 1977. Austr, J. Biol. Sci. 30, 229.

FACTORS AFFECTING THE VARIABILITY OF OVULATION RATES AFTER PMSG STIMULATION

J. Saumande, D. Chupin, J.C. Mariana, R. Ortavant, P. Mauleon
INRA - Station de Physiologie de la Reproduction
37380 Nouzilly France.

INTRODUCTION

In all reports dealing with hormonal induction of super-ovulation in cattle, variability of response is observed, whatever the criteria used. Many experiments have been carried out to clarify the reasons for these variations and ways to decrease them. This report deals with major and recent results.

In this paper, response to treatment is defined only as number of ovulations. Egg quality, which does not depend on number (Testart, 1975; Renard et al., 1977) has not been studied.

1. SOME FACTORS MODIFYING VARIABILITY OF OVARIAN RESPONSE TO PMSG

1. Relationship between mean ovulation number and variance

Many authors have shown that the mean number of ovulations increases with dose of PMSG, but it is rarely noted that variability increases also (Table 1). As variances are heterogeneous and the distributions of ovulation number are not normal, it is impossible to predict the response to an injection of PMSG.

Another model has been suggested to represent the variation as a continuous variable (Mariana, 1970). An individual threshold could exist for each number of ovulations and for each cow and, at the population level, one could define threshold such as, for example, 50% of cows respond by 'n' ovulations to a given dose of PMSG. This is the reason why all the following results are estimated not only by the mean and standard deviation but also by the distribution between some classes defined according to the aim of the experiments:

TABLE 1

THE DOSE RESPONSE RELATIONSHIP IN FPN COWS TREATED WITH PMSG

Dose of PMSG (iu)	Animal number	Ovulation rate		Percentage of animals with		
		m+sd	(range)	0-1 ovulation	2-4 ovulations	more than 4 ovulations
1600	62	2.46 + 4.26	(0 - 13)	61.3	22.6	16.1
2000	62	4.67 + 5.07	(1 - 19)	43.5	22.6	33.9
2500	60	5.24 + 6.15	(0 - 37)	41.7	15.0	43.3
3120	60	8.84 + 8.42	(0 - 40)	23.3	16.7	60.0

0 - 1, 2 - 3 - 4, more than 4 for the hormonal induction of twinning: except when the opportunity offers, one can consider that quadruplets is the maximum limit for a normal pregnancy in the cows.

0 - 1, 2 - 5, more than 5 for egg production in trans-plantation experiments.

This law of threshold response can be illustrated by two examples:

When Gordon et al. (1962) increased the dose of PMSG from 800 to 1000 iu, mean ovulation number increased from 1.43 to 1.77, but the distribution between the three classes defined here was not modified and the percentage of 0 - 1 ovulations did not decrease.

With doses of PMSG varying from 50 to 500 iu, Smith et al. (pers. comm.) obtained 25 to 29% of multiple ovulations at the end of an oestrus synchronisation treatment; with 750 iu, this percentage increased to 56%.

2. <u>Genetic origin of this variation</u>
   a) <u>Variations between breeds</u>
   We have reported here only the results of experiments planned to compare at the same time and with the same PMSG, the response of different breeds and suppressing the effects of many other factors such as for instance, year, location, nutrition, age, stage of lactation, interval since calving, type of treatments, quality of PMSG...

The results of some of these experiments are given in Table 2. The experimental conditions were as follows:

<u>Experiment 1</u>: Primiparous dairy cows (Friesian) or primiparous suckling cows (Charolais) having received, three months after calving, one im injection of PMSG on Day 16 of a normal oestrous cycle.

<u>Experiment II and III</u>: Double muscled heifers of several beef breeds of 18 months of age having received 1000 iu of PMSG

TABLE 2A

COMPARISON OF OVARIAN SENSITIVITY TO PMSG FOR DIFFERENT BREEDS. EXPERIMENTS I AND II.

Breed	Dose of PMSG (iu)	Animal number	Ovulation rate		Percentage of animals with:		
			m+sd	(range)	0 - 1 ovulations	2 - 4 ovulations	more than 4 ovulations
Experiment I							
Friesian	1600	15	2.47 ± 3.20	(0 - 13)	73.3	6.6	20.0
	2000	16	5.00 ± 4.30	(1 - 16)	43.8	12.5	43.8
Charolais	1600	17	6.17 ± 9.46	(1 - 41)	29.4	29.4	41.2
	2000	14	10.35 ± 5.86	(1 - 22)	7.1	7.1	85.7
Experiment II							
Charolais	1000	8	1.00 ± 0.76	(0 - 2)	75.0	25.0	0
Charolais x Blonde Aquitaine	1000	16	1.38 ± 0.81	(1 - 4)	75.0	25.0	0
Blonde Aquitaine x Charolais	1000	12	1.08 ± 1.08	(1 - 4)	83.0	17.0	0
Blonde Aquitaine	1000	11	1.46 ± 0.93	(1 - 3)	63.6	36.4	0

TABLE 2B

COMPARISON OF OVARIAN SENSITIVITY TO PMSG FOR DIFFERENT BREEDS. EXPERIMENTS III AND IV.

Breed	Dose of PMSG (iu)	Animal number	Ovulation rate		Percentage of animals with:		
			m+sd	(range)	0 - 1 ovulation	2 - 4 ovulations	more than 4 ovulations
Experiment III							
Maine-Anjou	1000	27	2.6 ± 2.6	(0 - 10)	55	26	19
Blonde Aquitaine	1000	10	2.5 ± 1.8	(1 - 6)	40	40	20
Blonde Aquitaine x Limousine	1000	10	2.0 ± 1.4	(1 - 4)	70	30	0
Limousine	1000	12	1.5 ± 1.2	(1 - 5)	83	8	8
Experiment IV							
Charolais	1600	35	2.94 ± 3.38	(0 - 16)	60	14.3	25.7
Normande	1600	43	2.60 ± 3.02	(0 - 10)	60.4	23.3	16.3

on the last day of a synchronisation treatment (Norethandrolone - 18 Days).

Experiment IV: Suckling cows (Charolais) and dairy cows (Normande) having received at two months postpartum 1600 iu of PMSG on Day 16 of a normal oestrous cycle.

These results show that some breeds are more sensitive to PMSG in regard to the number of ovulations: Maine-Anjou and Blonde d'Aquitaine breeds are the most sensitive, Limousine and Friesian breeds the least (Table 3). They are not related to the type of production (milk vs beef).

TABLE 3

CLASSIFICATION OF BREEDS ACCORDING TO THEIR SENSITIVITY TO PMSG

Sensitivity	Breed	Index of merit	Type of production
Lowest	Friesian	1	Dairy
	Limousine	2.19	Beef
	Normande	2.20	Dairy and beef
	Charolais	2.50	Beef
	Blonde d'Aquitaine	3.65	Beef
Highest	Maine-Anjou	3.80	Dairy and beef

b) Intra-breed variations

We have observed the differences in response to the same superovulation treatment (1000 iu PMSG, 1500 iu HCG) between six groups of Charolais half sisters (Table 4) (Mariana et al., 1977). Mean ovulation response varied from 1.47 (Group 1) to 3.62 (Group 6) and for these two groups, the proportion of cows responding with more than one ovulation was 31.6% vs 75.0% respectively.

These results demonstrate that it is difficult to get a really homogeneous group of cows for an experiment and the necessity to work on a large number of animals to obtain worthwhile conclusions.

TABLE 4

VARIATION OF SENSITIVITY AMONG HALF-SIBS OF THE CHAROLAIS BREED (1000 iu PMSG ON DAY 16)

Family number	Animal number	Ovulation rate		Percentage of animals with:		
		m±sd	(range)	0 - 1 ovulation	2 - 4 ovulations	more than 4 ovulations
3	19	1.47 ± 0.94	(0 - 4)	68.4	31.6	0
5	22	1.50 ± 0.95	(1 - 5)	68.2	27.3	4.5
1	21	1.71 ± 1.27	(1 - 6)	66.7	28.6	4.8
4	22	2.00 ± 1.66	(0 - 7)	50.0	40.9	9.1
6	16	2.25 ± 1.82	(1 - 6)	56.2	25.0	18.8
2	8	3.62 ± 2.64	(1 - 8)	25.0	37.5	37.5

3. Effect of ovarian status at the time of PMSG

   a) Ovarian appearance on Day 16

We have tried to correlate the appearance of the ovary observed by coelioscopy on the day of PMSG injection, with the ovarian response (Table 6). The ovaries were classified according to the number of small follicles (size less than 5 mm) and to the presence or absence of medium (6 - 8 mm) or large (more than 8mm) follicles (Table 5). Firstly, there is a great variability in follicular development or more precisely in the external follicular appearance between females at the same stage of the oestrous cycle. Secondly, the results of Dufour et al. (1972) show that we cannot determine on Day 16 how many follicles are going to ovulate. Indeed, many examples were found with only 2 or 3 follicles externally visible resulting in more than 20 or 30 ovulations post-injection.

However, it was hoped that the appearance of ovaries may reflect the level of ovarian activity at time of PMSG injection. For a preliminary and semi-quantitative analysis, the following criteria were considered; the ovarian response was judged as positive if there was a low percentage of non superovulation and as negative for the contrary ie a high percentage of non superovulated animals and also the presence of large follicles. Due to the high variability in the number of ovulations, the mean ovulation rate is a poor criterion for such a comparison. Moreover, endoscopy on Day 16 had influenced this number since it has prolonged the interval PMSG-oestrus (Table 7).

Finally, three facts are illustrated by Table 6, despite the breed difference for the degree of superovulation. Stimulation of an ovary having no follicles (Group 1) results in a high percentage of unsuperovulated responses. The ovaries with many small follicles are the more sensitive with highly positive responses. On the contrary, for the two other groups, ie with the presence of medium and large follicles, there were less positive responses and more negative responses. The presence of large follicles seems to inhibit the response to PMSG and shows a possible intra ovarian regulation.

TABLE 5

BOVINE OVARIAN CLASSIFICATION AFTER ENDOSCOPIC EXAMINATION AT DAY 16 OF THE OESTROUS CYCLE (DAY OF PMSG INJECTION)

Types of ovaries	Frequency of follicles (two ovaries) A size of			
	< 5 mm	6 - 8 mm	> 8 mm	
	Less than 5	No	No	Poor follicular development
	More than 5 (6 - 40)	Less than 5	No	Only small follicles
	Less than 5	Mean 2 (0 - 7)	Mean 2 (1 - 3)	Only medium and large follicles
	More than 5 (6 - 30)	1 (0 - 4)	1 (1 - 2)	Many small and few medium and large follicles

( ) Range

TABLE 6

OVARIAN APPEARANCE ON DAY 16 AND RESPONSE OF PMS

Type of ovaries	Number of treated cows[+]	Ovulation rate		Percentage of cows with		
		Mean	Range	No super-ovulation	ovulation and large unovulated follicles > 6 - 8 mm	> 10 mm
	6	1.5	1 - 3	66	0	0
	6	2.1	1 - 6	66	33	0
	12	5.5	1 - 33	25	50	0
	8	2.5	1 - 9	50	37	12
	12	5.6	1 - 24	33	50	25
	18	1.05	1 - 2	89	77	33
	4	4.7	1 - 9	25	75	75
	3	0.7	0 - 1	100	100	100

+ Two breeds studied: first one = Charolais; second one = Normande

TABLE 7

EFFECT OF OVARIAN EXAMINATION BY ENDOSCOPY ON SUBSEQUENT RESPONSE TO PMSG INJECTION

Breed	With endoscopy			Without endoscopy		
	n	$I - O^+$ m+sd	Number of ovulations m+sd	n	$I - O^+$ m+sd	Number of ovulations m+sd
Charolais	30	5.19 + 1.90* (a)	3.33 + 4.51	35	3.74 + 1.29 (b)	2.94 + 3.38
Normande	36	6.72 + 3.15* (c)	1.58 + 1.65	43	4.71 + 2.29** (d)	2.60 + 3.02

* 4 animals not detected in oestrus

** 1 animal not detected in oestrus

a and b: different with P < 0.01   c and d: different with P < 0.1

Other results not significantly different

+ injection - Oestrus interval

TABLE 8

EFFECT OF DAY OF TREATMENT ON RESPONSE TO 1600 iu PMSG (NORMANDE BREED)

Day of treatment	Animal number	PMSG injection -oestrus interval (days)	Ovulation rate		Percentage of animals with		
			m+sd	(range)	0 - 1 ovulation	2 - 4 ovulations	more than 4 ovulations
16	23	6.43 + 3.47	2.52 + 2.59	(0 - 10)	52.1	30.4	17.4
18	17	3.71 + 2.40	2.76 + 3.42	(1 - 12)	64.7	17.6	17.6

b)   Interval PMSG - oestrus

Usually superovulation treatments are carried out on a fixed
day of the oestrous cycle (chronological time) when females are not
at the same physiological stage of their cycle (different biological time). We have tried to adapt treatment according to
biological time.

Adaptation to individual oestrous cycle duration (Mauleon
et al., 1970$_1$):   62 dairy cows (Friesian) received 1600 iu of
PMSG four days before the expected oestrus (calculated according to mean duration of previous cycles). Results seem to
show that ovarian response is higher for an interval PMSG -
oestrus of 96 h, ie when the correction has been efficient.

In fact, statistical analysis shows that ovulation number
distribution is homogeneous despite the PMSG - oestrus interval.
It is only because the number of cows is greater for 96 h that
a higher variability appears (Figure 1).

This result is confirmed by those obtained with 2000 iu of
PMSG (Figure 2). Oestrous cycle length is modified and the
distribution of onsets of oestrus becomes bimodal. So is the
distribution of ovulations according to interval PMSG - oestrus.

For this type of treatment (PMSG on Day 16 - HCG on the
day of oestrus) there is no relationship between the interval
PMSG - oestrus and the number of ovulations. However, when
PMSG is injected too late (animals coming into oestrus, 24 -
36 h after treatment), superovulation cannot be induced.

Adaptation to breed oestrous cycle duration: in the
Normande breed, the oestrous cycle is longer by 1.5 days to that
in Charolais or Friesian breeds. We have compared the effect
of PMSG injected on Day 16 or Day 18 (Table 8). There was no
effect on oestrous cycle length or ovulation number. So, treatment can be carried out between Days 16 - 18 without variations
in ovulation rate.

c)   Treatment related to luteolysis

The pattern of progesterone levels during the oestrous cycle

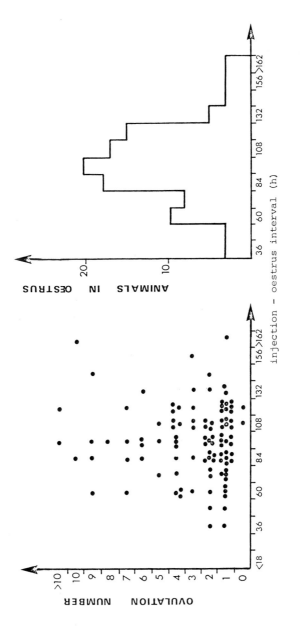

ig. 1. 1600 iu PMSG

208

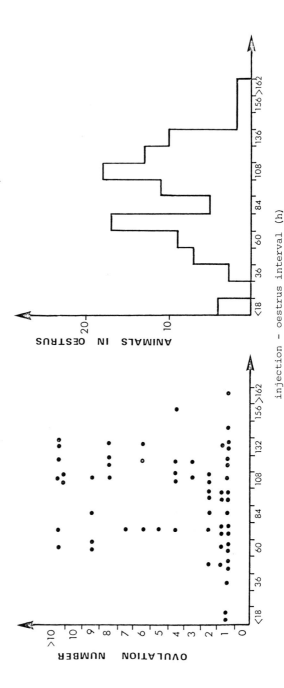

Fig. 2. 2000 iu PMSG

is different between breeds. It has been shown, at least for
Friesian, Normande and Charolais breeds, that the duration of
the luteal phase determines the duration of the oestrous cycle
(Yenikoye et al., 1977). So, it seems preferable to set the
treatment in relation to luteolysis instead of day of cycle.

To test this hypothesis, we have injected an analogue of
$PGF_{2\alpha}$ before, on the same day and after PMSG injection on Day
16. The best results are obtained when the two hormones are
injected on the same day (Table 9) whether the mean number of
ovulations or the percentage of superovulated animals is considered. So, at the end of the cycle, it is the same when
gonadotrophin stimulation and luteolysis occur at the same time.
It is to be noticed that when $PGF_{2\alpha}$ is injected with PMSG, the
treatment - oestrus interval is only 3 days; the theoretical
optimum interval of 96 h is observed when $PGF_{2\alpha}$ is injected
on Day 17 or Day 18 with the worst results of superovulation.

4.   Influence of various hormonal treatments
     a) PMSG injection on Day 16 and HCG on day of oestrus
     The dose (O, 750, 1500, 3000 iu) of HCG has no effect on
the mean ovulation response to four levels of PMSG (Mauleon et
al., 1970[1]). However, with 1500 iu and 3000 iu of HCG, a
decrease in the frequency of animals with only one ovulation
or with more than four ovulations was observed.

     b) Oestrous cycle control treatments
     One of the problems after PMSG administration is that
normal cycle length variation makes it difficult to control the
interval from PMSG to oestrus which is a disadvantage in egg
transfer. Recent oestrous control techniques (progestagens,
prostaglandin) may open up the possibility of a more precise
control of this interval. So, PMSG has been combined with such
treatments.

No significant modifications of the mean ovulation rate or
variability in response were observed after PMSG injection
combined with various progestagen treatments: FGA pessaries

TABLE 9

TIME OF PG ACCORDING TO PMSG INJECTION AND OVULATION RATE (FPN BREED; 1600 iu PMSG ON DAY 16)

Day of PG injection	Animal number	PMSG injection -oestrus interval (days) m+sd	Ovulation rate		Percentage of animals with:		
			m+sd	(range)	0 - 1 ovulation	2 - 4 ovulations	more than 4 ovulations
15	17	2.00 + 0.35	1.71 + 1.72	(0 - 7)	70.6	17.6	11.8
16	17	2.97 + 0.60	3.29 + 2.89	(1 - 9)	47.0	11.8	41.2
17	8	3.75 + 0.46	1.25 + 0.46	(1 - 2)	75.0	25.0	0
18	13	4.39 + 0.55	1.92 + 1.94	(0 - 7)	61.5	23.1	15.4

TABLE 10

NUMBER OF OVULATIONS OBTAINED AFTER PMSG INJECTION ON DAYS 6 AND 16 IN FRIESIAN COWS

Dose of PMSG (iu)		Animal number	Ovulation rate		Percentage of animals with:		
Day 6	Day 16		m+sd	(range)	0 - 1 ovulation	2 - 4 ovulations	more than 4 ovulations
800	800	19	1.84 + 0.96	(1 - 4)	47.4	52.6	0
800	1000	18	2.17 + 1.47	(1 - 6)	44.4	50.0	5.6

(Mauleon et al., 1970_2) or norethandrolone for 18-Day treatments (unpublished data) or SC 21009 for 9-Day treatments (unpublished data). Also, Beehan (1975; cited by Sreenan, 1977) found no significant differences in the mean ovulation response at three levels of PMSG in animals treated with progesterone or SC 21009.

Prostaglandin ($PGF_{2\alpha}$) synchronisation has allowed the use of PMSG at different stages of the oestrous cycle. It has been shown that treatment with PMSG on Days 8 - 12 of the cycle results in higher mean ovulation rates compared with treatments given earlier or later in the cycle (Philippo and Rowson, 1975). However, results for the treatment at the end of the cycle were not confirmed by others (Luhmann et al., 1975; Greve, 1976).

c) PMSG in split doses

Schilling and Holm (1963) have obtained superovulation with a low variability in PMSG-injected cows at Day 6 and Day 16 of the oestrous cycle. This result was repeated by some authors (Laster et al., 1971; McCaughey and Dow, 1977).

We have tried to verify such results - 800 iu of PMSG was injected on Day 6 and 800 or 1000 iu on Day 16 of the oestrous cycle of cows in autumn (Table 10). The results show that a high rate of superovulation was obtained with such a treatment although we never observed a superovulation response when 800 iu of PMSG was injected only on Day 16. As a matter of fact more than 50% of cows show a moderate superovulation without high responses. So it seems that such a treatment may be interesting for superovulation in cows. However, the same experiment in winter has not given as good results. Studies on folliculogenesis in the cow show that there are three important periods during the cycle: on Days 6, 12 and 16 (Mariana and Nguyen Huy, 1973) hence PMSG was injected on these 3 days but the results were poor (Table 11) perhaps because of the inadequacy of the dosage of PMSG. So, this type of treatment needs further investigation.

TABLE 11

OVULATION RATE AFTER INJECTIONS OF PMSG AT VARIOUS STAGES OF THE OESTROUS CYCLE OF FRIESIAN HEIFERS

Dose of PMSG (iu)			Animal number	Ovulation rate		Percentage of animals with:		
Day 6	Day 12	Day 16		$m \pm sd$	(range)	0 – 1 ovulation	2 – 4 ovulations	more than 4 ovulations
500	500	500	6	0.83 ± 0.41	(0 – 1)	100	0	0
500	500	1500	6	5.33 ± 5.96	(1 –16)	50	0	50
0	0	1500	6	17.00 ± 19.27	(1 –55)	16.7	0	83.3
500	500	–*	6	1.00 ± 0.00	(1)	100	0	0
0	500	–*	6	1.17 ± 0.41	(1 – 2)	83.3	16.7	0

* 500 µg of $PGF_2\alpha$ analogue on Day 14

TABLE 12

OVULATION RATE AFTER 8 H INFUSION OF PMSG AT DAY 12 IN FRIESIAN LACTATING COW

Total dose (iu)	Animal number	Ovulation rate		Percentage of animals with:		
		$m \pm sd$	(range)	0 – 1 ovulation	2 – 4 ovulations	more than 4 ovulations
8 h infusion						
800*	9	1.44 ± 1.01	(1 – 4)	77.7	22.2	0
1200	9	4.55 ± 4.74	(1 – 15)	44.4	22.2	33.3
1600	9	6.22 ± 3.35	(0 – 11)	22.2	0	77.7**
I.V. injection						
1200	9	1.88 ± 1.68	(0 – 6)	77.7	11.1	11.1

* In Friesian cows, 800 iu PMSG injected intramuscularly never induced superovulation
** With 1600 iu injected intramuscularly on Day 16, the percentage of animals with more than 4 ovulations never exceed 20%.

d) PMSG infusion on Day 12

With intravenous infusion of 800 or 1200 iu over several hours (Mariana, unpublished data), superovulation is obtained at a time when such doses are not superovulating. These results (Table 12) show that the mean ovulation response obtained with this last type of PMSG administration is higher than with one intramuscular injection even at Day 16 with the same dose. Therefore, the pattern of PMSG release could affect ovarian response.

5. Influence of repeated PMSG treatments on superovulatory response

For egg transfer, it is necessary to get a very high number of ova from one female within a limited time. Using a treatment of PMSG - HCG, the ovarian response has been found to decrease following repeated treatments (Willett et al., 1953; Hafez et al., 1965; Jainudeen et al., 1966; Mariana et al., 1970; Laster et al., 1971; Betteridge and Mitchell, 1974). In these experiments, treatments were repeated only 2 or 4 times, and for each animal the interval between injection and the dose of PMSG was not kept constant. Furthermore, we have no information for the PMSG - $PGF_{2\alpha}$ treatment which is now commonly used.

In the first experiment (Saumande and Chupin, 1977), we treated heifers every 7 - 9 weeks over a period of 14 months. Figure 3 shows that the mean ovulatory response regularly decreases for the first 4 or 5 treatments. Then it increases for the next 2 - 3 treatments. There are several possible explanations for such results:

Formation of antibodies, (Jainudeen et al., 1966): by radio-immunoassay, we cannot show anti-PMSG antibody formation

Seasonal influences

Exhaustion of follicular classes able to be stimulated.

The seasonal influence was tested in a second experiment. Five groups of 8 heifers were used and one group started treatment every three months. Each animal received every 6 weeks 2000 iu of PMSG on Day 8 and 500 µg of a $PGF_{2\alpha}$ analogue 48 hours

214

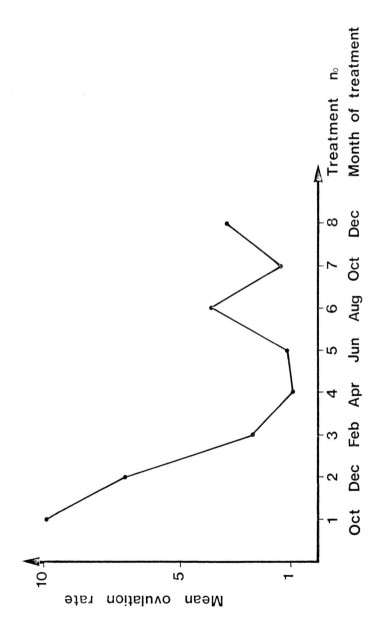

Fig. 3.

later.

The results (Figure 4) show that:

Again a regular decrease in the ovulation rate was observed.

An increase was generally observed in each group in February - April and July - August despite the season of the beginning of treatment (Table 13).

However, in each group, individual heifers showed various patterns of ovulatory response (Figure 5). Generally typical patterns were observed in some animals while others regularly had high or low superovulatory responses during the whole experiment.

So, it is difficult to analyse precisely the seasonal influence on superovulatory response.

Furthermore, several factors may interfere in the seasonal variations of ovulatory response at PMSG injection: ambient temperature, daily light duration. In this experiment, only the nutritional level was constant.

6. Influence of daily light duration on ovarian response at PMSG injection (Ortavant, Chupin, Locatelli, de Fontaubert, unpublished results).

Four groups of 4 Charolais heifers were subjected to the following light treatments: normal variations of daylight from September or May and either 8, 12 or 16 h light per day.

One month after the beginning of light treatment, heifers were synchronised with two injections of a $PGF_{2\alpha}$ analogue separated by 11 days; 20 days after the second injection of $PGF_{2\alpha}$ analogue, PMSG was injected intramuscularly at the dose of 1500 iu per 500 kg of body weight. The ovaries were examined 10 days after by endoscopy.

Finally, one month after the injection of PMSG, the light treatments of the different lots were modified so that each of

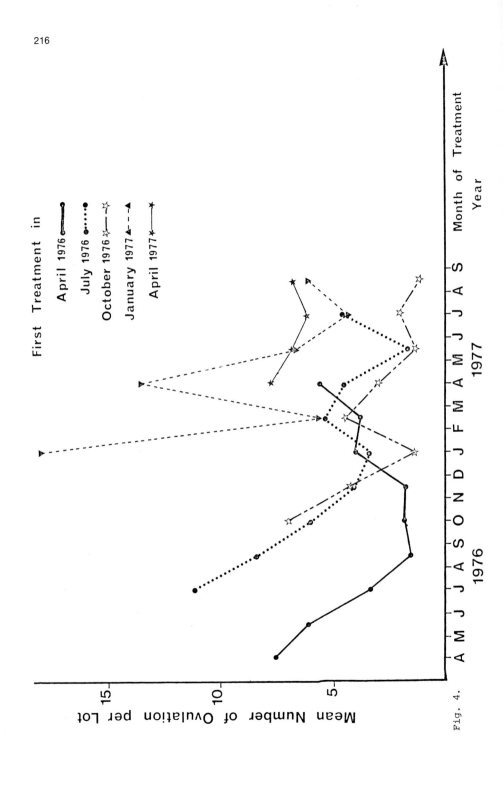

Fig. 4.

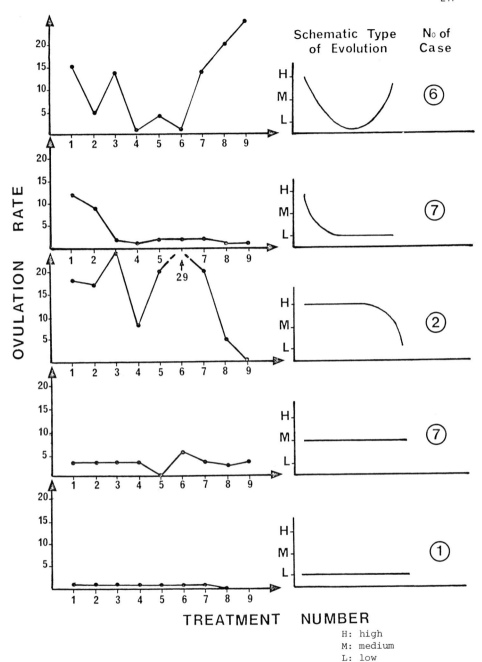

Fig. 5.   Individual responses to repeated superovulatory treatments

TABLE 13

TIME OF INCREASE FOR THE OVULATION RATE IN HEIFERS REPEATEDLY SUPEROVULATED

Animal treated for the 1st time in	Occurrence of a recovery in ovulation rate			
	Treatment number		Month	
Every 6 wk				
April 1976	7 - 8 - 9		Feb. - Apr.	
July 1976	6 - 7	9	Feb. - Apr.	July
October 1976	4 - 5	8	Feb. - Apr.	July
January 1977	3	6	Apr.	Aug.
April 1977		4		Aug.
Every 7 - 9 wk				
October 1977	6 - 7 - 8		Aug - Oct. Dec.	

TABLE 14

INFLUENCE OF DAILY LIGHT DURATION ON MEAN OVARIAN RESPONSE OF HEIFERS TO PMSG INJECTION $(m \pm sd)$ (n = 12 HEIFERS PER TREATMENT)

Light treatment	Mean ovulation rate (Ov.)	Follicles > 5 mm (Foll.)	Ov. + Foll.
Controls	2.2 + 0.41*	1.2 + 0.34	3.3 + 0.50
8 hours	2.2 + 0.56	1.7 + 0.45	3.8 + 0.36
12 hours	1.9 + 0.71	4.0 + 1.64	5.2 + 1.50
16 hours	3.3 + 1.18	4.3 + 1.67	7.5 + 1.67

them were subjected to each of the three light treatments in succession at the end of the experiment in the following manner; group A received firstly 16 h of light per day, then 8 h and then 12 h; group B: 12 h, 16 h, 8 h; group C: 8 h, 12 h, 16 h. The results were analysed according to light treatment thus:

Group 1: Normal variations of daylight from September to May

Group 2: 8 hours of light per day

Group 3: 12 hours of light per day

Group 4: 16 hours of light per day.

The results regarding the mean ovulation rate and the average number of follicles > 5 mm are reported in Table 14.

These two parameters were lower when animals were submitted to short daily light durations but the differences are not significant. However the sum of ovulations plus follicles is significantly lower in animals subjected to 8 h of daily light duration compared with those subjected to 16 h. Similar results are observed in the control group which was treated at the end of autumn and during winter. So it seems short days decrease the ability of follicles to be stimulated.

Menissier (1974) has found that the multiple birth frequency was lower in autumn, ie for ovulations occurring during winter.

Because of the low number of animals in this photoperiodic experiment, these preliminary results are now being retested in a new experiment with the same protocol.

2. EXPERIMENTAL RESULTS AND HYPOTHESIS ON THE ORIGIN OF VARIABILITY

So, we have more solutions and ideas for the control of the non-ovulatory response than for decreasing the variability. However, the biological law that variance increases with the mean in the case of induced superovulation results from an

incredible puzzle of intermingling factors. Recent progress
has been generated by new hypotheses as follows:

1. Follicular growth: primary cause of variability

From experimental results (Mariana, unpublished data), it
seems that an essential mechanism of variability is the passing
from the medium sized follicles (with mean diameter of more
than 238 μ in the rat) to preovulatory and ovulatory stage
(this threshold-size corresponds to antrum appearance).

One could consider two possible origins of variation of
the ovulation rate: one linked to the number of large follicles
and another linked to the aptitude of these large follicles to
go through to the preovulatory stage.

These two types of variation characterise the particular
sensibility of an ovary:

The first origin is considered appoximately as sensitivity
of follicles to FSH.

The second one would reflect sensitivity of large follicles
to LH and/or to other stimulations inducing ovulation.

If n is the number of large follicles among which are, for
a given animal, the preovulatory follicles and p the probability
for such a large follicle to become a preovulatory one, it will
be observed at the level of the animal population (supposing
that all these large follicles are independent in probability
to become preovulatory one) variances in the ovulation rates
$V(o)$:

a. n constant between animals, p variable
(such animals differ either in the magnitude of the
stimulus or the ability to respond to the stimulus).

$$V(o) = n\,\bar{p}\,\bar{q} + n_{(n-1)}\,\delta^2\bar{p} \quad \text{with:}$$

$\bar{p}$: mean value of p

$\delta^2\bar{p}$: variance of p

$\bar{q}$: $1 - \bar{p}$.

b. n variable in all animals and p constant
(in this case, animals have by nature comparable levels
of stimulation or comparable abilities to respond to
stimulation).

$$V (o) = \mu\ pq + p^2 \delta_\mu^2 \quad \text{with} : \mu : \quad \text{mean value of } n$$
$$\delta_\mu^2: \quad \text{variance of } n$$
$$q : 1 - p.$$

In these two cases, if n is increasing, the variance of
the ovulation rate increases also, ie an increase in the number
of antrum follicles involves an increase in the variance for
the ovulation rate (binomial variation) (Robertson, 1951).

2. Potential and induced variability

Ovulation is the outcome of sequential events which not
only involve classes of follicles and structures of ovary, but
also the activity of hypophysis and all the factors associated
with its function.

The study of follicular populations shows that this is
extremely variable for a given species and even among animals
of the same origin which, later, have by nature the same number
of ovulations.

It seems that the variability present at birth has been
eliminated by the mechanism of regulation inherent to the species.
The consequence of the equilibrium realised in the adult is that
the number of ovulations is comparable between animals, although
the levels in the classes are different in absolute value. However, the equilibrium is unstable. It is in a similar manner
that PMSG reveals the variability present in the animal. This
can be called the potential variability.

This potential variability, revealed through PMSG, is only
considered for the low levels of stimulation, when superovulation
is due to the protector effect of gonadotrophin against atresia
(Peters et al., 1975; Hay and Moor, 1977, unpublished data).
When the dose is increased, the number of ovulations obtained

cannot be elucidated in such a manner; superovulation involves other phenomena induced by hormonal injection and the variability will have other origins - this is the induced variability.

According to the aim of the treatment (induction of twinning or egg production for transplantation), it is probable that we provoke:

The potential variability in the first case (use of limitative dose of PMSG)

The potential variability plus the induced variability in the second case (much more important doses of PMSG).

CONCLUSION

From all these results, we draw five main conclusions:

- All experiments dealing with superovulation in cattle need a great number of animals as the response can be affected not only by environment, but also by breed and within a breed by origin of cows.

- It is insufficient to study results of treatments only from the mean number of induced ovulation.

- It is no longer evident that 96 h between PMSG and oestrus is the optimum interval.

- Changing treatments can affect the variability of responses, mainly in decreasing the occurrence of non superovulated animals.

- Among animals variability can be observed not only in the response to one superovulatory treatment, but also and independently, in the response to repetition of treatment.

It appears that the variability of responses is a consequence of the variability of sensitivity and follicular growth among animals. This explains why treatment cannot reduce variability; it stimulates us to find an early test of an animal's ability to be superovulated based on a better knowledge of folliculogenesis and its control.

ACKNOWLEDGEMENT

The authors wish to thank Dr. Leo Cahill for assistance with the preparation of the English manuscript.

REFERENCES

Beehan, D., 1975. In Sreenan and Beehan, 1976.

Betteridge, K.J. and Mitchell, D., 1974. Theriogenology 1, 69-82.

Church, R.B. and Shea, B.F., 1977. Can. J. Anim. Sci. 57, (1), 33-45.

Dufour, J., Whitmore, H.L., Ginther, O.J. and Casida, L.E., 1972. J. Anim. Sci. 34, (1), 85-87.

Gordon, I., Williams, G. and Edwards, J., 1962. J. Agric. Sci. 59, 143-199.

Greve, T., 1976. Theriogenology 5, 15-19.

Hafez, E.S.E., Jainudeen, M.R. and Lindsay, D.R., 1965. Acta Endocr. Suppl. 102, 5-43.

Jainudeen, M.R., Hafez, E.S.E., Gollnick, P.D. and Moustaja, L.A., 1966. Am. J. Vet. Res. 27, 669-675.

Laster, D.B., Turman, E.J., Stephens, D.F. and Renbarger, R.E. 1971. J. Anim. Sci. 33, (2), 443-449.

Luhmann, F.,Hahn,J. and Hahn, R., 1975. Zuchthygiene, 10, 77.

McGaughey, W.J., and Dow, C,, 1977. Vet. Rec. 100, (2), 29-30.

Mariana, J.C., 1970. Ann. Biol. anim. Bioch. Biophys. 10, (4), 575-579.

Mariana, J.C., Mauleon, P., Benoit, M., and Chupin, D. 1970. Ann. Biol. anim. Bioch. Biophys., Suppl. 1, 47-63.

Mariana, J.C. and Nguyen Huy, N., 1973. Ann. Biol. anim. Bioch. Biophys. 13, H.S., 211-221.

Mariana, J.C., Girard, P., and Chupin, D., 1977. Ann. Zootech. (in press).

Mauleon, P., Mariana, J.C., Benoit, M., Solari, A. and Chupin, D. 1970_1. Ann. Biol. anim. Bioch. Biophys. 10, 31-46.

Mauleon, P., Rey, J., Mariana, J.C., Benoit, M. and Benoit, M., 1970_2. Ann. Biol. anim. Bioch. Biophys. 10, 65-79.

Menissier, F. and Frebling, J. 1974. 25 th Ann. Meet. Europ. Fed. Zootech., Copenhagen.

Miller, W., Larsen, L.H., Nancarrow, C.D. and Cox, R.I., 1976. J. Reprod. Fert. 43, 384-385 (Abstr.).

Peters, H., Byskov, AG., Himelstein-Braw, R. and Faber, M., 1975. J. Reprod. Fert 45, 559-566.

224

Philippo, N. and Rowson, L.E.A. 1975. Ann. Biol. anim. Bioch. Biophys. 15, (2), 233-240.

Renard, J.P., Menezo, Y., Saumande, J. and Heyman, Y., 1977. EEC Conf. Galway, Sept. 27-30.

Robertson, A., 1951. Ann. Eugen. Bond. 16, (1), 1-15.

Saumande, J. and Chupin, D., 1977. Theriogenology 7, (3), 141-149.

Schilling, E. and Holm, W., 1963. J. Reprod. Fert. 5, (2), 283-286.

Smith, M.F., Burrell, W.C., Shipp, L.D., Sprott, L.R., Songster, B. and Wiltbank, J.N. Anim. Reprod. Sci., accepted for publication.

Sreenan, J.M. and Beehan, D., 1976. In 'Egg transfer in cattle'. Rowson L.E.A. Ed., 19-34.

Testart, J., 1975. Thesis CNRS no. AO 11631, Paris VI, pp. 113.

Willett, E.L., Buckner, P.J. and McShan, W.H., 1953. J. Dairy Sci. 36, 1083-1088.

Yenikoye, A., Mariana, J.C., Terqui, M., Ley, J.P. and Jolivet, E., 1977. Joint Meet. SSF-SNESF, Nottingham.

PRELIMINARY STUDIES ON THE HCG BINDING PROPERTIES OF BOVINE GRANULOSA CELLS

J.P. Gosling, Patricia Morgan and J.M. Sreenan*

Department of Biochemistry, University College, Galway.
*Agricultural Institute, Belclare, Co. Galway, Ireland.

ABSTRACT

^{125}I-HCG was prepared enzymatically, purified; and characterised using rat testis homogenate. Specific binding of ^{125}I-HCG to bovine granulosa cells was demonstrated Granulosa cells from bovine follicles of 3 different size ranges were prepared and their receptors specific for ^{125}I-HCG characterised. The association constants (K_a) of all these populations of receptors were in the range 9-11 x $10^9 M^{-1}$ but the number of receptors per cell increased dramatically as follicle size increased.

INTRODUCTION

Variability of response is the major problem with all superovulation treatments when applied to cattle. The number of follicles which become large is variable, but the percentage of large follicles which ovulate may be more variable. (Sreenan and Beehan, 1975). Therefore, an understanding of the characteristics of the normal, ripe, preovulatory follicle is desirable; so that treatments, or parts of treatments, tending to give rise to defective follicles may be identified, and modified if possible. Of course, part of the problem may be outside the follicle, for example the timing and size of the preovulatory LH surge (Sreenan, et al., this EEC Conference).

Several studies (Channing and Kammerman, 1973; Kammerman and Ross, 1975; Lee, 1976; Stauffer, et al., 1976) have demonstrated that, as porcine follicles increase in size the number of specific HCG-LH receptors per granulosa cell increases dramatically, with no detectable change in association constant (K_a). This increase is correlated with greater responsiveness of the adenyl cyclase system of the same cells to HCG stimulation. Zeleznik, et al. (1974) found that treatment of rats with FSH increased the binding of HCG to granulosa cells. Therefore, normal follicular development in the pig and the rat is apparently associated with an increased ability of the granulosa cells to respond to LH stimulation.

This paper shows that bovine granulosa cells specifically bind HCG, and that in the cow also, the number of receptors per cell increases as follicles increase in size.

MATERIALS AND METHODS

Gonadotrophins

The gonadotrophins employed were HCG, CR 119; bovine LH, NIH-LH-B9; ovine FSH, NIH-FSH-S10; bovine Prolactin, NIH-P-B4. We acknowledge NIAMDD for providing these free of charge.

Iodination

The reaction volume was 65 μl and contained 5 μg HCG, 0.5 mCi ^{125}ICI (Radiochemical Centre, Amersham), 1 μg Lactoperoxidase (Sigma, L-2005), 10 μg hydrogen peroxide and 5 μmoles Na PO$_4$, pH 7.1. Additional 10 ng lots of hydrogen peroxide were added at 2, 4, 6 and 10 minutes after starting. At 15 minutes 100 μl of 16% sucrose, 0.1% gelatin in PBS (0.14 M NaCl, 0.01 M NaPO$_4$ - NaOH, pH 7.0) was added and the complete mixture transferred to the top of a Bio-Gel P-60 column. This was equilibrated with 0.1% gelatin-PBS, which was also used as elution buffer.

Purification

Sepharose-5B concanavalin-A was obtained from Pharmacia and was used as described by Dufau et al. (1972).

Rat Testis Homogenate

This was prepared according to the method of Leidenberger and Reichert (1972) but with 1% egg albumin (Sigma grade II), 0.1 M sucrose, 5 Mm MgCl$_2$, 0.05 M tris-HCl, pH 7.5 as buffer.

Bovine Granulosa Cells

Ovaries were obtained fresh at the abattoir and placed immediately on ice. The animals were mostly heifers and non-pregnant. The granulosa cells were harvested by the method of Channing and Kammerman (1973) except that the buffer used was the same as for the rat testis preparation.

Experimental Proceedings

All incubations were performed on a shaking incubation bath at 37°C for 3 h with tests homogenate, and 6 hours with granulosa cells. Free label etc. was removed by three centrifugations and resuspension of the precipitate in fresh buffer. The final precipitate was counted in an automatic gamma counter (LKB-Wallac 80 000). Specific binding was estimated in all experiments by including tubes corresponding to each set of conditions which also contained 5 μg cold HCG. Mean counts

228

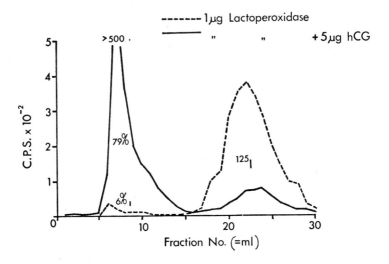

Fig. 1.

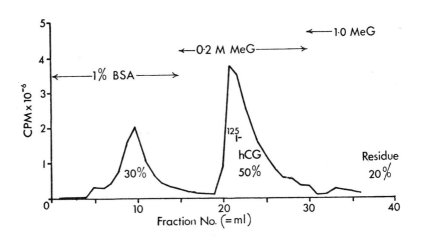

Fig. 2.

obtained in this way (NSB) were subtracted to give the count rate due to specific binding.

RESULTS

Preparation and Purification of ^{125}I-HCG

To prepare radioactively labelled HCG lactoperoxidase catalysed iodination was used. Figure 1 shows an elution profile from the Bio-Gel p-60 column used to separate the high molecular weight components of the reaction mixture from unreacted ^{125}I. Approximately equimolar amounts of HCG, ^{125}I and hydrogen peroxide are employed, and high specific activities in the range 50 - 80 μCi/μg are obtained. In the absence of added HCG little apparent self-iodination of the lactoperoxidase is found (Figure 1).

To increase the proportion of intact ^{125}I-HCG in the label, group-specific affinity chromatography on a column (140 x 5 mm) of Sepharose-4B concanavalin-A was used (Dufau,et al., 1972). ^{125}I-HCG with intact glycoprotein structure is eluted with 0.2 M methyl α-D-glucopyranoside (Figure 2).

Characterisation of ^{125}I-HCG

^{125}I-HCG purified as described above exhibits greatly reduced non-specific binding (NSB tube) to the glass incubation tubes and significantly greater specific binding to rat testis homogenate (Table 1). Non specific binding to glass is further reduced by prewashing the tubes with a solution of 1% egg white in buffer.

To investigate whether the label retains the biological activity of the original HCG, a rat testis radioreceptor assay system was used to construct displacement curves for HCG and ^{125}HCG (Figure 3). The estimated specific activity of 83.5 μCi/μg was close to the calculated specific activity of 73.9 μCi/μg (Greenwood, et al., 1963). On consideration of the errors in these types of estimations, this result indicates

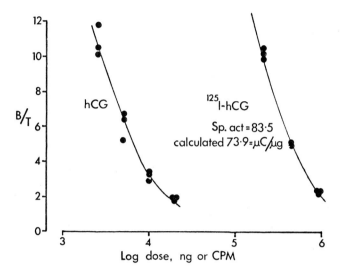

Fig. 3.

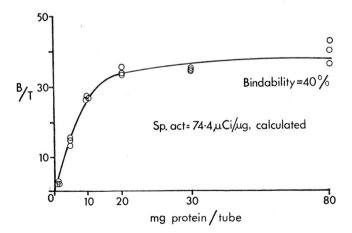

Fig. 4.

that full, or nearly full biologically specific binding
activity was retained; and that calculation of the specific
activity gave a good approximation of the true specific activity.

TABLE 1

EFFECT OF CON-A CHROMATOGRAPHY

| | Bound/Total %$\pm$ S.E., N = 3 | |
	Unpurified	After Con-A
NSB Tube (Not prewashed)	3.69 $\pm$ .17	3.02 $\pm$ .23
NSB Tube	3.19 $\pm$ .19	1.75 $\pm$ .05
NSB*+	3.09 $\pm$ .02	1.54 $\pm$ .06
B_o *	28.8 $\pm$ .47	32.4 $\pm$ .99

* With 5 mg rat testis homogenate.

+ With 5 µg HCG

Not even the purest available gonadotrophin preparations
contain 100% biologically active hormone, and the proportion
of active hormone in a typical ^{125}I-HCG preparation must be
determined if an accurate estimate of an affinity constant is
to be obtained. Incubation of a fixed amount of label with
increasing amounts of rat testis homogenate gives an estimate
of the maximum bindability of the label. In the experiment
of Figure 4, about 40% of the ^{125}I-HCG could be bound by the
testis receptors, and was therefore considered to be biologically active or at least bindable. In all experiments the
amount of label shown indicates only the bindable label present.

In order to check the experimental procedures to be employed with the bovine granulosa cells, a well characterised
LH-HCG receptor population was first investigated. Rat testis
homogenate (7 mg/tube) was incubated with increasing amounts
of ^{125}I-HCG and a hyperbolic saturation pattern obtained
(Figure 5). Fitting these points to a hyperbola by the programme of Cleland (1963) gave maximum binding of 2.7 x 10^{-12}
moles of ^{125}I-HCG per g of tissue. Catt and Dufau (1975) cite

232

a figure of 10^{-12} mole/g. The association constant, K_a was calculated to be 2.4 x 10^{10} M^{-1}. Catt and Dufau (1975) state that this K_a for HCG is 2.3 $\pm$ 1.3 (SD) x 10^{10} M^{-1}.

^{125}I-HCG Binding to Bovine Granulosa Cells

The time course of the binding of label to bovine granulosa cells was followed at 37° and at room temperature, 18° (Figure 6). At the lower temperature binding was low even after 10 hours. At 37° maximum binding was obtained after 6 - 10 hours. A similar curve was obtained by Lee (1976) with porcine cells. All subsequent incubations were performed at 37°C for 6 hours.

The specificity of ^{125}I-HCG binding to granulosa cells was also examined (Table 2). One thousand nanogram amounts of HCG or bovine LH abolished binding of label, but the same amount of ovine FSH or bovine prolactin had little or no effect.

TABLE 2

SPECIFICITY OF ^{125}HCG BINDING TO BOVINE GRANULOSA CELLS

	Relative Binding (B/B$_o$)			
Amount Added	HCG	bLH	oFSH	bPRL
0	1.00	-	-	-
100 ng	0.03	0.10	0.90	1.08
1000 ng	0.00	0.02	0.86	1.03

The capacity of granulosa cells from bovine follicles of different sizes to bind labelled HCG was then investigated. Cells from 182 small (2 - 6 mm diameter), from 42 medium (6 - 10 mm) and from 15 large (> 10 mm) follicles were pooled, washed, counted and aliquoted into incubation tubes. There were 3.0, 3.5 and 3.5 x 10^6 cells per tube from the small, medium and large follicles respectively. Ovaries from 31 animals were used. Each cell type was incubated with increasing concentra-

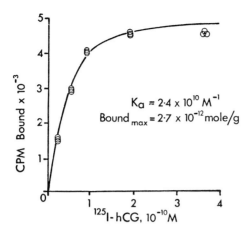

Fig. 5.

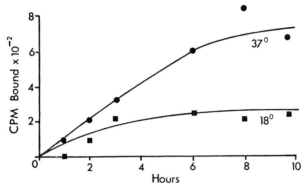

Fig. 6.

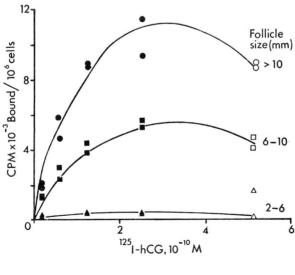

Fig. 7.

tions of ^{125}I-HCG, from 30 000 to 750 000 cpm per tube. Figure 7 shows cpm bound per 10^6 cells plotted against bindable label concentrations, and the binding capacity of each cell type is remarkably different: as in the pig, binding capacity increases with follicle size. However, binding appears to be inhibited at the highest label concentration in all cases, and these points (open symbols) here omitted when the data was fitted to curves (Cleland 1963, as described above). Table 3 shows the kinetic constants obtained. There was no significant difference between the 3 dissociation constants obtained and the association constant, K_a was in the range 9 - 11 x 10^9 M^{-1} for all three types of cell. However, the apparent number of receptors per cell was distinctly different for each class of cell, increasing 20-fold, from 140 to 2063 to 4025 sites with increasing follicle size. An earlier experiment of this kind gave similar, though less precise results.

TABLE 3

[125]I HCG BINDING TO BOVINE GRANULOSA CELLS

Size of Follicles	$K_a(M^{-1})$	K_d^* $(\times 10^{-10}M) \pm$ S.E.	Binding Sites per cell $\pm$ S.E.
2-6 nm	10.9 x 10^9	0.92 $\pm$ 0.90	140 $\pm$ 50
6-10 nm	9.17x 10^9	1.09 $\pm$ 0.20	2063 $\pm$ 164
> 10 nm	9.26x 10^9	1.08 $\pm$ 0.31	4025 $\pm$ 510

$*K_a = 1/K_d$

DISCUSSION

An adequate description of follicular development at the molecular level is a long way in the future. However, significant advances toward this have been achieved in the past five years. One such advance is the discovery that in the pig the number of LH-HCG receptors per cell and the sensitivity to LH stimulation of granulosa cells increase as follicles mature, probably under the influence of FSH (Richards and Medgley, 1976). The results presented in this paper suggest that a similar process occurs in the cow.

Human chorionic gonadotrophin was chosen for this study because of its greater robustness and resistance to inactivation during iodination. In the study of binding specificity, only LH of the hormones tested competed with HCG for the granulosa cell receptors.

The association constant for HCG and bovine granulosa cells was estimated to be $9 - 11 \times 10^9$ M^{-1} which is equivalent to a dissociation constant of $0.9 - 1.1 \times 10^{-10}M$ or about 2.5 $\times 10^{-10}$ if it is not corrected for non-bindable ^{125}I-HCG. Dissociation constants (mostly uncorrected) reported for porcine granulosa cells and HCG are $1.7 \times 10^{-10}M$ (Kammerman and Ross, 1975); $10 \times 10^{-10}M$ (Lee, 1976); and $2.9 \times 10^{-10}M$ (corrected) (Stauffer et al., 1976). Papaionannou and Gospodarowicz (1975) reported a value of $5.3 \times 10^{-10}M$ for the dissociation constant between HCG and isolated bovine luteal cells.

The increase in the number of receptors per granulosa cell found to be associated with follicular enlargement was dramatic and accords well in magnitude with the increases reported for porcine follicles (Kammerman and Ross, 1975; Lee, 1976, Stauffer et al., 1976). For the largest follicles (> 10 mm diameter) the number of receptors detected per cell was 4 000 which is considerably lower than the figure of 50 000 per cell reported by Papaionannou and Gospodarowicz (1975) for bovine luteal cells prepared from the corpus luteum of mid-pregnancy. The inhibition of HCG binding to granulosa cells at the highest HCG concentration is not easy to explain at this time. Perhaps it depends on the length of the incubation period. An equivalent depression in binding, though smaller, is apparent from the curve with rat testis homogenate.

An ideal superovulation treatment would cause the development of a precise number of follicles which would all ovulate in a predetermined period. Consequently, any agent or proced-

ure tending to promote the development of inadequate follicles
must be avoided. Therefore, we intend to monitor the development of HCG binding capacity of granulosa cells from superovulated ovaries and to examine the effects of particular agents
on such development.

ACKNOWLEDGEMENTS

We are grateful to the Agricultural Institute for a grant-
in-aid, and to Professor P.F. Fottrell for his interest and
support. We thank Mr. Eddie Casey of the Galway Abattoir for
his co-operation and the butchers for their patience and
interest.

REFERENCES

Catt, K.J., and Dufau, M.L. 1975. Gonadal receptors for luteinizing Hormone and chorionic gonadotrophin. In: Methods in Enzymology Vol. XXXVII, Part B; Eds. B.W. O'Malley and J.G. Hardman. Academic Press, New York.

Channing, C.P. and Kammerman, S. 1973. Characteristics of gonadotrophin receptors of porcine granulosa cells during follicle maturation. Endocrinology 92, 531-540.

Cleland, 1963. Nature 198, 463-465.

Dufau, M.L., Tsuruhara, T. and Catt, K.J. 1972. Interaction of glycoprotein hormones with agarose-concanavalin A. Biochim. Biophys. Acta 278, 281-292.

Greenwood, F.C., Hunter, W.M. and Glover, J.S. 1963. The preparation of ^{131}I-labelled human growth hormone of high specific radioactivity. Biochim. J. 89, 114-123.

Kammerman, S. and Ross, J. 1975. Increase in numbers of gonadotrophin receptors on granulosa cells during follicle maturation. J. Clin. Endocrinol. Met. 41, 546-550.

Lee, C.Y. 1976. The Porcine Ovarian Follicle: III Development of chorionic gonadotrophin in receptors associated with increase in adenyl cyclase activity during follicle maturation. Endocrinology 99, 42-48.

Leidenberger, F., and Reichert, L.E., Jr. 1972. Evaluation of a rat testis homogenate radioligand receptor assay for human pituitary LH. Endocrinology. 91, 901-909.

Sreenan, J.M. and Beehan, D. 1976. Methods of induction of superovulation in the cow and transfer results. Proc. EEC. Conference 'Egg Transfer in Cattle'. CEC., Eur. 5491 pp. 19-34.

Stauffer, R.L., Tyrey, L. and Schamberg, D.W. 1976. Changes in ^{125}I labelled human chorionic gonadotrophin (HCG) binding to porcine granulosa cells during follicle development and cell culture. Endocrinology 99, 516-525.

Zeleznik, A.J., Midgley, A.R.,Jr., and Reichert, L.E. Jr. 1974. Granulosa cell maturation in the rat: Increased binding of human chorionic gonadotrophin following treatment with follicle stimulating hormone in vivo. Endocrinology 93, 818-825.

P. Mauleon *(France)*

May we have some questions for Dr. Schams please.

K. Betteridge *(Canada)*

Dr. Schams, I noticed in the profiles for progesterone, you had one animal in which the level declined after six days. What proportion of animals do you find with this premature regression of the corpora lutea?

D. Schams *(West Germany)*

There was only one case and it agrees very well with the morphological picture during egg removal. There must be something happening to these corpora lutea because they looked very pale and small although we don't know the reason why. But there was only one case.

K. Betteridge

We have been recovering embryos later than some people and we see this regression in about 10% of animals.

S. Willadsen *(UK)*

If you use prostaglandin to synchronise oestrus in sheep which have been treated with PMSG, you get very high proportions of prematurely regressing corpora lutea - of the order of 30 to 40%. I think this possibly has something to do with the build-up or release of the endogenous prostaglandin which, of course, you deprive of the possibility to act when you come in with your prostaglandin from an exogenous source. It is quite frequent to get this in the sheep.

D. Schams

It is quite interesting to know this and it may have some influence on the quality of the eggs - I don't know.

S. Willadsen

Well, if you get premature regression, the sheep usually

comes back on heat between Days 5 and 8. This is the case with the animals we have been working on where we have been trying to recover eggs on Day 7. If there is regression of the corpora lutea you very rarely find any eggs - they will have been expelled.

K. Betteridge

The same is true of cows: I think this might be very important where people are trying to recover embryos at about the stage when regression is occurring.

D. Schams

In this case recovery was about Day 7 or 8. Perhaps Dr. Schneider has some comment on the one animal I mentioned. If my memory is correct I believe the quality of the eggs was also very poor.

U. Schneider (West Germany)

Yes, I can't remember the exact quality but probably they were stopped at the 2-cell stage; we find that quite often.

R. Newcomb (UK)

We looked at progesterone and oestrogen levels after oestrus in a series of cattle. It did appear in one animal that we were getting premature regression of the corpora lutea. As you know, you get a peak of oestrogen at Day 6 after oestrus in all or most superovulated cattle, and in this particular animal we got a very high level of oestrogen which led to an immediate fall in pro-gesterone level. It could be that you have a luteolytic effect of oestrogen at this stage in some animals, but it doesn't seem to happen very commonly at the stages at which we are looking for eggs, which is about Day 7.

H. Papkoff (USA)

I would like to make a comment about the disappearance studies relating to the PMSG that is used. I think that in a crude PMSG preparation there are different molecular species of PMSG containing varying amounts of sialic acid and these are usually fractionated out when we obtain what we consider to be

the most purified. I think this might explain the bi-phasic
nature of the occurrence, that some of the low sialic acid containing materials would have a shorter half-life compared to the
form with much greater sialic acid. I think it would be interesting if we could isolate a species with a lesser amount of
sialic acid and make some comparisons.

D. Schams
     Well, if you calculate half-life I guess it is very common
that you never have just one component, you always find at least
two.

H. Papkoff
     It seems to me you had a very quick decrease. Your first
component was about a two day half-life ....

D. Schams
     Yes, but the peptide hormones also have a very short half-
life compared with PMSG, for instance. LH average is about 30
minutes and PMSG is 5 days.

P. Mauleon
     Thank you all very much. We must move on now to Dr. Sreenan

D. Schams
     One question on your last point, Dr. Sreenan, I didn't
really understand because you pointed out that with a good responding animal with a high ovulation rate, the time interval
was shorter.

J.M. Sreenan (Ireland)
     Yes, the time between the onset of oestrus and the peak of
LH.

D. Schams
     But then you postulated that the output of LH was also
greater and that is what I don't understand.

J.M. Sreenan

The shorter interval between oestrus and LH peaks were
associated with higher ovulation rates. Also these shorter
intervals were associated with greater total LH output. Total
LH output here is not the peak value reached for each animal but
rather the output over a six hour period including the peak; in
other words, the peak value plus the values at -2 and +2 hours.

P. Mauleon

I would like to recall the inverse results existing in the
sheep breeds which naturally have a difference in ovulation rates.
You will remember in the Romanoff breed, for example, a breed
with a high level of ovulation, there is a long interval between
oestrus and LH peak, it is exactly contrary.

S. Quirke *(Ireland)*

If I may comment on your last point. That is in a natural
oestrus situation but if you give PMS to sheep there is evidence
that you advance the timing of LH peak, certainly relative to the
onset of oestrus, so I think the same applies to the sheep as
the cow in response to PMSG.

B. Hoffmann *(West Germany)*

Do you have an explanation as to why you have a lower response on Day 12 of the oestrous cycle to your PMSG treatment.

J.M. Sreenan

We think it may possibly be associated with follicular
development, normal follicular growth at this time, but we are
not sure about it. We are looking at days in more detail now.

R. Newcomb *(UK)*

In the series of animals that we did, there was no difference in ovulation rate between Days 9, 10, 11 and 12, but we did
get this same significant increase between Days 8 and 9.

J.M. Sreenan

I think we would agree that this is the case because the

numbers of animals involved per treatment were much higher as
we went from Days 7 and 8, to Days 9 and 10, until we came to
Day 12 where we had a smaller group.

J.P. Renard *(France)*

You say that there is no difference in superovulation at
any particular time of the year. Have you studied the quality
of the eggs, and do you find a modification of this quality in
respect to season?

J.M. Sreenan

Yes, what we are showing here are simply ovarian responses
to PMSG. I think we have a poorer quality of eggs during the
winter time but this is confounded by the fact that at this time
of the year, while ovarian response does not differ, the oestrus
response does. So procedures for breeding the animals are more
difficult in the winter months, particularly in January and February, in our situation. So egg quality would be poorer at this
time but this would be a consequence of the animal not coming
in standing oestrus and therefore time of insemination may not
be correct.

J.P. Renard

You mean the donor coming into oestrus?

J.M. Sreenan

Yes, the donor coming in standing oestrus. The response
is lower in winter.

S. Quirke

Dr. Sreenan, have you observed any premature regression of
the corpora lutea as was observed by Dr. Willadsen in the sheep?

J.M. Sreenan

We have, but on very rare occasions - possibly as low as
1 - 2%.

P. Mauleon

    Thank you, Dr. Sreenan. Can we have questions for Dr.
Seidel now?

J.C. Mariana *(France)*

    Have you a significant reduction of standard deviation with
FSH against PMS?

G.E. Seidel *(USA)*

    I honestly haven't looked at it with this recent data. In
the very first experiment there was not, in terms of the number
of embryos produced.

S. Willadsen

    Have you had a chance to look at the steroid profiles of
the ones which presumably you would say are marginally closer
to the normal situation?

G.E. Seidel

    Yes, we looked very intensively at steroid profiles several
years ago but only up to Days 4 or 5 after oestrus. It seems
to me that with the bad luck we are having in trying to recover
good embryos, non-surgically, from PMSG treated donors, this may
be related to something going wrong after that time. At Days 4
and 5 there seems to be very little difference but I think as you
go later the situation deteriorates with the PMSG. Unfortunately,
we have not looked at the hormonal levels on Days, 6, 7 and 8,
when things may be going badly.

L.E. Rowson *(UK)*

    In your earlier slides you showed a number of ovulations
and a number of abnormal eggs, and it seemed to me that you have
got a very high proportion of abnormal eggs - something like 33%
of the eggs were abnormal. Can you explain this?

G.E. Seidel

    A lot of them were just unfertilised. Those figures
included unfertilised plus embryos sufficiently abnormal that we

would not transfer them. So I don't think about 30% is too bad.
However, it seems that later we are doing better. When one is
recovering non-surgically, especially late with the FSH treatment, it is possible that the abnormal embryos are difficult to
find or disintegrating by that time and therefore it is not a
true picture compared to what would have been seen several days
earlier.

L.E. Rowson
    Were they inseminated with frozen semen?

G.E. Seidel
    On the field type of data like the last experiment we have
to use the semen that the owner of the donor requests. About
90% is frozen semen. Every one of those last ten were frozen
semen and that is what we did over the last month.

J.P. Renard
    My question is the same as Mariana's except that it is in
respect of embryo quality. Do you find for the FSH and PMS
treatments the same variation in the proportion of good eggs?

G.E. Seidel
    There is no clear difference. One would like to be able
to ascribe the findings to a specific phenomenon but it seems
that there are a few more embryos, a few more normal, a few more
fertilised and perhaps a little bit higher pregnancy rate. In
the final analysis we get more pregnancies with the FSH but it
cannot be ascribed to one specific factor, or at least, I have
not been able to sort it out.

H. Papkoff
    Have you any idea how low a dose of FSH you can give and
still get effective ovulation? I am asking this because in
making some rough calculations as you were talking, I have a
feeling that you are giving much less FSH, as it is, than you
are giving PMSG, on the basis of purified molecules.

G.E. Seidel

There have been some studies done with other objectives
in mind, for example, producing twins, and 10 mg given in just
the right way will produce two or three ovulations. There are
so many permutations and combinations of ways of going about
this that it is almost frustrating. The answer to your question
is, no, as one lowers the FSH to half of what I presented here,
the response declines significantly and if one doubles it one is
asking for trouble - the ovary blows up and one gets terrible
responses, as happens with PMSG if one goes to 4 000 or 5.000 iu
in our experience.

P. Mauleon

Now we move on to the paper by Saumande for your questions
and comments.

B. Hoffmann

By treating the animal with this massive dose of PMSG, I
think you are changing the whole system, you are producing more
follicles. Don't you think that to some extent you are also
raising the basal level of progesterone during the phase after
injecting PMSG so that you would expect some higher levels than
in a cow which had not been treated with PMSG? I think the follicles have the capacity to synthesise more progesterone which
you might pick up in peripheral plasma.

J. Saumande *(France)*

It is possible, but what must be remembered is that the
level of progesterone during the LH peak is not higher in the
animal with the highest ovulation rate. So I think if follicles
can secrete progesterone they are not responsible for all the
progesterone secreted at the time of the LH peak.

D. Schams

For your prostaglandins, which kind of preparation did you
use, analogue or natural?

J. Saumande

Analogue.

D. Schams

We have observed these kinds of patterns with the natural $F_2\alpha$ and it was that something had happened to the preparation itself.

J. Saumande

I hope not.

P. Mauleon

Are there any other comments? The results seem to be very similar to those of Dr. Sreenan.

K. Betteridge

Do you know how repeatable the profiles are? Have you repeated the treatments in the same animals and found the profiles repeatable in successive treatments?

J. Saumande

I don't think so; I have no information on hormonal levels in such animals, only for LH and LH is about the same. But as progesterone and oestradiol are related mainly to the number of ovulations, and the number of ovulations decreases with repeated treatments, it can't be reproduced.

K. Betteridge

So you don't think there is any chance of using oestrogen production to identify a good responder for future treatments?

J. Saumande

I do not think that is the best approach. At the present time we are trying to stimulate prepubertal animals and relate their sensitivity at this time to when they become pubertal.

R. Newcomb

In answer to Dr. Betteridge's question, I am afraid we

haven't done any hormonal work but we have treated a large number of cows for superovulation and carried out non-surgical recovery. There does seem to be a within animal consistency; if an animal responds well initially it does tend to respond well in subsequent ovulations. Those that don't respond well initially tend to give a poor response subsequently. But we haven't carried out any hormonal work though I think it would be interesting, from the point of view of identifying animals which are particularly good for superovulation.

J. Saumande

This subject has been referred to in the Mariana paper.

P. Mauleon

Well, I think that this is good time to ask for questions on that paper.

A.J. Breeuwsma

What, in your opinion, determines exactly ovulation time, the PMS injection, the PMS dosage, or the prostaglandin injection which shows a luteolytic effect?

J. Saumande

I don't know but I think it is the prostaglandin injection; I have no definite information.

A. Brand (Netherlands)

Do you think these responses would be the same in lactating dairy cattle or do you think that the results will be different?

J. Saumande

We have no experience on this.

I. Wilmut (UK)

Could I just ask you to confirm a point? Are you suggesting from the endoscopy experiments that the follicles which are approximately 5 mm in diameter when you give the PMS injection, are the follicles which will ultimately ovulate?

J. Saumande

No, there is an experiment published in 1972 which shows that the follicles which will ovulate can only be detected three days before ovulation. Secondly, in these experiments we found some animals which had only 10 or 20 small follicles of about 5 mm diameter, and yet more than 40 ovulations after treatment with PMS.

I. Wilmut

In answer to that, surely there would be even fewer large follicles than there are follicles of 5 mm? So what we mean then, is that follicles of from 5 mm upwards, all of them are the follicles which ovulate. Is that the right answer?

J. Saumande

Some which were smaller than 5 mm ovulated.

K. Betteridge

Am I right in saying that there is a big difference of opinion here? Ray Newcomb is saying that you can get repeatability and now you are telling us that you don't get repeatability ....

J. Saumande

What we say is that animals which superovulate at the first treatment can be expected to superovulate at the next treatment, but animals which do not respond to the first treatment do not respond later.

K. Betteridge

But nevertheless, the response in a group of animals tends to go down.

J. Saumande

Yes, but with some animals the ovulation rate is constant throughout all the experiments.

D. Schams

Have you looked at the same animal by means of endoscopy
for ovarian activity in separate cycles and has the same animal
shown the same ovarian activity or was there a change? Let us
say, in one cycle you observed only two or three small follicles, is it possible that in another cycle you would observe
10 or more small follicles in this animal? I am asking generally because if good responding animals show a very high ovarian
activity all the time, ie, always have a pool of small follicles,
and the weak responding ones have only one or two small follicles
available during each cycle, this might be the explanation why
certain animals respond very well and others do not.

J. Saumande

We have no information on that.

M.P. Boland *(Ireland)*

Have you looked at the fertilisation rate?

J. Saumande

No, all these figures are concerned only with the number
of ovulations.

S. Hanrahan *(Ireland)*

You talked about the differences between your families –
your half-sister groups. Were those differences significant?
The variations shown on your slide are difficult to assess as to
whether they are real or are sampling variations. Did you test
that?

J. Saumande

There were significant variations in the mean number of
ovulations and also in the distributions.

S. Hanrahan

We have been looking at a lot of ovulationary data in the
Finnish Landrace breed which has got an ovulation rate of around
4, ranging from 2 to 9 in natural situations. The repeatability

is very high, somewhere between 0.7 and 0.8 for the correlation. So, in the natural situation there is a very high repeatability, both within the same season and from one year to the next. Whatever it is that gives a certain ovulation rate is quite repeatable.

H. Coulthard *(UK)*

I wonder if you have looked at the interval between repeated operations or collections and what effect this has. We have had limited results on 16 animals but it does seem that if you extend the interval between operations above about 90 days, you get a drop in the response. Also, if you try to re-operate or re-collect too soon, again you have a drop in response.*

P. Mauleon

And now may we have some question on Dr. Gosling's paper?

R.J. Ryan *(USA)*

I would just like to congratulate Dr. Gosling for doing a very nice job; I am particularly pleased to see this data for some other species besides the rat and the pig. There is one problem that continually plagues us in dealing with rats and pigs and I wonder if you have had success in dealing with it with cows. That is, how do you estimate the number of binding sites that are already occupied by endogenous hormones? You might, in fact, be underestimating the number of sites in the large follicle because you might have relatively high concentrations of endogenous LH at that point in time, occupying some of these sites. Have you looked at that problem?

J.P. Gosling

No, at this stage it is early days for us yet and we have not attempted to look at this. The methods that have been used for this problem have been to incubate the receptors at low pH in an attempt to elute the bound hormone, and then to measure this by radioimmunoassay or some similar method. I don't see

---

* See appendix at conclusion of this Discussion Period

how this would work with intact cells, I don't think it would
be satisfactory - there would be too many things happening. To
my knowledge it has only been applied to studies of receptor
preparations. I would be interested to know whether Dr. Ryan
has come across any such studies which indicate that the approach
that Patricia Morgan and I have been taking is invalid. However,
if I may make a personal comment, I feel that the one criterion
is, does it make sense and does it fit in with everything else.

R.J. Ryan

I'm obviously biased; to my mind there is no question about
it, I think it makes perfect sense. But I think that there is
one thing that you have to bear in mind. There are really four
criteria for saying that something is a receptor: its hormonal
specificity: its tissue specificity: its affinity being high
enough to be applicable to the concentrations of hormone that
exist in blood, and lastly, (and most difficult), that it relates to a biologic effect. Now, unfortunately, at the present
time, all we can do is to make correlations between receptor
binding and biologic effect. We don't know all the intervening
steps and until we know them all we are not going to be able to
answer that question definitively. I think what needs to be done
in your system, as in others, is to make as many correlations
with biologic responses and receptor concentrations or receptor
numbers, as you possibly can.

J.C. Mariana *(France)*

Have you controlled the histological picture of your follicles ....

J.P. Gosling

No, we have simply taken follicles of various sizes from
animals at slaughter. This is a statistical survey.

W.C.D. Hare *(Canada)*

This is probably in the literature but I am interested to
know how you estimate the number of cells.

J.P. Gosling

Patricia Morgan is the expert on this. It is done by the standard haematological method, using a slide and counting the number of cells between each group and applying the appropriate correction.

P. Mauleon

Thank you Dr. Gosling. Before opening the general discussion, I would like to make some comments.

For a better comparison of our research results, I think we must choose a similar method of expression of the effect of hormones on the ovary. I would like to focus your attention on the necessary criteria which showed variations. Mariana said to me that many gentlemen were drowned in a river which had an average depth of 1 metre; I don't know if we are not sometimes in the same condition.

In order to have a better discussion I propose that we divide the topics into groups, first, the treatment and what is the best treatment for PMS injection - once, twice or more? What do you think about repeat treatments, for example,and how is it possible to predict the response to the treatment? Finally, with regard to the contradictory research on seasonal variation, I propose to look at an experiment of the effect of season on ovarian stimulation.

Well now, does anyone wish to put a question about the best treatment?

S. Willadsen

Now that it is possible to recover eggs non-surgically, how often and how soon can you do your recoveries? We have had a little information on that from Coulthard*but do any of the other groups have any information on that matter?

---

* See appendix at conclusion of this Discussion Period

P. Mauleon

   I think egg recovery will be discussed tomorrow.

S. Willadsen

   This is with reference to repeating the superovulatory
regime and really doesn't affect egg recovery.

P. Mauleon

   Dr. Saumande can you answer this?

J. Saumande

   I think last year, Testart tried to repeat treatment in
animals four times and eggs were recovered by non-surgical
methods and he found a similar decrease in the recovery rate as
we have found in the ovulation rate.

S. Willadsen

   This was in successive cycles?

J. Saumande

   No.

Y. Heyman *(France)*

   The interval between successive treatments was six weeks.

W.W. Lampeter *(West Germany)*

   I had a group of 20 experimental animals which I have
treated up to six times now. I stimulated them and after doing
the flushing I gave them prostaglandin injections; they came
into heat approximately 10 days later. That was an artificial
heat and I waited for the natural cycle and again repeated the
stimulation which adds up to 30 or 40 days. The more I stimulated, the more the results declined. Sometimes I had a cystic
reaction or sometimes I could palpate only very small ovaries.

G.E. Seidel

   We have superovulated perhaps 100 or 150 donors at about
60 day intervals. Our standard scheme has been to randomly use

FSH the first time and PMSG the second, and then vice versa. We
have noticed no decline in response by doing things that way.
We have a limited amount of data on the third superovulation
where we go from FSH to PMS to FSH, or PMS to FSH to PMS, and
there seems to be very little effect there as well. Now, we are
switching hormones and we are waiting 60 days between treatments.
We too are wondering how frequently one can superovulate and we
are starting a very thorough experiment on that. We did some
preliminary work to see how much we could get away with. For
example, we superovulated donors and on Day 10 of the superovulation cycle we tried again, so that would be about a 14 day
period - and it doesn't work. It is pushing the system too hard
it seems; I think we had five or six donors in this preliminary
study and we recovered good embryos from one of them. The others
came into heat at the wrong time, or didn't come into heat, or
had a poor response. So, I can tell you what doesn't work but
there must be something in between that will work as well. We
think that one can superovulate at about 35 or 40 day intervals
successfully.

R. Newcomb

I think this is going to be one of the subjects where the
practical men are going to give the answer before the scientists.
I think there will be tremendous differences in response to repeat superovulation depending on the routine used. One paper
in Theriogenology reported considerable decrease in response
after the first time but this isn't our experience at all. Where
we looked at 2 000 and 1 000 iu of PMSG there does tend, the
second time, to be a drop in the 1 000 unit group. In other words,
the threshold to PMSG may have been raised but there is no difference when one uses 2 000 iu. Our results tend to be quite
repeatable. If anything there is a drop the second time but
thereafter the response levels out at the same level that it was
in the first superovulation. The routine we use is PMS mid-
cycle, followed by the prostaglandin and insemination at oestrus.
Then prostaglandin again on Day 24 because we have found that
most superovulated animals return to oestrus 27 days after the
induced oestrus, instead of the usual 21 days. They seem to take

their repeat cycle from Day 6 when they have the oestrogen peak.
So we inject on Day 24 in order to get a return to heat on Day
27. We then superovulate again on Day 10 of the subsequent cycle.
This brings it round to roughly a six week interval. I think
there is something involved in the method of repeated superovulations which will be responsible and when we are able to define how different methods affect it, then, perhaps, we can start
looking at the hormonal or follicular reasons why repeat superovulation isn't good with some routines and is very effective
with others.

R.J. Ryan

In listening today I was reminiscing a little about the
attempts to induce ovulation in women that went on 8 or 10 years
ago. I know that women are different from cows! I also know
that the object of the treatment with women was not to superovulate
them but to produce one ovulation. However, in effect, many of
them were superovulated. A few of the factors which were involved
there I can cite to you and you may find them relevant or not
relevant. You have to make that judgement.

Firstly, it required both FSH and LH to induce responsiveness. Secondly, it generally required a minimum of nine days of
treatment before you got a response in terms of inducing an ovulation. Thirdly, they generally started treating the women with
very small doses of both LH and FSH and would treat them for one
cycle; it they got a failure they would go the next cycle and
double the dose. Now, in general, if you give too much FSH and
LH during that first 9 - 10 day period, you are more apt to get
superovulations but the thing that was most related to superovulations, or the development of acute large cysts in the ovary,
was the dose of HCG that was used to ovulate. Finally, the system that was evolved was to start with low doses and to monitor
urinary oestrogen; when you got a urinary oestrogen of around
100 micrograms/day, you had a responsive system; if you got more
than 200 micrograms/day, you had too responsive a system and
you backed off. Now, I don't know whether that is applicable to
cows or not but maybe you can see some points to it.

K. Betteridge

To pursue this human analogy, has anyone been trying any
of the anti-oestrogens in cows as a way of increasing their
ovulation rate? Secondly, related to this, in sheep, someone
working in Edinburgh has been using an antiandrostenediol and
doubling the ovulation rate. Has this approach been tried by
anyone here?

J. Saumande

I have no superovulatory results from tests but from progesterone levels it is evident that we can find the dosage which
will increase LH and probably also FSH secretion. But no animals
come into oestrus after this treatment.

K. Betteridge

Why?

J. Saumande

I don't know.

K. Betteridge

Maybe the anti-oestrogenic effect is too long?

J. Saumande

Perhaps. I have no information on that.

J.P. Renard

I would like to make one point about the comparison of
different kinds of treatment for donors. People make these comparisons very often by just the number of ovulations shown but
very seldom by the quality of the embryo. In fact, there is
great variability in one treatment between donors of the number
of good embryos. Recently we have done an experiment and we used
a modified treatment of PMS and prostaglandin. We used not two
but three prostaglandin injections to get synchronisation of the
donors and then superovulation. I mean two prostaglandin injections to get synchronisation then PMS on Day 10 and then
another prostaglandin. What was surprising was that out of about

20 to 25 donors treated, the number of ovulations was the same
with this three prostaglandin treatments in comparison with two
prostaglandin treatments carried out at the same time. The number of ovulations was about the same but what was surprising
was that the number of good embryos recovered at Days 8, 9 and
10, was much lower from the treatment with three prostaglandin
injections.

T. Greve *(Denmark)*

I have some comments on this repeated superovulation. For
us, working in Denmark on superovulation in high yielding dairy
cows, this repeated superovulation has more of an academic interest than a practical interest because usually we only do it
one time around and that's it. Secondly, I would be very con-
scientious about some of these high yielding dairy cows because
we know that even one superovulatory treatment very often turns
them cystic. We know from plasma progesterone that it very often
takes a long time for the cow to return to normal. Dr. Newcomb
mentioned that it takes about 27 or 28 days before they come back
but for dairy cows it takes a much longer time, so I think you
must be very careful. Also, there may be some genetic influence
in response to superovulation with regard to cysts. We know, for
example, that in certain lines of dairy cows within the Friesian
breed which we have tried to repeat superovulate, we have always
had a poor response; each time they tended to develop cysts and
the egg quality was very poor. We had one cow under treatment
and each time we tried superovulation, even though we had two or
three months in between, we had very poor egg quality. We have
now examined the daughters and so on to see if there is any genetic trait here. Then we have also repeated these superovulations
with several Charolais heifers and I just wanted to say that even
though the number of ovulations has been very constant, the quality of the eggs has been dropping. This is by the non-surgical
egg recovery method. In other cases we have seen a very significant drop-off and we cannot support the idea of seasonal
variation in this superovulation. We have examined it very carefully over the last two years. There is no seasonal variation
either in breed or dairy cows. It's the same thing.

What I am getting at is that the individual variation is
so tremendous that just to examine, for example, the time from
calving to the initiation of superovulation or the lactational
status, or endocrine status, or ovarian status, wouldn't be of
great importance compared with the individual variation. We are
trying to develop some patterns but it is very confusing right
now.

S. Hanrahan

I would like to offer a comment on genetic types of variation in sheep. There was a report some years ago about surprising differences between haemoglobin genotypes in some Indian
sheep. We checked these out in Finn sheep that we were working
with, comparing the heterozygotes and one of the homozygotes
for haemoglobin genotype; there was a two-fold difference in response to a standard dose of PMSG, six corpora lutea for one
genotype and somewhere between 13 and 14 for the other. So when
you start looking for genetic differences amongst your cows, some
information like this might very simply answer the genetic variation that you might be seeking.

S. Willadsen

I would like to comment on that. I am not saying that we
have any information that disproves that this might be the case
in the Finnish Landrace and also in these Indian sheep, but we
have also looked at some haemoglobin types in the Welsh mountain
sheep and there is no difference whatsoever, so you can't say
that this is specifically associated with haemoglobin types, or,
for that matter, with any other genetic traits that we can think
of at the moment. It could be an observation in one line of
sheep but it wouldn't necessarily apply to cattle.

There is one other thing I would like to discuss. It seems
to me that it is very much a standard treatment that people are
using to superovulate cattle, that is to give a compound with
FSH and LH activity on some predicted day during the middle of
the cycle and then give prostaglandin afterwards. I would like
to know what the cause of failure is. In other words, why do

some cows become cystic and would there be very simple measures
you could take which would allow you to avoid, say, cystic cows?
Let's not worry about recruiting new follicles in the ovary but
just worry about the treatment that we have at the moment. Is
that one sufficient and could it be modified in such a way that
it would cater for cows that have a tendency to become cystic?
Would it help to give a compound with LH activity? This seems
to help in the sheep.

J.M. Sreenan

I would like to comment first on the question of method
of treatment. The data we presented indicates that treatment
with progestagens, given at any stage of the cycle with PMSG
prior to removal, results in ovarian responses similar to PMSG -
Prostaglandin treatment at mid-cycle. As to the reason for cystic follicles arising, we do know that donor animals that come
into oestrus very early after PMSG treatment have a high proportion of cystic or unovulated large follicles. Perhaps these
follicles have not enough binding sites for LH at this stage -
they are still maturing and while there may be an LH surge, the
follicle is unable to bind the LH.

S. Willadsen

Well, have you tried to give these cows ovulating hormones?

J.M. Sreenan

It is not possible in advance to predict animals that are
going to have cystic follicles and therefore one would have to
use releasing hormones or HCG routinely. Our treatment has always been PMSG 48 hours prior to prostaglandin injection or
progestagen removal and therefore we do not have a high number
of early oestrus animals. Perhaps someone else here has used
releasing hormones or HCG?

S. Willadsen

This would relate to the sheep but in the sheep you can
give very high doses of PMSG. All our sheep are inseminated
surgically and during one period we looked at all the ovaries

before we inseminated the sheep. In the sheep, ovulation normally occurs at the end of the oestrous period whereas in these ones we frequently found an ovulation at the beginning of oestrus. Now, in those we routinely gave a shot of HCG. It is not a very large sample but it would be some 20 or 30 sheep,I think. There was no difference in the final ovulation rate in these ones and the ones which were not treated with HCG. In other words, it looked to us as if we were, in fact, ovulating sheep which otherwise might not have ovulated more than the single ovulation. I think somebody has already mentioned that PMSG has this LH activity and this is probably what ovulates the sheep in the first place.

A. Brand

I believe when Dr. Sugie from Japan published his first results with the non-surgical collection of embryos, he was treating his donor animals with a cocktail of hormones, oestrogens, HCG, etc. Maybe I am wrong but Dr. Allen told us this morning that oestrogen can increase the number of receptor cells in follicles for FSH and maybe also for HCG. Is there somebody who has treated the donor animals just the day before that they are expected to come in heat and would that be a possibility to increase ovulation rate?

H. Papkoff

I would just like to make a plea that somewhere along the line you people who are involved in the superovulation studies try to do at least a confirmatory study using pure, chemically defined hormones. I realise how difficult these may be to obtain but I think until that is done there will always be a degree of ambiguity as to what the hormonal requirements are.

S. Willadsen

If you could supply the purified material we will be quite happy to use it.

P. Mauleon

I would suggest that we finish this session by a presentation of the effect of season on ovarian stimulation. Perhaps

Dr. Ortovant can present his slide.

R. Ortovant *(France)*

Yes, I think there are various opinions about the seasonal variations of ovarian responses after PMSG stimulation.

Some people think that there is no seasonal variation and others think there are seasonal variations. So, I think there are several factors which can interfere with seasonal variation, feeding, ambient temperature, daily light duration, and I think it is necessary to control each of these factors. We have started an experiment in a climatic chamber where feeding, ambient temperature and daily light duration are controlled. For this, four groups of heifers have been used - a control group which was subjected to normal variation of daily light, a second group which received eight hours of light per day, a third group which received twelve hours of light per day and finally a fourth group which received sixteen hours of light per day. Each group, except the control group, were subjected systematically to each light treatment by rotation. They received a moderate stimulation by PMSG, that is to say, 1 400 iu per animal, at the sixteenth day of the cycle. The results indicate that there is no difference between groups in terms of mean ovulation rate. The number of follicles greater than 5 mm is lower in the control group and in the group which received eight hours of light per day than it is in the other two groups. This experiment was done from October to April and thus the control group received light during winter.

When we add the ovulated follicles and the other follicles even these follicles do not belong to the same generation as indicated by Dr. Saumande, nevertheless they indicate a general follicle stimulation in these animals. We can see for the control group subjected to shorter days and for the group which received eight hours of light per day, the general follicle stimulation is lower than in the other groups.

W. Lampeter

I have a private interpretation of the different results
several people have concerning seasonal differences in reaction
to PMSG stimulation. The Danish group, for example, say that
they don't see it. I agree because the seasonal difference between their winter and their summer is not as great as in other
regions. I am thinking, for example, of Canada, where I have
been for a couple of months. The winter in Canada is relatively
hard and relatively long. They close down their operation in
January and February because of poor results and other regions,
such as Ireland, do not have such seasonal differences as we
experience in the more northern parts of Canada. So I explain
it simply with this.

J.M. Sreenan

I was just wondering whether it would be fair to ask
Dr. Betteridge to make some comments because he has got some data
from his Canadian group that I suspect shows no seasonal variation.

K. Betteridge

This is true: we have found no seasonal variation. I have
to say that our interpretation of the data is retrospect, it
wasn't an experiment designed to show (or not to show) seasonal
variation. However, our seasonal variation is very severe but
this is confounded with some animals being inside during treatment and, so, again, you can't even define that. I think one
can say that the closing down of units in winter is not only
because of ovulation rates. In fact, I think it is more related
to the poorer results obtained in these periods which are not
necessarily related to a reduced ovulation rate.

M. Thibier (France)

I would like to go back to Dr. Ortovant's data. I understood that the cows you were talking about in the control group
were submitted to normal daylight from November to April which
means that during that time the daylight period was increasing
from about 6 to 7 hours in December to 12 hours in April, and

yet the results you have shown are quite similar to the eight
hours of the treated group. Would you suggest that there could
be an interaction between the duration of the daylight and the
increasing duration of the daylight?

R. Ortovant

     The control group was subjected to the normal variation –
first decreasing duration of daylight from approximately eleven
to eight hours from October to December. After that there was
an increase of daylight from eight hours to approximately thir-
teen hours at the end of April. However, I have not given all
the details of this experiment. Each group received three
treatments in succession and the control group received two and
these figures represent the average of the three treatments.
But, concerning the variation of light, the difference between
constant daylight and the variation of daylight I cannot answer
now in cattle. There are many experiments in sheep about the
action of photo period on ovarian response and until now it is
not possible to analyse the action of the daily light duration
and the action of the variation between the increase and the
decrease. It seems that the difference of the hormone we are
studying, for example, for prolactin secretion, it is necessary
to give a very long daily duration, 16 hours of light, whilst
for other hormones like LH, the number of pulses of LH per day
is also subject to the normal variation of daylight but with a
very large duration between eleven hours and more than fifteen
hours. However, again, it is not possible to differentiate the
total duration and the decrease or increase of daylight. I
cannot say more than that.

P. Mauleon

     I think now it is time for us to stop. May I thank all
the speakers and all the contributors to the discussion.

APPENDIX

A CONTRIBUTION TO THE DISCUSSION

H. Coulthard

It has appeared to us since the advent of more frequent
non-surgical collections, that the interval between superovulations is important. Below are our results so far on 28
such repeat superovulations. (Table 1).

Repeat superovulation was carried out on all occasions
with a batch and dose of PMS which we would have expected to give
a similar or better response, this being on most occasions
identical to the previous dose and batch and given on the same
day of the cycle approximately (ie day 11 $\pm$ 1 day).

Thus we have concluded that repeat superovulation is best
carried our during the 41 to 100 days period and now aim to re-
collect during a 6 to 7 week period as far as possible.

An apparent drop in response when an interval of 90+ days
between superovulation stimulations was used, was also seen
during the period in which we were collecting surgically.

TABLE 1

Interval between superovulations	Similar response to that immediately preceding (ie ± 2 ovulations)	Better response than that immediately preceding (ie 3+ ovulations more)	Worse response than that immediately preceding (ie more than 3 ovulations fewer than before)
Less than 30 days	0	0	5
31 - 40 "	2	0	0
41 - 60 "	9	3	1
61 - 80 "	2	1	0
81 - 100 "	1	0	1
101 - 120 "	0	0	1
121 - 130 "	0	0	2

SESSION 2

OVUM RECOVERY AND TRANSFER

Chairman:

L.E.A. Rowson

METHODS OF OVUM RECOVERY AND FACTORS
AFFECTING FERTILISATION OF SUPEROVULATED BOVINE OVA

G.E. Seidel, Jr., R.P. Elsden, L.D. Nelson, and J.F. Hasler

Animal Reproduction Laboratory, Colorado State University,
Fort Collins, Colorado 80523 USA.

ABSTRACT

Non-surgical embryo collection is now widely used because it can be done on the farm and can be repeated indefinitely without damaging the donor. For consistent success, manipulative skill is required in addition to proper instruments.

Fertilisation rates of ova recovered non-surgically were 89% with unsuperovulated donors and 70% with superovulated donors ($P < .05$). The pregnancy rate after transfer of normal, unsuperovulated embryos (73%) was not significantly higher than with superovulated embryos (64%).

More embryos were recovered surgically when gonadotrophin treatment was begun in mid-cycle followed by prostaglandin $F_2\alpha$ ($PGF_2\alpha$) than when administered on Day 15 of the cycle without $PGF_2\alpha$ (6.6 vs 2.6 normal embryos per donor, $P < 0.5$). More normal ova were recovered 3 days than 6 - 8 days after oestrus (5.2 vs 4.0 in one trial and 7.3 vs 3.9 ($P < .05$) in a second). More normal eggs were recovered with unfrozen than with frozen semen in one experiment (7.1 vs 4.0, $P < 0.5$) but no significant differences were found in another.

Previously infertile cows can often reproduce with embryo transfer but success rates are low. From 78 embryos recovered from 14 of 25 cows with extended periods of infertility, 37 pregnancies were obtained. The percentage of embryos developing into pregnancies was similar to that obtained with embryos recovered at corresponding stages of the oestrus cycle from donors without known fertility problems.

NON-SURGICAL RECOVERY OF EMBRYOS

The most striking and universal change in embryo transfer
methodology over the past few years has been the rapid transition from surgical to non-surgical methods of embryo collection.
Brand and Drost (1977) have provided a thorough review of the
literature on non-surgical methods beginning with the report
of Rowson and Dowling (1949). Since their review, Greve et al.
(1977) have described slight modifications of previous methods
and confirmed experiences of others.

We, as well as most other researchers working in this
field, have been interested in developing non-surgical methods
for some time. Our early attempts (Seidel, 1970), like those
of others (Brand and Drost, 1977), employed methodology which
was very similar in principle to that which is used successfully now. However, until recently, only Sugie et al. (1972)
were able to achieve consistently satisfactory results. Success does not depend on the instruments alone, for success has
been achieved with a wide variety of collection devices.

In addition to a workable instrument one must also have
(1) ova in the uterus,

(2) excellent manipulative skills,

(3) large volumes of fluid (although there are unpublished
reports that volumes as low as 50 ml per horn are adequate
with certain techniques) together with equipment and personnel to examine the fluid adequately,

(4) adequate confinement of the donor and good epidural
anaesthesia, and

(5) practice and confidence.

Even when these conditions are satisfied, one must recognise
limitations such as the inability to recover ova from the
oviduct, the possibility of infecting the uterus unless extreme
care is taken, and the inability to pass the instruments through
the cervix of all cattle. These are minor problems, however.

We have performed over 400 non-surgical collections from more than 200 donors, representing a variety of breeds and ages, although most were virgin heifers. There have only been six instances when a Foley catheter could not be passed into the uterus by first using a cervical expander as described by Elsden et al. (1976). Under some conditions, the incidence of this problem may be greater, such as with breeds of cattle of Brahman extraction.

The advantages of non-surgical collection have been enumerated by Brand and Drost (1977). In our opinion, the single greatest advantage is the virtual elimination of the formation of adhesions which permits repeated recovery of embryos without damaging the donor. We have flushed one donor 16 times and plan to continue indefinitely. It should be pointed out that the risk of adhesions is not completely eliminated by trans-vaginal methods such as those described by Testart and Godard-Siour (1975). Repeated surgery can lead to infertility as suggested by some of our unpublished data: 12 of 12 donors conceived after one or two inseminations following one surgery, whereas only 8 of 13 conceived after one or two inseminations when two surgical recoveries were performed. We have seen a number of cows that were sterile after surgical recovery of embryos, probably because of the resulting adhesions, however, it is very difficult to get proper control data for such comparisons.

Only limited data are available comparing the efficacy of surgical and non-surgical methods. We are of the opinion that the differences would be small and that such an experiment is not justified because of the adhesions which would probably result in the surgical group. Further, comparisons of two specific methods would still leave questions concerning dozens of other possible comparisons. Surgical methods of embryo recovery have been reviewed by Elsden (1977) and Elsden and Betteridge (1977).

COMPARISONS BETWEEN SUPEROVULATED AND UNSUPEROVULATED DONORS

Non-surgical methods of ova recovery enable comparisons between unsuperovulated and superovulated ova from the same population of donors. Elsden et al. (1976) recovered ova in 36 of 51 attempts (71%) from unsuperovulated donors without known fertility problems and in 24 of 26 attempts with super-ovulated donors. Calculation of the percentage of ova recover-ed from superovulated donors is not very meaningful because the number of corpora lutea (CL) is only estimated by rectal palpation. Numbers of CLs are also frequently underestimated when embryos are recovered surgically. The accuracy with which numbers of CLs are determined may account for some of the discrepancies in embryo recovery rates found in the literature.

The fertilisation rate for the group of donors studied (Elsden et al., 1976) without superovulation was 89% (32 of 36) but only 70% (125 of 179) with superovulation (P < .05) despite the fact that superovulated donors were inseminated three times instead of once and with six times as much semen. Hasler et al. (unpublished) have examined 286 unfertilised eggs from superovulated donors with bright-field and Normarski optics. Eleven of these (4%) contained one or more spermatozoa embedded in the zona pellucida and another 11 (4%) had sperma-tozoa adhering to the outer surface of the zona. The remaining 92% had no spermatozoa associated with them. This may mean that spermatozoa never reached the oocytes.

In Table 1 pregnancy rates from transferring unsuperovulated and superovulated embryos from the same population of donors are presented. Methods of superovulation and recovery were as described by Elsden et al. (1976). The percentage of retarded embryos from superovulated donors (18%) was not significantly higher than from unsuperovulated donors (10%), although this difference may become significant with larger numbers. Simil-arly, there were no significant differences in pregnancy rates between embryos transferred surgically from unsuperovulated or superovulated donors whether they were normal or retarded.

Again, some of these differences may become significant as more
data are accumulated. Much higher pregnancy rates (P< .01)
resulted after the transfer of normal embryos than of retarded
embryos from both superovulated and unsuperovulated donors.

TABLE 1

PREGNANCY RATES WITH EMBRYOS FROM SUPEROVULATED AND UNSUPEROVULATED
DONORS*

	Unsuperovulated donors		Superovulated donors	
	No. of embryos transfer-red	% Developing to 90-day pregnancies	No. of embryos transfer-red	% Developing to 90-day pregnancies
Normal embryos	52	73	239	64
Retarded embryos	6	17	53	21
% retarded	10		18	

* Elsden et al., unpublished.

FACTORS AFFECTING FERTILISATION OF SUPEROVULATED OVA

Summaries of unpublished experiments (Seidel et al.) where
embryos were collected surgically are presented in Tables 2 and
3. In the first experiment (Table 2) 41 Hereford-Angus cross-
bred heifers that failed to become pregnant after having received an embryo surgically were assigned to treatments that
consisted of the factorial combinations of:

1) Three gonadotrophin treatments (1800 iu of PMSG; 1000
iu of PMSG on each of two consecutive days; or 10 injections at halfday intervals of 5,5,4,4,3,3,2,2,2, and 2 mg
Armour FSH plus 20% Armour LH)

2) Initiation of treatment at two stages of the oestrous
cycle (Day 15 with no prostaglandin $F_2\alpha$ or Day 10 with
25 mg $PFG_2\alpha$ on Day 12 followed by 12 mg $PGF_2\alpha$ 4 hours
later; the 1000 - 1000 iu of PMSG treatments were initiated

TABLE 2

SURGICAL RECOVERY OF EMBRYOS EXPERIMENT I[+]

Treatments	No. of heifers	No. of ovulations	No. of normal embryos	Total ova	% Recovery	% Normal embryos
Gonadotrophin On Day 15	20	15.1	2.6*	4.9*	36	51
Gonadotrophin On Day 10 + PG	21	17.6	6.6	9.5	54	67
Embryo recovery On Day 3	23	13.0	5.2	7.2	53	65
Embryo recovery On Day 6	18	19.7	4.0	7.1	37	53
Frozen semen, 2 inseminations	15	21.0	5.2	9.2	51	58
Frozen semen, 4 inseminations	13	14.1	4.4	6.7	51	68
Fresh semen, 2 inseminations	13	13.9	4.2	5.6	32	50

+ Seidel et al., unpublished.

* $P < .05$

274

on days 9 and 14)

3) Embryo collection either 3 or 6 days after the beginning of oestrus (heifers not showing behavioral oestrus
were given HCG on Day 16 or Day 21 of the cycle and inseminated assuming oestrus began at that time)

4) Three insemination treatments (frozen semen 12 and
24 hours after oestrus was first observed, frozen semen
8, 16, 24, and 32 hours after oestrus, and unfrozen semen
12 and 24 hours after oestrus).

Each insemination dose of unfrozen semen contained 50 million progressively motile spermatozoa and each dose of frozen
semen contained 60 million progressively motile spermatozoa
prior to freezing.

The gonadotrophin treatments have been described elsewhere in these proceedings (Seidel et al., 1977). From Table 2
it can be seen that significantly more total ova and normal
embryos were recovered with mid-cycle gonadotrophin treatments
plus $PGF_2\alpha$ than when gonadotrophins were administered toward
the end of the cycle. This was probably due to a more constant
interval between initiation of gonadotrophin treatment and
luteal regression with the $PGF_2\alpha$ treatment. These results
are similar to those reviewed by Betteridge (1977).

Although there were no significant differences between
embryo recovery rates on Days 3 and 6, recovery tended to be
higher on Day 3. The data in the table are least-squares
means which represent the means of individual donors. If one
calculates the percent recovery on the basis of total CLs and
total ova per treatment, the recovery rate was 55% on Day 3
which is significantly (P<.01) different from 36% on Day 6.
These data are in agreement with the conclusions that Elsden
and Betteridge (1977) reached after reviewing the literature.

No significant differences were found among the three
insemination treatments (Table 2). If one calculates the

TABLE 3

SURGICAL RECOVERY OF EMBRYOS, EXPERIMENT II[+]

Treatments	No. of heifers	No. of ovulations	No. of normal embryos	Total ova	% Recovery	% Normal embryos
PG, 22.5 mg	38	11.0	6.2	7.9	68	73
PG, 45 mg	38	12.0	5.0	6.8	57	73
Embryo recovery on Day 3	38	12.4	7.3*	9.6**	72*	70
Embryo recovery on Days 6-8	38	10.6	3.9	5.1	53	75
Previous surgery	48	10.7	4.0	5.3*	59	71
No previous surgery	28	12.3	7.1	9.4	66	74
Unfrozen semen	40	12.7	7.1*	8.8	68	72
Frozen semen	36	10.3	4.0	5.9	57	73

+ Seidel et al., unpublished.

*p < .05

**p < .01

percentage of normal embryos by dividing the number of normal embryos by the total ova (compare to individual percentages of each donor averaged in the Table), a slightly different picture emerges with 57, 66, and 75% normal embryos respectively for the 2X frozen, 4X frozen, and 2X unfrozen treatments. Nearly all of the abnormal eggs were unfertilised.

Based on the results of the experiment just described, another experiment (Table 3) was done comparing FSH, FSH-LH, and PMSG treatments begun on Day 10 of the oestrous cycle for all donors (described elsewhere in these proceedings, Seidel et al., 1977). The gonadotrophin treatments were in factorial combinations with

1) Two doses of $PGF_2\alpha$ (15 mg on day 12 followed by 7.5 mg 9 hours later versus double these amounts)

2) Two embryo recovery times (Day 3 or Days 6 - 8 of the cycle)

3) Two seminal treatments (the 2X unfrozen and the 2X frozen treatments of the previous experiment).

Heifers not observed in oestrus by 3 days after $PGF_2\alpha$ treatment were inseminated as if they had been in oestrus 2½ days after $PGF_2\alpha$ treatment. Within each subclass there were two Hereford-Angus crossbred heifers which had not become pregnant after receiving an embryo surgically and one or two heifers that had not had surgery.

There were no differences among the gonadotrophin treatments (Seidel et al., 1977) or between the two $PGF_2\alpha$ treatments (Table 3). However, the low prostaglandin dose may not be sufficient for extremely large cows.

In this experiment significantly more total and normal ova were recovered 3 days after oestrus than 6 - 8 days after oestrus. The percentage of ova recovered was also higher at 3 days. This confirms and extends the data of the previous experiment.

Significantly more ova (P <.05) were recovered from donors
that had not had previous surgery. The reason for this is
obscure. Only a few donors had adhesions thought to impair
reproductive function and these were not included in the
experiment. Since donors with previous surgery had not become
pregnant after receiving a normal embryo, they may have been
selected for low fertility.

In this experiment, significantly (P <.05) more normal
ova were recovered from donors inseminated with unfrozen than
with frozen semen. The percentage of normal embryos (essentially the percentage fertilised) was not higher with unfrozen
semen when averaging individual donors. However, if one
determines the percentage normal on an overall basis, the unfrozen semen resulted in 13% more normal embryos. Possibly a
higher percentage of ova were originally unfertilised with the
frozen semen treatment and selective loss of the unfertilised
ova occurred. Additional experiments are required to sort
this out. There seem to be few other reports where fresh and
frozen semen from the same bulls were compared directly. Onuma
et al. (1970) found a clear advantage with unfrozen semen for
superovulated calves and Sreenan and Beehan (1976) presented
data with a similar trend, although spermatozoal numbers were
confounded with treatments.

EMBRYO TRANSFER WITH SUB-FERTILE DONORS

From the experiment just described, it appears that the
previous history of donors may be related to the ability to
recover normal embryos. Elsden et al. (1976) were able to
recover only four ova in 38 non-surgical attempts (11%) with
unsuperovulated donors with known fertility problems.

Bowen et al. (1978) studied 25 infertile cows in detail
at laparotomy, usually after superovulation. Ten of these

had idiopathic infertility. Sixteen attempts were made to recover embryos surgically from these 10 donors on days 4 - 6 of the oestrous cycle. From these 16 attempts, 36 transferable embryos were recovered resulting in 18 calves.

The remaining 15 infertile donors had adhesions (usually from Caesarian section), oviductal obstructions, or chronic metritis. From 18 recovery attempts with these donors, 42 transferable embryos were obtained resulting in 19 calves.

Although only 37 pregnancies resulted from 14 of these 25 infertile cows, this represented a considerably improved reproductive rate since most had not had a calf for at least 2 years. The embryos recovered had normal potential for development, especially considering the stage of the oestrous cycle at which recovery and transfer were made.

Another major value of surgical embryo recovery from infertile cows is that a more accurate diagnosis and prognosis is possible in many cases (Archbald et al., 1976); also, certain lesions may be corrected surgically. Non-surgical methods of embryo recovery are also useful for working with infertile cows (Elsden et al., unpublished) and surgery should probably be used only after non-surgical methods have failed. Any of these procedures should probably be limited to overcome infertility due to age, injury, or disease so as not to propagate genetic defects.

ACKNOWLEDGEMENTS

This work was supported in part by grants from Select Sires, Inc., Plains, City, Ohio and the Colorado State University Experiment Station through Regional Project W-112. The THAM Salt of prostaglandin $F_2\alpha$ was donated by the Upjohn Co., Kalamazoo, Michigan through Dr. James Lauderdale. Excellent technical assistance was provided by J. Bowen, R. Bowen, L. Case, R. Carter, N. Homan, L. Roesener, and S. Seidel.

REFERENCES

Archbald, L.F., Warren, J.M. and Sworts, D. 1976. Use of embryo transfer
    technique in the diagnosis of bovine uterine-tube pathology. Vet.
    Med. Small Anim. Clinician, 71: 208-209.

Betteridge, K.J. 1977. Techniques and results in cattle: Superovulation.
    In: Embryo Transfer in Farm Animals, ed. K.J. Betteridge, Canada
    Department of Agriculture Monograph 16. pp. 1-9.

Bowen, R.A., Elsden, R.P. and Seidel, Jr. G.E. 1978. Use of embryo trans-
    fer techniques for infertile cows. J. Am. Vet. Med. Assn. In Press.

Brand, A. and Drost, M. 1977. Embryo collection by non-surgical methods.
    In: Embryo Transfer in Farm Animals. ed. K.J. Betteridge. Canada
    Department of Agriculture Monograph 16. pp. 16-19.

Elsden, R.P. 1977. Embryo collection by surgical methods. In: Embryo
    Transfer in Farm Animals, ed. K.J. Betteridge. Canada Department
    of Agriculture Monograph 16. pp. 10-13.

Elsden, R.P. and Betteridge, K.J. 1977 Effects of different
    surgical collection methods on yield of ova. In: Embryo Transfer
    in Farm Animals, ed. K.J. Betteridge. Canada. Department of
    Agriculture Monograph 16. pp. 14-16.

Elsden, R.P., Hasler, J.F. and Seidel, Jr. G.E. 1976. Non-surgical rec-
    overy of bovine eggs. Theriogenology 6: 523-532.

Greve, T., Lehn-Jensen, H. and Rasbech, N.O. 1977. Non-surgical recovery
    of bovine embryos. Theriogenology. 7 : 239-250.

Onuma, H., Hahn, J. and Foote, R.H. 1970. Factors affecting superovulation,
    fertilisation, and recovery of superovulated ova in prepubertal cattle.
    J. Reprod. Fertil. 21 : 119-126.

Rowson, L.E.A. and Dowling, D.F. 1949. An apparatus for the extraction of
    fertilised eggs from the living cow. Vet. Rec. 61 : 191.

Seidel, G.E., Jr. 1970. Superovulatory response, viability of ova, and
    luteinising hormone and progesterone levels in gonadotrophin treated
    dairy calves. Ph.D. Thesis, Cornell University. pp. 72-73.

Seidel, G.E., Jr., Elsden, R.P., Nelson, L.D. and Bowen, R.A. 1977. Super-
    ovulation of cattle with pregnant mare's serum gonadotrophin and
    follicle stimulating hormone. EEC Conference on Control of Reproduct-
    ion in the Cow. In press.

Sreenan, J.M. and Beehan, D. 1976. Methods of induction of superovulation
    in the cow and transfer results. In: Egg Transfer in Cattle, ed.

L.E.A. Rowson. Commission of the European Communities. Luxembourg. EUR 5491. pp. 19-34.

Sugie, T., Soma, T., Fukumitsu, S. and Otsuki, K. 1972. Studies on the ovum transfer in cattle, with special reference to collection of ova by means of non-surgical techniques (in Japanese; English summary). Nat'l. Inst. Anim. Ind. Bull. 25: 27-34.

Testart, J. and Godard-Siour, C. 1975. Transvaginal recovery of uterine eggs in the cow. Theriogenology 4 : 157-161.

RECOVERY AND TRANSFER OF EMBRYOS BY NON-SURGICAL PROCEDURES IN LACTATING DAIRY CATTLE

A. Brand, M.H. Aarts, D. Zaayer*and W.D. Oxender**

Clinic for Veterinary Obstetrics, Gynaecology and AI., University of Utrecht, The Netherlands.

* Sandery 11, Delden, The Netherlands.

** Department of Large Animal Surgery and Medicine, College of Vet. Medicine, Michigan State University, East Lansing, USA.

ABSTRACT

Non-surgical recoveries and transfers of bovine embryos were performed on farms and at the clinic in lactating dairy cattle. One hundred and ten gynaecologically normal lactating dairy cows, varying in age from 3 to 16 years and in daily milk production from 6 to 40 kg, were superovulated by a regime of 3000 iu PMSG administered between Day 8 and 13, followed by prostaglandin 48 h later. Eighty-two cows (74%) responded with 3 or more corpora lutes. Twenty-eight cows responded poorly with less than 3 corpora lutea or with varying numbers of unovulated follicles and none or one to two corpora lutea. In 10 cows the cervix could not be penetrated by the embryo collector. Seventy-two cows were flushed either at Days 6 to 8 or 11 to 15. In 61 (55%) of the 110 superovulated cows embryos were recovered. The average recovery rate per cow flushed was 3.1 normal and 0.7 degenerate embryos and/or unfertilised ova.

Eighty-three of the non-surgically recovered embryos were transferred non-surgically to recipient heifers by means of a modified Cassou apparatus used for insemination of sheep. Twenty-seven recipients (32%) became pregnant, diagnosed by rectal palpation between Days 40 and 50 of pregnancy. Three of these recipients aborted between Days 60 and 90 of pregnancy. There was no difference in pregnancy rate following transfer at Days 6 to 9, or at Days 11 to 15. Twenty embryos, recovered at Days 13, 14 and 15, were sexed. Five recipients became pregnant following non-surgical transfer of a sexed Day 13 or 14 embryo.

INTRODUCTION

Development of non-surgical recovery techniques of bovine embryos was exceedingly slow before 1976 but rather expansive thereafter (Drost et al., 1976; Elsden et al., 1976; Rowe et al., 1976; Greve et al., 1977; Lampeter, 1977; Brand et al., 1977). Most workers make use of totally flexible one-way balloon catheters and a reverse flush. Although promising recovery rates up to 6.9 embryos and ova have been reported (Elsden et al., 1976; Lampeter, 1977), pregnancy rates following non-surgical transfer are rather low and most embryos transferred non-surgically were recovered by surgical procedures.

Moreover most experiments for non-surgical embryo recovery in cattle have been done with (beef) heifers and information on lactating dairy cows is scarce (Greve et al., 1977; Brand et al., 1977).

The aim of this paper is to describe the non-surgical recovery results in 110 superovulated lactating dairy cows both at farms and at the clinic and pregnancy rates in 83 heifers following the non-surgical transfer of embryos which were recovered non-surgically.

MATERIAL AND METHODS

A total of 110 lactating dairy cows of the Friesian and Meuse-Rhine and Isel breeds were used as potential donors. They were kept on various farms or belonged to the dairy herd of the clinic and varied in age from 3 to 16 years, but most cows were in their 3rd to 5th lactation period. The interval from the last parturition to the time of PMSG treatment varied from 40 to 270 days. Seventy percent of the donors had an interval shorter than 90 days at the time of PMSG treatment. The annual milk production varied from 4000 to 9000 kg and the daily production from 6 to 40 kg. All cows were examined clinically before treatment either by us or by local veterinary practitioners. Cows with recognisable abnormalities in their reproductive organs were excluded from treatment.

All cows were injected intramuscularly with 3000 iu PMSG (Folligon Intervet) between Days 8 and 13 (Day 0 = day of oestrus). If the last date of oestrus was not exactly known PMSG was administered at the time when a well developed corpus luteum could be detected by rectal palpation. Luteolysis was induced by either 500 μg Cloprostenol or 25 mg $PGF_2\alpha$, 48 h after PMSG administration. The donors were inseminated twice, 12 to 18 hours apart, with a single or double dose of deep frozen semen. Sixty per cent of the cows were flushed at the farms and the rest at the clinic.

The recovery apparatus consisted of a one-way Foley catheter* (18 Ch) with a 10 cm long extended tip as described by Frost et al.(1976), Brand et al. (1977). Before flushing the rectum was emptied and an epidural anaesthesia was accomplished with 5 ml Xylocain (2%). After the tail was tied out of the way, the vulva and surrounding area was cleaned with water followed by 70% alcohol. A sterile vaginal speculum 4 cm in diameter was always introduced into the vagina in order to examine the vagina for the presence of abnormal secretion and to prevent contamination of the catheter during passage through the vulva and vagina. Flushing of the uterine horn was performed by using 500 ml of Dulbecco's phosphate buffered saline, enriched with 4 mg of bovine serum albumin and 100 iu sodium penicillin G and 0.05 mg dihydrostreptomycin sulphate per ml. The fluid was infused into the uterine horn, then squeezed out by rectal manipulation and finaly collected in conical glasses which were kept at a temperature between 30 and $37^{\circ}C$. After collection 3 to 4 lots of 10 ml samples were aspirated from the bottom of the conical glasses with a sterile pipette and placed in a concave dish and examined for the presence of embryos under a stereomicroscope.

* Willy Rush Stuttgart, W. Germany.

Heifers, 15 to 16 months old, were used as recipient animals for non-surgically recovered embryos. Most recipients were synchronised with the donor by administration of 25 mg of PGF2α or 0.5 mg Cloprostenol. Prostaglandin was administered half a day earlier in recipient heifers than in donor cows. Between recovery and transfer the embryos were stored at a temperature varying from 30 to $37^{\circ}C$ and for periods ranging from 60 to 300 minutes. Transport of embryos (sometimes over 150 km) occurred in 5 ml vials filled with 2 ml PBI at $37^{\circ}C$. Before transfer,the recipient animals were examined rectally and the site of the corpus luteum was determined. Per recipient a single embryo was transferred into the horn ipsilateral to the corpus luteum. All embryos were transferred by means of an extended Cassou apparatus used for insemination of sheep. The embryo was sucked up into a ministraw which was inserted into the Cassou gun. The Cassou apparatus was then covered by a sheath. A second sheath with a wider diameter which was cut open over the whole length served as a vaginal speculum. The outer sheat was withdrawn from the instrument as soon as the tip of the Cassou apparatus had reached the ostium externum of the cervix. All the equipment had been sterilised before hand. The recipients were given an epidural injection before transfer. The anaesthetic and hygienic measures were performed as in the donor cows. Recipient heifers were examined for pregnancy by rectal palpation 40 to 50 days after their last oestrus.

RESULTS AND DISCUSSION

The non-surgical recovery results are presented in Table 1. Of the 110 cows treated with PMSG and prostaglandin, 82 (74%) responded with more than 2 ovulations. The other cows had only 1 or 2 corpora lutea or had varying numbers of unruptured follicles and no corpora lutea. Some of these latter cows might have ovulated which was however not possible to diagnose rectally.

In 10 (9%) cows it was not possible to pass a 18 Ch or

smaller catheter through the cervix. Four of these cows had
had dystocia at their last parturition. Three Holstein Friesian
cows had a long (10 cm) cervix with a diameter of less than
3 cm. All these three cows were daughters of the same bull.
In three other cows no abnormalities could be diagnosed. One
of the latter cows was slaughtered and showed a distinct kink
in the cervix about 2 cm from the internal os.

TABLE 1

NON-SURGICAL EMBRYO RECOVERY RESULTS FROM SUPEROVULATED LACTATING
DAIRY COWS

Day of embryo recovery	No. of Cows			No. of attempts	
	Treated	Responding $\geq$3 CL	Cervix not pass able	Unsuccess- ful	Success- ful
6 - 8	44	32	2	5	25
11 - 15	66	50	8	6	36
Total	110	82	10	11	61
	No. of normal embryos			No. of degenerate embryos or unfertilised eggs		
	Total	Range	Mean	Total	Range	Mean
6 - 8	87	0-17	2.9	35	0-4	1.1
11 - 15	136	0-8	3.2	18	0-4	0.4
Total	223	0-17	3.1	53	0-4	0.7

No expanders were used in very valuable cows as no data are
available on fertility following the use of cervical expanders.

Seventy-two cows were flushed either at Days 6 to 8 or 11
to 15. In 61 (55%) of the 110 superovulated cows, flushing of
the uterus was successful. Two hundred and twenty-three normal
embryos and 53 ova or degenerate embryos were recovered from
the 72 cows flushed which is 3.1 normal embryos and 0.7 ova or
degenerate embryos per cow flushed.

Our embryo recovery rates are lower than those obtained by Elsden et al. (1976), Rowe et al. (1976) and Lampeter (1977) but comparable to those of Greve et al., who also worked with lactating dairy cattle.

Differences may be due to breeds, age of the donor (heifers or cows), lactational status, superovulation response and or instruments used. In 10 (9%) cows the catheter was blocked by a rather sticky string of mucus, 5 to 10 cm in length, causing loss of fluid and embryos along the balloon during manipulation of the uterus.

After placement of a new catheter fluid recovery was normal. Blocking of the catheter by mucus was not related to a certain stage of the cycle.

In eleven cows no embryos or ova were recovered. In three of these cows the recovered fluid was cloudy due to an endometritis which might have caused early degeneration and resolution of embryos and or ova. In two other cases of cloudy fluid collection however, normal embryos were recovered.

The overall recovery rate of fluid was 90% which is lower than reported by Greve et al. (1977) who also worked with lactating dairy cattle. Most cows flushed in our study were older than 4 years. The length of the uterus in slaughtered cows, aged 3 to 8 years, varied from 22 to 48 cm with a mean of 33 cm (Zaayer, 1977). The best fluid recovery rates in our study were obtained in cows with relatively small uteri.

The number of ovulations was not determined by laparoscopy as was done by Rowe et al. (1976). Compared with the average recovery rate of embryos and ova in 32 comparable slaughtered dairy cows at the end of their lactation period (Brand et al., 1977) 1.1 less normal and 1.2 less degenerate embryos and or ova per cow were recovered in this study. This means that the average recovery rate in this study was 74% of the recovery rate after slaughter.

Recovery of bovine embryos at Day 11 to 15 has the advantage of a more rapid recovery from the flushing fluid. Day 12 embryos can already be recognised by the naked eye. Another advantage is that the quality of these later stage embryos can be determined much better than that of unhatched embryos.

In Table 2 the pregnancy results are shown following non-surgical transfer of a single embryo to 83 recipient heifers. All transferred embryos were recovered non-surgically.

TABLE 2

PREGNANCY RESULTS FOLLOWING NON-SURGICAL TRANSFER OF A SINGLE EMBRYO TO HEIFERS SYNCHRONISED TO WITHIN 1 DAY OF THE DONOR.*

Day of cycle	No. of embryos transferred	Pregnant No. (%)	No. Aborted
6 - 8	32	12 (37)	-
11 - 12	29	10 (34)	1
13 - 14	16*	5 (31)	2
15	6*	0 (0)	-
Total	83	27 (32)	3

* Sexed embryos

In total 90 recipient heifers were selected but in 7 (8%) heifers it was not possible to pass the Cassou gun through the cervix. An overall pregnancy rate of 32% was obtained. Of the 27 pregnancies obtained following non-surgical transfers, 11 recipients gave birth to a healthy calf, 3 aborted between Days 60 and 90 of pregnancy and 13 recipients are still pregnant.

There was no significant difference in pregnancy rates following transfer of embryos of different stages of development. This is in agreement with results obtained by Renard, et al. (1977). From a review of results since 1969 by Brand and Drost (1977) it can be concluded that following non-surgical transfer of a single embryo (257 recipients) an average

pregnancy rate of 28 (0 to 43) was obtained. An average pregnancy rate of 56.5 following surgical transfer of a single
embryo and of 74 following transfer of two embryos can be calculated from a review of results since 1969 by Betteridge (1977).
Recently Renard et al. (1977) obtained pregnancy rates of 37%
following non-surgical transfer of a single embryo and 60%
after bilateral transfers.

Bilateral transfers therefore seem to favour pregnancy
rates in cattle and embryonic death seems to occur more frequently following non-surgical than following surgical transfer.
Many factors may play a role in embryonic death such as the
quality of the embryo to be transferred, the degree of synchrony between donor and recipients, the uterine environment,
the time of production and release of luteotrophic factors in
relation to luteolytic factors etc. but all these factors
affect surgical as well as non-surgical transfer equally. The
only difference is the way and site of transfer which for unknown reasons may lower the pregnancy rate following non-surgical transfer.

One of the reasons may be that during non-surgical
transfer occasionally some uterine trauma with subsequent
bleeding occurs, caused by the Cassou apparatus. The released
blood (serum) may kill the embryo because according to Chang
(1949) bovine serum which is not heat treated is embryotoxic.
This could account for some of the pregnancy failures both in
surgical and non-surgical transfers. In surgical methods,
however, bleeding is very much under control.

With regard to the transfer of later stage embryos,
Betteridge, Mitchell, Eaglesome and Randall (1976) showed that
bovine embryos can be successfully transferred surgically up
to Day 16. Our results show that bovine embryos recovered non-
surgically at Days 13 and 14 can develop normally after a non-
surgical transfer by means of a Cassou apparatus. So far no
pregnancies have been obtained following transfer of Day 15

embryos (0.5 - 4 cm in length) which were all damaged during recovery by an 18 Ch catheter. However this technical failure can be overcome by using catheters of a bigger diameter.

Of 21 embryos recovered non-surgically on Days 13, 14 and 15, 20 were sexed according to the method of Mitchell et al. (1976). Five recipients became pregnant following non-surgical transfer of a sexed Day 13 or Day 14 embryo. (Bosma, 1977). However two recipients aborted (Table 2) 60 to 90 days after their last oestrus.

In both donors and recipients the method used for the non-surgical recovery and transfer of embryos did not inter-fere with subsequent fertility.

Fertility rate at the first oestrus after flushing proved to be higher than 60%. From this study and from others it may therefore be concluded that for embryo transfer in lactating dairy cattle non-surgical procedures are preferable to surgic-al methods with their inherent surgical risk. Fertility rates after non-surgical transfer however have to be improved.

REFERENCES

Betteridge, K.J., Mitchell, D., Eaglesome, M.D. and Randall, G.C.B. 1976.
Embryo transfer in cattle 10-17 days in oestrus. VIIIth Intern.
Congr. Anim. Prod. and A.I. Krakow. Vol. III 237-240.

Betteridge, K.J. 1977. Summary of factors affecting success rates in
surgical embryo transfer. In: Embryo transfer in farm animals.
A review of techniques and applications p. 29-31. Ed. Betteridge,
K.J. Canada Department of Agriculture, Monograph No. 16.

Brand, A. and Drost, M. 1977. Embryo transfer by non-surgical methods.
In: Embryo transfer in farm animals. A review of techniques and
applications. p. 31-34, Ed. Betteridge, K.J. Canada Department of
Agriculture, Monograph No. 16.

Brand, A., Trounson, A.O., Aarts, M.H., Drost, M.H. and Zaayer, D. 1977.
Superovulation on non-surgical embryo recovery in the lactating
dairy cow. Anim. Prod. in press.

Chang, M.C. 1949. Effects of heterologous sera on fertilised rabbit ova.
J. Gen. Physiol. 32, 291-300.

Drost, M., Brand, A. and Aarts, M.H. 1976. A device for non-surgical
recovery of bovine embryos. Theriogenology, 6, 503-507.

Elsden, R.P., Hasler, J.F. and Seidel, G.E. 1976. Non-surgical recovery
of bovine eggs. Theriogenology, 6, 523-531.

Greve, T., Lehn-Jensen, H. and Rasbech, N.O. 1977. Non-surgical recovery
of bovine embryos. Theriogenology, 7, 239-248.

Lampeter, W.W. 1977. Erfahrungen mit der unblutigen Gewinnung von Rinder
embryonen. Zuchthyg. 12, 8-13.

Mitchell, D., Hare, W.C.D., Betteridge, K.J., Eaglesome, M.D. and Randall,
G.C.B. 1976. Sexing and transfer of bovine embryos. VIIIth Intern.
Congr. Anim. Prod. and A.I. Krakow. Vol. III, 258-261.

Rasbech, N.O. 1976. Nicht-chirurgische Gewinnung und Übertragung von
Rinderembryonen unter Praxisverhältnissen. Dtsch. Tierärztl. Wschr.
83, 524-526.

Renard, J.P., Heyman, Y. and du Mesnil du Buisson, F. 1977. Unilateral
and bilateral cervical transfer of bovine embryos at the blastocyst stage. Theriogenology, 7, 189-193.

Rowe, R.F., Del Campo, M.R., Eilts, C.L., French, L.R., Winch, R.P. and
    Ginther, O.J. 1976. A single cannula technique for non-surgical
    collection of ova from cattle. Theriogenology, 6 471-483.
Zaayer, D. personal communication.

Bosma, A.A. 1977. in press.

THE NON-SURGICAL RECOVERY AND TRANSFER OF BOVINE EMBRYOS

R. Newcomb, W.R. Christie and L.E.A. Rowson

ARC Institute of Animal Physiology, Animal Research Station,
Cambridge CB3 0JQ, UK.

ABSTRACT

A system for non-surgically recovering embryos from the bovine uterus has been devised using a three lumen PVC catheter which is passed through a cervical introducer to the tip of the uterus. Using this system and with animals anaesthetised in left flank recumbency an embryo recovery rate has been obtained which closely approximates the success of surgical recovery.

A within animal comparison of the pregnancy rate obtained after transferring embryos surgically with the transfer of embryos non-surgically has revealed large differences in the success rates of the two methods. In one group of heifers embryos were transferred to the tip of the ipsilateral uterine horn surgically and non-surgically to the contralateral horn. In a second group the position of surgical and non-surgical transfers was reversed. When embryos were transferred to the ipsilateral uterine horn surgically the pregnancy rate was high despite a much poorer survival of non-surgically transferred embryos in the contralateral uterine horn. However, where embryos were transferred non-surgically to the ipsilateral uterine horn a low pregnancy rate resulted with both surgically and non-surgically transferred embryos.

In view of these findings the effect of position of surgically transferred eggs was investigated with inconclusive results.

INTRODUCTION

It is an essential prerequisite for the widespread application of egg transfer techniques that non-surgical means of egg recovery and transfer are perfected. It is desirable that the present success of surgical recovery and transfer methods is achieved non-surgically.

In order to recover eggs non-surgically a consistently reliable method of passing a catheter into the uterus via the cervix is required and the technique must dislodge eggs from their location at the ovarian extremity of the uterine horns.

The transfer of eggs non-surgically to the bovine uterus has resulted in consistently poorer success than surgical methods. Although infection, egg rejection and unsuitable placement have all been considered as causes of this poorer success there is an incomplete understanding of the mechanisms involved.

In order to investigate these problems further three experiments were devised.

1) The non-surgical recovery of Day 7 bovine embryos using a three-way PVC catheter introduced into the uterus through a metal cannula and passed to the ovarian extremity of the uterus.

2) The effect of transferring Day 7 embryos surgically and non-surgically to the same animal.

3) The effect of site of transfer of Day 7 bovine embryos.

MATERIALS AND METHODS

Experimental animals were predominantly Hereford cross-bred 15 - 18 month old bulling heifers. In all experiments anaesthesia was induced with thiopentone sodium and maintained with a halothane/nitrous oxide/oxygen gas mixture following a period of 24 - 48 hours starvation.

Experiment 1

Forty-two heifers were divided into 2 groups and superovulated with either 1000 iu PMSG (Group 1.1) or 2000 iu PMSG
(Group 1.2) injected i/m between Days 9 - 12 of the oestrous
cycle (oestrus = Day 0) followed 2 days later by 1000 µg Cloprostenol (ICI). The heifers were inseminated with liquid sem-
en at the induced oestrus and embryos were recovered on Day 7.

The animals were secured in left flank recumbency with the
head tilted downwards. The vulva was disinfected with surgical
spirit and a metal speculum passed into the vagina up to the
cervix (Figure 1.1.). The central core of the speculum was
removed and a metal introducer (consisting of an outer metal
cannula and a central insert which at its leading end forms a
smooth tapered outline) passed through the speculum (Figure
1.2). The speculum was then withdrawn and the introducer passed through the cervix (using the rectal technique) and inserted into the appropriate uterine horn as far as could be easily
achieved without causing trauma (Figure 1.3). The central insert was then withdrawn and a sterile PVC three lumen catheter
passed through the metal cannula and manipulated gently as far
as possible toward the utero-tubal junction (UTJ). The cuff
of the catheter was inflated until the uterus was sufficiently
distended to avoid movement of the catheter, but insufficiently
inflated to rupture the uterus (Figure 1.4). Several modifications of the catheter were used in this experiment but basically the catheter consisted of a 14FG 3-lumen PVC extrudate
with a 20 ml latex cuff. One lumen carried flushing medium to
an inlet hole at the tip of the catheter. The largest lumen
collected the flushings from an outlet hole just in front of
the cuff and the smallest was used to inflate the cuff. At
the trailing end of this catheter means were provided for the
introduction and collection of fluid and introduction of air.

The method used to recover eggs was consistent. After

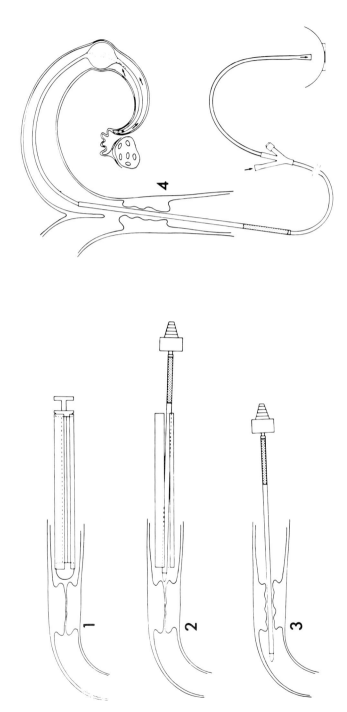

Fig. 1   Diagramatic representation of the method used to recover embryos non-surgically.

1. Speculum passed into the vagina up to the cervix.

2. Central core of speculum removed and introducer passed to cervix.

3. Introducer passed through the cervix into a uterine horn.

4. Central insert of introducer removed and P.V.C. catheter passed to the tip of the uterus and the cuff inflated.

the catheter was correctly positioned a total of 300 ml flushing medium (PBS - Whittingham, 1971) was passed into the uterus
through the inlet hole at the tip of the catheter. Initially
fluid and entrained embryos were prevented from passing into
the oviduct by pinching the UTJ between thumb and forefinger.
The first 100 ml was recovered into glass egg collecting dishes
(Camlab) and at this stage the UTJ was released and the uterus
gently manipulated by squeezing and releasing as the remaining
200 ml was collected into 50 ml boiling tubes.

The collection dishes were placed directly under a stereo-
dissecting microscope to locate embryos. The boiling tubes
were allowed to stand for 20 min when all but 10 ml of fluid
was carefully withdrawn with a syringe and long needle. The
remaining fluid was decanted into further collection dishes
for microscopical examination.

Assessment of the number of ovulations was made initially
by rectal estimate followed by visual observation of the
exteriorised ovaries through a conventional mid-line laparotomy
incision.

This same group of 42 animals was superovulated on a
second occasion some 6 - 8 weeks later, allocating animals
randomly to 1000 or 2000 iu PMSG treatment groups. Embryos
were recovered surgically on Day 7 using the method described
by Newcomb and Rowson (1975). Only animals completely free of
adhesions of the fimbria were included in the data.

Experiment 2
A supply of embryos for this experiment was recovered
surgically or non-surgically as described in Experiment 1.
Recipients were selected after a natural or prostaglandin
induced (500 μg Cloprostenol) oestrus which occurred synchronously with the donor oestrus.

A group of 40 recipients was anaesthetised and one embryo
transferred surgically to the tip of the uterine horn (Rowson

et al.,1969). After surgical repair of the incision the animal
was placed in left flank recumbency and one egg placed in the
opposite uterine horn using a sterile Cassou insemination pipette sheathed with sterile nylon film. The pipette was inserted into the vagina up to the cervix and then the nylon film
was drawn back and the pipette passed into the uterus.

Embryos were placed either surgically into the uterine
horn ipsilateral to the corpus luteum and non-surgically contralaterally (Group 2.1, n=20) or were placed non-surgically
ipsilaterally and surgically contralaterally (Group 2.2, n=20).

Animals seen to return to oestrus were palpated 35 - 42
days after oestrus and were diagnosed non-pregnant on this
basis. All other animals were slaughtered between 42 and 65
days after oestrus and the embryos dissected from the uterus.

Experiment 3

A group of 20 heifers was anaesthetised and one Day 7
embryo was surgically transferred either to the tip or to the
base of the ipsilateral uterine horn of each heifer. Transfer
was achieved by using a blunted miniature trochar and cannula,
consisting of an 18 gauge needle and metal insert, to puncture
the uterus. The metal insert was then withdrawn and a nylon
cannula (length 30 cm, O/D 0.63 mm) containing the embryo was
passed through it either to the tip or base of the uterus.
Two lengths of trochar were used

i) 8.5 cm in length inserted into the uterine horn at
the position of the intercornual ligament and directed
toward the cervix and

ii) 5 cm long and inserted 7 cm from the UTJ and directed
towards it.

Pregnancy was assessed by heat detection and rectal palpation 35 - 42 days after oestrus.

RESULTS

Experiment 1

The results of experiment 1 appear in Table 1. A total
of 216 embryos was recovered. Of these 166 (76.9%) were in
the first 100 ml (ie before any manipulation was commenced),
42 (19.4%) in the 2nd 100 ml and 8 (3.7%) in the 3rd 100 ml of
fluid recovered.

In donors receiving 1000 iu of PMSG (Group 1.1) 62.6% of
embryos were recovered by non-surgical means compared with
79.2% surgically. The equivalent results for donors receiving
2000 iu PMSG (Group 1.2) were 55.2% and 63.9%. There was no
difference in recovery rate (P< 0.05) between surgical and non-
surgical recovery for groups 1.1 or group 1.2 but the differences between the surgical and non-surgical recovery rates for
the 1000 and 2000 iu groups combined was significantly different (P< 0.02). There was no significant difference (P< 0.05)
in ovulation rate between the first (non-surgical recovery)
superovulation and the second (surgical recovery) for donors
receiving 2000 iu PMSG, but there were significantly (P< 0.01)
fewer ovulations on the second occasion for donors receiving
1000 iu PMSG.

In the 1000 iu PMSG group the mean percentage recovery
non-surgically was greater from the left than from the right
uterine horn (67.6 $\pm$ 8.6 cf 50.2 $\pm$ 10.3). This was also true
of the 2000 iu PMSG group of donors (58.1 $\pm$ 1.1 cf 53.2 $\pm$ 6.9).
However, these differences were not significant (P< 0.05). The
fluid recovery from the left and right uterine horns did not
differ (93.6 $\pm$ 1.2% left cf 95.6 $\pm$ 0.8% right horn of group 1.1
and 96.7 $\pm$ 1.1 left cf 93.7 $\pm$ 1.7% right horn of group 1.2) nor
the mean position of the catheter tip (4.5 $\pm$ 0.4 cm from the
UTJ left cf 4.4 $\pm$ 0.4 cm right in group 1.1 and 4.4 $\pm$ 0.4 cf
5.0 $\pm$ 0.4 cm right in group 1.2). The correlation between
actual and rectal assessment of numbers of ovulations was 0.98
(P< 0.001).

TABLE 1

THE EFFICIENCY OF NON-SURGICAL COMPARED WITH SURGICAL RECOVERY OF DAY 7 BOVINE EMBRYOS

	Non-Surgical recovery (1st superovulation)			Surgical recovery (2nd superovulation)		
	Left horn	Right horn	Total	Left horn	Right horn	Total
1000 iu PMSG (Group 1)						
No. of ovulations	2.0 ± 0.3	2.7 ± 0.5	(n=20) 4.7 ± 0.7	1.1 ± 0.3	1.1 ± 0.3	(n=17) 2.2 ± 0.5
No. of embryos	1.3 ± 0.2	1.5 ± 0.4	2.8 ± 0.5	0.9 ± 0.3	0.9 ± 0.2	1.8 ± 0.4
% recovery	67.6 ± 8.6	50.2 ± 10.3	*62.6 ± 8.1	71.7 ± 13.2	90.9 ± 6.1	*79.2 ± 8.4
2000 iu PMSG (Group 2)						
No. of ovulations	7.9 ± 1.4	6.4 ± 0.9	(n=21) 14.3 ± 2.1	4.9 ± 0.9	6.2 ± 1.3	(n=17) 10.7 ± 1.9
No. of embryos	4.6 ± 1.1	3.1 ± 0.6	7.6 ± 1.6	3.0 ± 0.5	3.2 ± 0.7	6.2 ± 1.0
% recovery	58.1 ± 7.1	53.2 ± 6.9	§55.2 ± 6.0	68.5 ± 7.9	48.5 ± 8.0	§63.9 ± 6.2

* and § differences are not significant (P< 0.05) values are means ± SEM.

Experiment 2

The results of experiment 2 are given in Table 2. A high proportion of the heifers became pregnant (85%) when one embryo was transferred to the ipsilateral uterine horn surgically and one embryo transferred non-surgically to the contralateral uterine horn. When one embryo was transferred non-surgically to the ipsilateral horn and surgically to the contralateral horn the pregnancy rate was by contrast low (35%).

In group 2.1 despite the high pregnancy rate the embryo survival rate of non-surgically transferred embryos was poor (7/20) compared to the surgical survival rate (17/20). In group 2.2 both surgical and non-surgical embryo survival rates were poor (6/20 cf. 4/20).

TABLE 2

COMPARISON OF TRANSFERRING ONE DAY 7 EMBRYO SURGICALLY TO THE UTERINE HORN IPSILATERAL TO THE CORPUS LUTEUM AND ONE EMBRYO NON-SURGICALLY TO THE CONTRALATERAL UTERINE HORN, OR NON-SURGICALLY TO THE IPSILATERAL AND SURGICALLY TO THE CONTRALATERAL UTERINE HORN

Group 1 Surgical ipsilateral			Group 2 Non-surgical ipsilateral		
No. of heifers	No. of pregnancies		No. of heifers	No. of pregnancies	
	Ipsi. (S)	Contra. (NS)		Ipsi. (NS)	Contra. (S)
20	17	7	20	4	6
Total pregnant (%) 17 (85)			Total pregnant (%) 7 (35)		

S = Surgical

NS = Non-surgical

Experiment 3

The results of experiment 3 are given in Table 3. Very similar results were obtained after transfer to the tip and to the base of the uterus.

TABLE 3

COMPARISON OF TRANSFERRING ONE DAY 7 EMBRYO SURGICALLY TO THE TIP OR TO
THE BASE OF THE UTERINE HORN IPSILATERAL TO THE CORPUS LUTEUM

Transfer to tip		Transfer to base	
No. of heifers	No. pregnant (%)	No. of heifers	No. pregnant (%)
10	6(60)	10	5(50)

DISCUSSION

The recovery rate of embryos using the non-surgical method described in experiment 1 was lower than the recovery rate
using surgery on the same group of animals on a second occasion.
A contributing factor towards the improved recovery rate with
surgery was undoubtedly the lower ovulation rates in both 1000
iu and 2000 iu groups on the repeat superovulation. In both
surgical and non-surgical groups a higher recovery rate was
obtained in the low PMSG dosage groups. The non-surgical recovery rate compares very favourably with the success claimed
by other workers (Elsden et al., 1976; Rowe et al., 1976) and
is probably more reliable since ovulation rate was assessed by
exteriorisation of the ovaries and not rectal palpation or laparotomy. Part of this success may be attributed to passing the
catheter to the tip of the uterus where it has previously been
shown that most eggs are situated (Newcomb et al., 1975). It
is interesting in this respect that 76.9% of all embryos recovered were in the first 100 ml of fluid recovered and before
any manipulation was commenced. A further contribution towards
the success of this technique may have been the excellent control and relaxation achieved by using anaesthesia It may be
that the difference in recovery rate between right and left
uterine horns in this non-surgical series was attributable to
the positioning of the animal because neither fluid recovery

nor the positioning of the catheter would appear to be responsible as they were similar for both horns.

The transfer results in experiment 2 support previous work
by Brand et al. (1975_1) and Seidel et al. (1975) that if adequate precautions are taken infection would not appear to be the
reason for the lower success of non-surgical compared with
surgical transfer and that rejection of embryos is probably
not implicated at Day 7 (Brand et al., 1975_2; Newcomb and Rowson, 1976).

The transfer of an embryo to the contralateral uterine
horn immediately after a surgical transfer to the ipsilateral
horn did not disturb the surgical pregnancy rate which remained
very high (85%).

The number of embryos surviving non-surgically in group
2.1 was low, but the survival rate was nearly double that when
the position of surgical and non-surgical transfer was reversed.
This may reflect some supportive role by the embryo within the
ipsilateral uterine horn which was possibly exerting a stronger
luteotrophic action than an embryo placed within the ipsilateral
uterine horn non-surgically.

The low pregnancy rate obtained in group 2.2 from embryos
transferred surgically to the contralateral uterine horn strongly
supports previous work by Newcomb and Rowson (1976) and Sreenan (1976) which indicated that an embryo placed in the contralateral uterine horn had only a poor chance of survival.

These results suggest also that unless a high pregnancy
rate is obtained in the ipsilateral uterine horn it is unlikely
that a high pregnancy rate will be achieved in the contralateral uterine horn. The pregnancy rate after AI tends to be
lower than after egg transfer and if non-surgical methods become as efficient as is currently achieved surgically it is
probable that the method used by some workers (Boland, et al.,
1975) of transferring one egg non-surgically to the contralat-

eral horn of a previously inseminated recipient, would be less efficient that transferring one egg to each uterine horn.

The low pregnancy rate achieved after the non-surgical transfer of embryos in this experiment suggests that either the site of the non-surgical transfer or some feature of passing the pipette through the cervix is responsible for the poor results. The result of placing embryos either at the tip or at the base of the ipsilateral uterine horn surgically in experiment 3 is inconclusive, but position of the embryo may be less important than passage through the cervix. The pregnancy rate resulting from transfer to the tip was lower than that achieved in experiment 2, possibly a consequence of using a different method of transfer. The effect of site of transfer remains an open question although both in this experiment and in Sreenan's (1976) study slightly more pregnancies were obtained after transfer to the tip of the ipsilateral uterine horn than to the base.

REFERENCES

Boland, M.P., Crosby, T.F. and Gordon, I. 1975. Twin pregnancy in cattle established by non-surgical egg transfer. Brit. vet. J. 131, 738-740.

Brand, A., Gunnink, J.W., Drost, M., Aarts, M.H. and de Bois, C.H.W. (1975$_1$). Non-surgical embryo transfer in cattle. II. Bacteriological aspects. In: 'Egg Transfer in Cattle'. EEC Symposium, Cambridge 10-12 December 1975, pp. 57-66.

Brand, A., Taverne, M.A.M., van der Weyden, G.C., Aarts, M.H., Dieleman, S.J., Fontijne, P., Drost, M. and de Bois, C.H.W. 1975$_2$. Non-surgical embryo transfer in cattle. I. Myometrial activity as a possible cause of embryo expulsion. In: 'Egg Transfer in Cattle'. EEC Symposium Cambridge 10-12 December 1975, pp. 41-56.

Elsden, R.P., Hasler, J.F. and Seidel, G.E. Jr. 1976. Non-surgical recovery of bovine eggs. Theriogenology 6, 523-532.

Newcomb, R and Rowson, L.E.A. 1975. A technique for the simultaneous flushing of ova from the bovine oviduct and uterus. Vet. Rec. 96, 468-469.

Newcomb, R and Rowson, L.E.A. 1976. Aspects of the non-surgical transfer of bovine eggs. VIIIth Int. Congr. Anim. Reprod. & A.I., Krakow. Proceedings Vol. III, pp. 262-265.

Newcomb, R., Rowson, L.E.A. and Trounson, A.O. 1975. The entry of superovulated eggs into the uterus. In: 'Egg Transfer in Cattle'. EEC Symposium, Cambridge 10-12 December 1975, pp. 1-15.

Rowe, R.F., Del Campo, M.R., Eilts, C.L., French, L.R., Winch, R.P. and Ginther, O.J. 1976. A single cannula technique for non-surgical collection of ova from cattle. Theriogenology 6, 471-483.

Rowson, L.E.A., Moor, R.M. and Lawson, R.A.S. 1969. Fertility following egg transfer in the cow; effect of method, medium and synchronisation of oestrus. J. Reprod. Fert. 18, 517-523.

Seidel, G.E. Jr., Bowen, J.M., Homan, N.R. and Okun, N.E. 1975. Fertility of heifers with sham embryo transfer through the cervix. Vet. Rec. 97, 307-308.

Sreenan, J.M. 1976. Egg transfer in the cow: effect of site of transfer. VIIIth Int. Congr. Anim. Reprod. & A.I., Krakow. Proceedings Vol. III, pp. 269-272.

Whittingham, D.G. 1971. Survival of mouse embryos after freezing and thawing. Nature, Lond. 233, 125-126.

NON-SURGICAL RECOVERY OF BOVINE EMBRYOS UNDER FARM CONDITIONS

W.W. Lampeter

Institut für Tierzucht und Tierhygiene der Universität München
Lehrstuhl für Tierzucht, München, W. Germany.

INTRODUCTION

The great advantage of embryo transfer for the breeders
lies in the possibility of multiplying the offspring of a
superior cow as compared to the number of offspring produced
by natural mating or artificial insemination. (Foote and
Onuma, 1970), Gordon (1975), Morcan (1972) and Rowson, (1974).
An intensive preselection of donor animals is essential for an
effective breeding programme. The most promising goal for
such programmes within the scope of the present techniques of
embryo transfer seems to be working with donors that are potential producers of AI bulls. In West Germany, there are AI
breeding programmes for all breeds in all regions including
almost 90% of the national herd, and data are computerised.

At the Munich University, a pilot project was carried
out over a two months period in early 1977 to examine the
technical possibilities of embryo transfer working with highly
selected superior cows under farm conditions. Morcan, (1972).
From a 230 000 cow population, 2% were computer selected as
potential donors strictly on performance data. The following
criteria were set for the selection:

a)   Potential AI bull dam

b)   High performance data (milk, butterfat)

c)   Six years and older

d)   Dual purpose breed (Fleckvieh)

e)   In milk production

MATERIAL AND METHOD

A region of Northern Bavaria was picked out for the project. There are approximately 230 000 cows in this area, in
three state operated breeding institutions. These institutions
screened, strictly on performance data, 50 potential donors
for the project out of their data pools. A second selection
was carried out by ourselves. The only criteria for our own
selection were:

f) No reproductive irregularity

g) No signs of vaginal, cervical or uterine infection

h) At least 40 days of 2 heats after calving.

Due to these selection criteria, only 21 animals out of
the 50 could be used for the experiment. The others were excluded for various reasons, mainly because they were again in
calf or had just received AI service. In some cases we detected infections.

Only a few farmers did not want to co-operate. The animals were leased from the farmer for the duration of the experiment, but were left in their own barns where stimulation and
collection of embryos was performed. In addition to the fee,
the farmer will receive one female calf out of the offspring of
his donor cow.

All animals selected were purebred German Fleckvieh.
Animals with reported breeding difficulties (repeat breeders)
were included. They ranged in age from 6 - 13 years, in
weight from 500 - 700 kg and in milk production from 4700 -
7200 l per 305 days lactation period. For oestrus detection,
we were fully dependent on the farmers' observations. All
animals were in a good nutritional and health condition and
came from different farms.

Over the period of two months (1st February to 31st March

1977) the 21 animals were stimulated. Stimulation was carried
out on Days 9 to 11. The stimulation agent was PMSG and 3000 iu
were given per injection. Prostaglandin was injected 48 hours
later, followed by three AI. Embryo collection was carried out
non-surgically and mostly on Day 7 after induced heat (Table 1).

TABLE 1

RESULTS OF STIMULATION AND COLLECTION

Animals stimulated		Animals reacted to stimulation	Animals flushed	Animals whose embryos were transferred
No.	21	16	11	9
%	100	76.2	52.4	42.9

The embryo collection team consisted of three persons,
two veterinarians and one assistant. A VW van was equipped with
the necessary instrumentation to function as a mobile embryo
collection unit. The total time for one collection varied from
45 - 60 minutes which included handling of instruments. Most
time was needed for microscopic work to search the medium for
embryos. This was done either in the mobile unit or in the
embryo transfer station itself.

The Catheter used for flushing was a two-way Foley Catheter
stiffened by a stainless steel stylet. A metal head at the tip
of the Catheter allowed for smooth passage through the cervix.
Both horns could be flushed with the same Catheter. By pulling
back the Catheter caudally to the bifurcation, after having
flushed one horn, it stiffened again inside the uterus. The
correct position of the stylet inside the Catheter was indicated
by a metallic clicking noise when the stylet slipped into the
metal flushing head. The Catheter was moved into the second
horn under rectal control.

Flushing was by means of a 60 ml syringe. To reduce the
amount of flushing medium and consequently to minimise time
consuming microscopic work, a filter was added to the flushing

Catheter. A three-way valve added to the filter allowed re-
use of the flushing medium several times and reduced the amount
to 120 ml per uterine horn. The first syringe containing 60 ml
medium was used to check if the Catheter was in the right position. Then the second 60 ml was flushed (at least 8 times) in
and out of the uterine horn via the filter. By flushing the
filter itself in the opposite direction, the medium containing
the embryos was collected in glass flasks and stored in a thermos
block.

The embryos were located with a Wild M7 sterodissection
microscope. Finally the embryos were transferred to storage
medium (PBS and 20% foetal calf serum and Gentamycin) and into
a thermos container ready for transport to the Embryo Transfer
station where transfer was performed surgically on the same
day or in a few cases the day after. (Table 2). The recipient
cows were selected from a pool of 220 Fleckvieh heifers and
had been synchronised with $PGF_2\alpha$. Their ages ranged from 16 -
20 months.

TABLE 2

RESULTS OF TRANSFERS

No. of embryos transferred	No. of recipients	Recipients palpated at 90 days pos.	% Pregnancy rate
34	31	10	32.25

RESULTS

The summary of results of the two month study is presented in Table 3. Five of the 21 donors (23.8%) did not react
to the stimulation. In one case the cervix could not be passed with the Catheter. In another case endometritis was observed and therefore no flushing was done. Curiously enough, this
donor ended up with a twin pregnancy. In two cases stimulation

TABLE 3

SUMMARY OF RESULTS - 1ST FEBRUARY - 31ST MARCH 1977

Donor	Stimulation	Palpation right	Palpation left	No. of embryos collected	No. of embryos transferred	Pregnancies
1	+	7-10	6-7	11	8	4
2	+	6	4	4	3	1
3	+	4	3	5	3	-
4	+	7	3	5	4	-
5	+	2	0	1	1	-
6	+	3	2	1	1	1
7	+	5-6	3-4	7	5	1
8	-	0	0	no flushing	-	-
9	+	2	1		-	-
10	-	1	0		-	-
11	-	0	0		-	-
12	-	0	0		-	-
13	-	0	0		-	-
14	+	2	1		-	-
15	+	2	2		-	-
16	+	4	6	5	-	-
17	+	6-7	5-6	--/*	-	-
18	+	3	2	8	6	2
19	+	8-10	5	4	3	1
20	+	3	3	0	-	-
21	+	6	6	7	-	-
21	16 x + 5 x -	71-78	52-55	58	34	10

/* Cervix could not be passed.

was poor (less than 4 CL), therefore no flushing was carried
out either.

Eleven cows were flushed successfully and a total of 58
embryos could be collected.   34 of these were transferred
into 31 recipients, the rest were either unfertilised, in the
2 - 4 cell stage, or lost. Ten pregnancies could be palpated
60 - 90 days after transfer which means a pregnancy rate of 32%.

CONCLUSION

There are some promising possiblities to incorporate embryo
transfer into geneticists' work. One is to multiply rare
animals with exceptional characteristics, or to produce AI
bulls by working with bull dams - which was the subject of
this study - or to follow other breeding concepts to accelerate
genetic improvement. (Cunningham, 1975;   Hill and Land, 1975).
In all cases, selection has to be done based only on performance
data of the potential donors. These animals are at least 3
years old and in most cases under heavy production stress.

If embryo transfer is to become a worthwhile tool to geneticists, the rate of failure should not be higher than 10% in
any group of donors. Stimulation, ovulation and recovery rates
have to be increased and fertilisation of eggs has to be improved.
It should be possible to produce 6 - 7 calves per donor per year
carrying out only two flushings consecutively. As long as it is
not possible to get satisfactory results on each selected
individual donor, embryo transfer as it stands today cannot yet
be recommended for breeding programmes under field conditions.

REFERENCES

Foote, R.H. and Onuma, H. 1970. Superovulation, ovum collection, culture and transfer - a review, J. Dairy. Sci., 53, 1681.

Gordon, I. 1975. Problems and prospects in cattle egg transfer. Irish Vet. Journal, 29, (2, 3) 21.

Morcan, L. 1972. Intensive cattle breeding with ova transplantation. NZ. J. Agric. 125, 15.

Rowson, L.E.A. 1974. The role of research in animal reproduction. Vet. Rec., 95, 276.

Drost, M., Anderson, G.B., Cupps, P.T., Horton, M.B., Warner, P.V. and Wright, R.W. 1975. A field study on embryo transfer in cattle. J.A.V.M.A.

Kräusslich, H. 1975. Application of superovulation and egg transplantation in AI breeding programmes for dual purpose cattle. EEC Agric. Res. Sem., EUR 5491, 333.

Smidt, D. 1973. Eitransplantation und Zyklussynchronisation beim Rind Forschungsaufgaben und Anwendungsmöglichkeiten. Vortrag DGFZ und Haustiergenetik, Giessen.

Cunningham, E. 1975. The use of egg-transfer techniques in genetic improvement EEC Agric. Res. Sem., EUR 5491, 333.

Hill, W.G. and Land, R.B. 1975. Superovulation and ovum transplantation in genetic improvement programmes. EEC Agric. Res. Sem., EUR 5491, 355.

I. Wilmut *(UK)*

Dr. Brand, I think you are suggesting two entirely different mechanisms for failure, one is the possibility of releasing
toxic mechanisms from the uterus and the other is the possibility
that the embryos are not able to prevent luteolysis. I think
we have got an easy way to distinguish between these and that is
to look at embryos two or three days after transfer, or five
days after transfer, because in one case the toxic factor will
have killed the embryo and in the other case the embryo will be
developing or it may perhaps be retarded. So my question is,
have you, or anybody else, got information on the development of
embryos after non-surgical transfer?

A. Brand *(Netherlands)*

I don't think it has been done in cattle but R.A.S. Lawson
carried out these experiments in ewes. It hasn't been published
yet, but I agree it is still very important.

M. Boland *(Ireland)*

You mentioned an abortion rate of about 20% in your non-
surgical transfers. At what stage of gestation did the abortions
occur and what, in your opinion, are the contributing factors?

A. Brand

Of the 27 animals I mentioned, 12 had already calved and
5 of the remainder aborted. Just at that time we had a viral infection in our animals and that may have been the reason. On the
other hand, other people have had a high abortion rate as well.
It is very difficult to establish what has been going on a few
days or weeks earlier because you are always too late to examine
it.

M.T. Kane *(Ireland)*

I might make a comment on the question of the toxicity of
serum. The original work that Chang did showed that the serum of
one species was toxic to the eggs of another. I know there are

suggestions in the literature that cow serum is toxic to cow eggs unless it is first heat treated. I don't know that it is definitely the case that serum should be toxic.

J.M. Sreenan

My question is not really directed to Dr. Brand but rather to Dr. Schneider or Dr. Boland, whether they might comment on this abortion problem. It is something which is significant; yet it is something which we haven't seen very much in our laboratory - we have a low rate of abortion, and it is not different after non-surgical transfer, surgical transfer or normal breeding. I wonder whether the French group might also comment.

Y. Heyman (France)

It is difficult for us to comment because following surgical transfer our control was slaughtered between Days 45 and 120 of pregnancy. We never had infection after surgical transfer.

A. Brand

I think maybe it is just sub-clinical infection. We have have taken care of asepsis too by sterilising and scrubbing and disinfecting everything. It is only this year that we had a problem. Last year we didn't have one abortion, so I don't know what the cause is

U. Schneider (West Germany)

I will also present some data showing that we find a certain number of abortions. It seems to me that in one year you find more and in another year, somewhat less. Similar data was presented by Gary Anderson this year in Madison where, in the 1975/76 transfer year, they had a lot of abortions. They tried to study this during that year and the occurrence was very low. So I don't think anyone knows exactly what causes the abortions.

L.E.A. Rowson (UK)

I think we must stop at that point and perhaps take it up again in the general discussion. I would like to have some questions on Ray Newcomb's paper right now.

I. Gordon (Ireland)

On the point that you made about the success rate following surgical transfer with single eggs, that there is a possibility of a higher success rate by using bilateral transfer, you said that the pregnancy rate that you would get in your recipients would be higher is you were using egg transfer than with the artificially inseminated animal. I am not too sure that I would agree with that. Of course, it all depends upon the fact that the mated recipient has a fertilised egg there. On the basis of the evidence on fertilisation rates in cattle, then the finding of insemination or fertilisation is favourable, that most of the eggs, in fact, are fertilised in the cow. I would have thought that the chances of having a decent egg on the ipsilateral side would be higher if you are dealing with the inseminated cow than if you are dealing with the animal which has an egg transferred to the ipsilateral side.

R. Newcomb (UK)

Well, this is certainly true at the moment. What I was saying was that if you want to obtain a high pregnancy rate in the contralateral horn by non-surgical transfer, the evidence there (and other published evidence) strongly suggests that you must have a very high pregnancy rate in the ipsilateral horn. Now, the pregnancy rate which has occurred after artificial insemination, as far as I am aware, has never approached the sort of figures that have been quoted after surgical transfer of eggs. Would you agree with that?

I. Gordon

I'm not sure that I would on the basis of the published data.

R. Newcomb

Well, obviously that is a matter for dispute but what is indisputable is that it seems very necessary that you have got to have a high pregnancy rate in the ipsilateral horn if you are going to get a high pregnancy rate in the contralateral horn. One hears figures of 40 - 63% pregnancy rate after AI which means that twinning rates are going to be quite low by egg transfer if you can't get above that sort of pregnancy rate in the ipsilateral horn.

I. Gordon

I still can't get over the point that you can get a better pregnancy rate in the ipsilateral horn by doing egg transfer than by having the cow left to its own devices in a normal fertilisation situation.

R. Newcomb

Well, certainly it is our experience that we get a much higher pregnancy rate after egg transfer than we do with AI.

J.M. Sreenan

I think there is probably a slight misunderstanding here. I think perhaps Professor Gordon may not have understood that what you are suggesting is if we could get non-surgical pregnancy rates comparable with our best current surgical pregnancy rates, that then you will have a pregnancy rate higher than artificial insemination results. I think Mr. Newcomb is probably thinking of the pregnancy rates obtained from artificial insemination of large numbers of animals in a breeding season and the resulting calving rates here would be in the region of 50% after a single insemination. However, I think Professor Gordon is also right in that if you select your animals for breeding and the timing of insemination is correct, as he says, the published evidence is that fertilisation rate is in the region of 90%. Therefore, if you can adjust factors in the normal breeding situation to increase that pregnancy rate also, then I think it is a viable proposition to transfer a single egg to a bred recipient because you have less manipulation. In terms

of donor animals I would also favour this because you get twice
the spread of eggs - or you can transfer to twice the number of
recipient animals. But the pregnancy rates that are obtainable
at the moment would favour what Mr. Newcomb is saying, that
bilateral transfer gives a higher pregnancy rate.

M. Boland

In relation to your bilateral non-surgical transfers,
have you carried out transfers using the Cassou Gun? If so,
what sort of pregnancy rates are you getting?

R. Newcomb

We haven't carried out any bilateral transfers; we have
been looking here at what we think are basic problems and then
hypothesising from that. The results of the non-surgical transfer which Mr. Rowson and Alan Trounson did about a year ago have
not been published yet but perhaps I could give them here.

On single egg transfers to the ipsilateral horn, at Day
6, the pregnancy rate was 25%; at Day 7, 40%; at Day 8, 50%;
and at Day 9, 59%.

M. Boland

Have you any experience with the two egg transfer?

R. Newcomb

Not with the Cassou Gun - no.

J. Saumande (France)

You have shown that pregnancy rarely occurs when eggs
are transferred to the contralateral horn. As far as twinning
is concerned, did you try to transfer eggs in animals moderately
superovulated in order to have corpora lutea on both ovaries?

R. Newcomb

No, we haven't done that. I believe some work has been
done along these lines in sheep.

L.E.A. Rowson

If there are no more questions I think we ought to move
on to Dr. Lampeter's paper.

I. Wilmut

If you are attempting to induce superovulation about
seven or eight weeks after calving, this is the time when these
animals will have almost reached their maximum lactation. Do you
think there would be any advantage in trying to get superovulation,
say six months after calving, or at some other time away from
the peak of production?

W.W. Lampeter *(West Germany)*

Yes, but if we do it under field conditions we have to
consider the calving time; we can't keep them open for six months,
the farmers wouldn't agree to that.

I. Wilmut

Not even if you were hoping to produce an AI bull?

W.W. Lampeter

Not even then because you see they are registered in the
herd book system and have to have a calving interval of less
than 400 days.

U. Schneider

May I make a comment on this. We have very preliminary
data that superovulation response decreases if the calving interval is longer than 100 days.

L.E.A. Rowson

At this point may I turn the session over to general
discussion and bring in all our speakers. I would like to re-
mind you that there was no time earlier for Dr. Seidel to answer
questions immediately following his paper so perhaps we could
fire some at him to start with.

M. Boland

You mentioned in one of your first slides, Dr. Seidel, that you had a very poor recovery rate of eggs following gonadotrophin treatment on Day 15. Have you any ideas why this should be so?

Secondly, how soon after the failure of your recipients to become pregnant did you treat them for superovulation?

G.E. Seidel (USA)

On the first question, I didn't present any data on that but I talked about it. We have done some recoveries of PMS treated cattle at Day 15 and it is very variable. I have the suspicion that PMSG may be causing problems later in that a lot of the problems with non-surgical recovery are not related to the method but simply to the fact that there are no embryos there. I don't want to say any more about that; I don't have the data with me.

On the question of the failed recipients, they were done at least two cycles and usually three, after failing to become pregnant.

Y. Heyman

I assumed you had some problems with fertilisation rates in your donors with three inseminations. I want to know at what time after the prostaglandin injection you made your inseminations because with two inseminations at 56 h and 72 h after prostaglandin we had about 90% fertilisation.

G.E. Seidel

Again, this depends on the treatment I think. We inseminate 12, 24 and 36 h after the cows first come into heat for commercial cows. For experimental donors, we simply do 12 and 24 h as a standard regimen. With the non-surgical recovery from FSH treated donors that I presented yesterday, we have an 80% fertilisation rate. These recoveries however, are done 7, 8 and 9 days after oestrus and perhaps we are simply not recovering

the unfertilised ova. Maybe the unfertilised ova are select-
ively lost - this is something we don't know. However, with
surgical recovery the fertilisation rates are usually 65 - 70%
with PMSG; with FSH they may be a little higher. I'm not sure
how to attack this question of selective loss of unfertilised
eggs.

K. Betteridge *(Canada)*

This question of selective loss of unfertilised eggs
keeps cropping up. We don't know whether it happens sometimes
but it certainly doesn't happen all the time because one can recover unfertilised eggs as late as Days 16 and 17.

R. Newcomb

The only comment I would like to make is that I think
the quality of the semen which one uses is particularly important
in the superovulated animal. In the work which we have done,
both with liquid and frozen semen, our fertilisation rate is similar to that of the French group, about 90%. However, we have
done some work in a more commercial situation where we have been
using semen which people have said we ought to use, ie, frozen
semen. In that situation we ended up with quite a lot of unfertilised eggs until we put our foot down and used a bull which was
of a very high fertility. I think that a lot of the work which
Dr. Seidel is doing is with semen which people tell him ought to
be used and not necessarily semen with the highest fertility.

G.E. Seidel

This is definitely true; these people are paying $1,500
per pregnant recipient and we must use the semen they give us.
We do look at the semen very carefully and if it is not acceptable, we will not use it. We have had plenty of experience of
using bad semen and very often one finds out afterwards.

It reminds me of a case where a farmer insisted we use
a certain semen. He said it was good and it looked pretty good
under the microscope but we got all unfertilised eggs and we
found out later that he had inseminated 10 cows and got one

pregnant! So, you are absolutely right - poor or mediocre quality semen will cause problems.

H. Coulthard (UK)

I am in a private situation rather like Dr. Seidel and
I have used both methods of collection, Mr. Elsden's and Ray
Newcomb's. I think the conclusion that we are coming to now,
although it is based on only about 50 animals, is that some cows
are suitable for one method and some are suitable for the other.
What we are setting out to do is to use the Newcomb method with
a normal animal with a normal tract. If there has been any surgical interference at all, or any distortion of the uterus, then
the Elsden method certainly seems to be the best one.

There is one thing which hasn't been brought out very
strongly and it is a thing which we found made a tremendous difference to us. That is the positioning of the cow. I noticed
one of Ray Newcomb's illustrations showed his flush with the
front of the cow tipped up - this is in the standing cow. We
have found this makes a tremendous difference both to the ease
of passing the catheter and to the efficiency of the collection.

I would like to ask Dr. Seidel one question about his
infertile cows. Like him we have a tremendous influx of old,
infertile, Charolais cows which are difficult to deal with. But
we have found that when we have collected a number of what appeared to be fertile embryos on the first occasion, we have had
a low pregnancy rate from those embryos but subsequently we have
had reasonable pregnancy rates. I just wonder whether this has
been Dr. Seidel's experience with the infertile cows.

G.E. Seidel

If you get good embryos, the recipients get pregnant.
That's our experience.

J.P. Ozil (France)

I would like to ask a question about the collection. How
many times do you massage and is there any resorption of fluid

when massage is prolonged? Do you notice any difference between cows and young heifers?

G.E. Seidel

Let's start at the last question first. I should say that I do not do the collections so this is second-hand information, but I'm there a lot and I get the feeling that one feels a lot more comfortable collecting from heifers than cows - they have a much smaller uterus. However, our collection rate is just about as good with cows, provided they are not abnormal cows. I would like to emphasise that we decide the cows are normal before we do the collection, not afterwards. In other words, this is not a question of selecting the data based on the results. When the cow comes in we put her into either the fertile or infertile categories. We have two different numbering systems to keep them straight. Infertility is of all sorts.

So, as far as the massaging is concerned, each filling of the uterine horn, maybe 40 or 50 ml, is massaged a little bit. In cows it is a little different in that the uterus will expand more readily with repeated fillings and therefore while there may be 40 or 50 ml in the first filling of the uterus, we increase it to as high as 200 or 250 ml in, let's say, an eighteen year old cow.

On your question about fluid resorption by the endometrium, there is undoubtedly some of this but the fluid is in there for such a short length of time that we get better than 90% of it back. So I don't think this is a problem. This business about fluid going up the oviduct - we worry about it, we hold off the oviduct when the horn is under pressure. I am sure it would be a problem in a small percentage of cows but I am not convinced that with the pressures we use it would be a problem but we still take that precaution.

L.E.A. Rowson

Can we now widen the discussion to cover all the papers?

J.S. Perry *(UK)*

Dr. Brand made an interesting suggestion when he was talking about the place in the uterus where you deposit the elongated embryo. He suggested that an antiluteolytic or a luteotrophic substance emanating from that embryo is spread around the uterine lumen by uterine movement. I have always been impressed by, literally, the lengths to which the embryos go in the three domestic species to extend the embryo in such a way that there is trophoblast in contact with endometrium throughout the length of the uterus. I wonder, if this is the case, then the degree to which the recovered embryo is tangled up, or tied in knots, when you put it back in the uterus. Is this possibly a factor in the antiluteolytic, luteotrophic effect of the embryo, wherever you put it?

A. Brand

Inside the uterus maybe that is true and that is the meaning of del Campo. However, in my opinion, there is another factor in that maybe intraluminal fluids may be transported from the contra- to the ipsilateral, or vice versa, just by movement of the uterus. It is just a suggestion but I can show you some data presented by del Campo. What he did was to sever one uterine horn and also the intercornual ligaments. He injected his recipient animals from Day 13 to Day 24 with progesterone. Now, progesterone does not interfere with the lifespan of the corpus luteum at all. Then he transferred two embryos to the contralateral side. In this case there was no chance that the luteotrophic substance could be transported through the body of the uterus into the ipsilateral horn; luteolysis occurred here. However, in this case, the intercornual ligament was severed. He got more maintained corpora lutea of a heavier weight at Day 24, in this case, without severance of the intercornual ligament. He proposed that the intercornual ligament maintains the equilibrium. If you sever it then the horns act independently of each other and may be that is the reason that interluminal fluids and also membranes can pass much more easily to the body. However, I don't think that at Days 16 and 17 the embryo has already reached that size; it is discovering the whole

endometrium so there must be another factor and it may be just
the interluminal fluid.

There is another point which has already been mentioned.
I think that if you transfer an embryo to the contralateral horn
or just to the body of the uterus, non-surgically, you need excellent embryos - the fittest of all the embryos because only
the fittest will survive. That is our reason for our transfer
at Day 13 or Day 14 because then you can distinguish between an
embryo which has developed and you get such a big variation between embryos which are only small and embryos which are already
large and I think that is a very good criterion to select your
embryos, because if they are still in the zona pellucida you can
only say, "It is poor looking" or, "It is good looking" but you
don't have good criteria.

R. Newcomb

Dr. Brand, you have quoted del Campo's work and I wonder
how you explain it in relation to the theory which you have been
propounding because del Campo has found, at Day 24, no difference
between putting embryos in the contralateral horn compared to the
ipsilateral horn. There was equal survival of embryos and there
was no difference between corpora lutea weight.

A. Brand

There was a difference between corpora lutea weight.

R. Newcomb

It was not significant.

A. Brand

There was no difference in survival rate but he was injecting his animals with progesterone so the corpus luteum of
that animal itself showed luteolysis but the embryos were still
alive because there was an extra corpus luteum induced in them.

L.E.A. Rowson

I wonder if I could mention that where the ipsilateral

horn is severed, if you take that horn out and transfer to the
other horn you get a high pregnancy rate. This has been done
with sheep and also the cow although it has not been published
in the cow.

I. Wilmut

Dr. Brand, how do you relate your theory to the data
presented by Newcomb just a few minutes ago with the surgical
transfer into one horn and the non-surgical into the other?

A. Brand

I think there is an influence from the ipsilateral horn;
embryos transferred in the ipsilateral horn can help the embryo
transferred in the contralateral horn - the HCG may help because
you are then postponing luteolysis, giving your embryo much more
time to develop and produce luteotrophic substances. If you
transfer an embryo to the contralateral horn I think only one
fifth of all these embryos will survive.

I. Wilmut

But unless I misunderstood his data, having an embryo
transferred into the ipsilateral horn surgically did not help.

A. Brand

I think the results of transferring embryos just to the
base of the uterine horn are a little bit contradictory. Boland
has some results which show a difference; Sreenan and Newcomb
have results showing no difference. I think experiments should
be carried out on more animals. Maybe there are other unknown
factors which are important.

T. Greve (Denmark)

You said that there was a natural selection of the embryos at Day 7 or Day 8, and we just heard from Israel some
time ago that it seems that Day 7 is a very critical day for
the embryos. OK, so you have this selection and then on Day 12
you have more normal embryos which is what you would expect.
Now, I want to ask you because I don't quite remember the

figures, what is the pregnancy rate of Day 12 embryos, it ought
to be very high since you have got rid of all the poor embryos?

A. Brand

I don't think I can give you an answer at the moment
because up until now we haven't found any difference. As you
know when you start an experiment such as this, there are so many
factors involved. Maybe I can answer that question next year.

W.W. Lampeter

We are using many different terms: 'good' embryo, 'poor'
embryo, 'fit' embryo, 'embryo which will make it' and 'embryo
which won't make it'. Is there anybody in the room who can give
us a precise definition of a 'good' embryo? Is it colour, the
cell stage - what is it?

A. Brand

I can't tell you. That is the reason why I like to use
Day 13 embryos.

W.W. Lampeter

We have also found, surgically, embryos at Days 4 and
5 which ended up as a pregnancy. So it is not definitely Day
13 or 12 or any other day.

A. Brand

Well, I can't say just by looking into a microscope at
Day 8 that a particular embryo is good. I know that it doesn't
look as good as another embryo but that is all I can say.

T. Greve

Have you done any progesterone profiles or oestrogen
profiles, any endocrinological studies in recipients, because
I think that is a very appropriate subject for ipsi- or contralateral transfers, to see how it is? I think you are suggesting
that if you are injecting recipients with progesterone or with
HCG which would extend the cycle,then you would get a higher
degree of survival rate of the embryos. Is it necessary for the

progesterone level to be at a certain concentration in order for
the recipient to maintain the pregnancy?

A. Brand

I don't think so; I think you need a certain level of
progesterone. It is known that there are pregnant animals with
higher and lower levels but it is the continuation that is important. I just hope that the presence of HCG will give the
embryo more chance to produce its luteotrophic substance and
develop. Del Campo has shown that just administering progesterone keeps embryos alive.

G.E. Seidel

I have got some data on that point. We looked at progesterone levels on the day of transfer in 530 heifers and there
was no difference in levels between those which get pregnant and
those which do not, unfortunately. We also looked at them in
mid-cycle and again there was no difference. I have got a slide
of some data that relates to natural endocrinology. This data
worried me for a long time:

Cycle length (days)	Number recipients	Pregnant (%)
17 - 19	126	41
19.5 - 21.5	338	56
22 - 24	76	39

This was a large pool of recipients that have different
cycle lengths. This is cycle length prior to transfer. We
divided the recipients into three groups. All of them had cycle
lengths between 17 and 24 days. This relates to transfers mostly
at Days 4 and 5 and, as you probably would expect, those with
the normal cycle length had the higher pregnancy rate but we did
not expect this large difference which is statistically significant ($P < .001$). This data also provides information for some
of the synchronisation work. I don't know what is going on; it
is another situation where the data are a little too conclusive
to ignore. We are currently looking at the data from transfers
on Days 7 and 8 and seeing if it still holds. I would be

interested in anybody's interpretation of what is going on here.
I have been sitting on this data for quite a while because I want
some more just to make sure it is correct.

T. Greve

I think one of the main problems is fertility following
non-surgical recovery of embryos. Have you any data on this
issue, for instance, the length of the cycle, when do they come
in heat normally and what is the conception rate and maybe the
number of inseminations after the non-surgical recoveries?

G.E. Seidel

You say conception rates after non-surgical recovery -
we have not seen any difference and we have recovered embryos
from a lot of animals six times, seven times, in a row, without
superovulation for example. Sometimes they short-cycle. On the
average we get a two day shorter cycle after non-surgical recovery.
The cattle form the two categories, those that short-cycle -
maybe 20%, and those that do not. This may be an advantage.
Sometimes we recover an embryo, the cow comes into heat 10 days
after she has been in heat, or three days after recovering an
embryo, we breed her again and get another good one - we have
done this a number of times so it is not necessarily bad that
they short cycle and it's not unexpected because this amount of
manipulation probably causes prostaglandin release and so on.

A. Brand

The farmers in our country are very satisfied with the
non-surgical procedure. When we first started we used the second oestrus after we had recovered embryos and now we use the
first oestrus after superovulation. Most of the animals become
pregnant after the first insemination; I think that is quite
nice because if you use the non-surgical method you are not inter-
fering with subsequent fertility at all.

W.W. Lampeter

I am not referring to the study I have presented but in
other cases after doing non-surgical collection, we induced the

following heat by administering prostaglandin and then using the first heat for insemination and we did not get the best results. If we didn't induce heat by administering prostaglandin, that is, if we waited for the next natural heat, we had quite normal results after AI. So I think there is a difference, if we use prostaglandin after stimulation the conception rate seems to be a little bit lower compared to waiting for the next natural heat.

R. Newcomb

I wonder if I could ask Dr. Seidel whether he has noticed any difference in the transfer success rate of eggs which have been recovered either surgically or non-surgically - whether they seem to have the same success rate on surgical transfer?

G.E. Seidel

There's no difference but it is a little bit confounded because we tend to recover the non-surgical embryos later and therefore there is a little more selection. However, we see absolutely no difference although we haven't done the experiment expressly to do that.

I. Gordon

I would like to ask Dr. Seidel whether he takes particular precautions to guard against temperature drop in his non-surgical collection? I watched him on the slide he put up, sitting there waiting for the fluid to come down. That's OK on a nice day, but ....

G.E. Seidel

We needed the sunshine to take the picture! Usually we do it inside. The ambient temperature is usually about $20^{o}C$ and we don't worry about it particularly. It may go up to about $25^{o}C$. We use warm fluid - warmed to body temperature. It takes a while for that to cool down and then we put it in an incubator.

I want to make one other point on the infertility. One of the uses of non-surgical methods has been with infertile cows

and this is more of a clinical expression than scientific fact
but we have a feeling that there is a certain amount of therapeutic value in doing non-surgical recoveries - putting 800 ml
of fluid in an infertile cow, especially with certain types of
chronic endometritus and so on. We have had a number of cases
where a farmer hasn't been able to get a cow pregnant for a year
or two, we do a non-surgical recovery, he breeds her and ends up
with two pregnancies in a 20-day period. So there are cases
where, in fact, it appears to be valuable rather than detrimental.

L.E.A. Rowson

I am afraid that we have to stop at that point. Thank
you all very much for your participation.

CERVICAL EMBRYO TRANSFER AT DIFFERENT STAGES IN CATTLE

Y. Heyman, J.-P. Renard, J.-P. Ozil and F. du Mesnil du Buisson
INRA - Station centrale de Physiologie animale
78350 Jouy en Josas, France

ABSTRACT

*Ninety cattle embryos at various developmental stages between Day 7
and Day 13 were transferred by the cervical route into each of the uterine
horns of 45 recipient heifers. No significant difference between the days
of transfer was found; 57.7% of the recipients were pregnant and 43.3% of the
embryos implanted. After bilateral transfer, implantation rate did not
significantly differ in the horns ipsilateral and contralateral to the corpus
luteum. The choice of an optimum day for embryo transfer in cattle is
discussed.*

INTRODUCTION

Embryo transfer by the cervical route at early stages (Day 3 to Day 5) gave poor results (Lawson et al., 1975). However, Sreenan (1975) obtained 50% pregnancy with transfers at Day 7.

We have shown that hatched blastocysts can be successfully transplanted (Renard et al., 1977). Betteridge et al. (1976) have demonstrated that by the surgical route the chances of successful transfer diminish after Day 14. In the present report, we compare blastocyst survival after cervical transfer at different developmental stages between Day 7 and Day 13.

MATERIALS AND METHODS

Seventy-three Charolais or crossbred Charolais heifers (28 donors and 45 recipients) 18 to 24 months old were used. Heat was detected twice daily with a vasectomised bull.

Donors

We used the superovulation treatment proposed by Phillippo and Rowson (1975); 2 100 iu PMSG (Intervet, batch V81) was injected on Day 10 of the cycle, followed by a 500 µg injection of a prostaglandin analogue*. The animals were inseminated twice with frozen sperm at 56 and 72 hours after Cloprostenol injection: the embryos were recovered after slaughter of the donors 7 to 13 days after the onset of heat.

Recipients

The recipient cycle was synchronised with that of the donors by two 500 µg injections of Cloprostenol given at an 11-day interval. The recipients were divided into 3 groups and were all implanted with two embryos, one in each uterine horn, using the cervical transfer technique described by Renard et al. (1977).

    Group 1   17 recipients were given 34 embryos recovered
                  between Day 7 and Day 9. The transplanted

* Cloprostenol ICI 80996

embryos were at the unhatched blastocyst stage
and had a diameter of less than 200 μ.

Group 2   15 recipients were given 30 embryos recovered
between Day 10 and Day 11. These embryos were at
the hatched blastocyst stage and were still
spherical or just beginning to elongate; their
diameter was between 200 and 500 μ.

Group 3   13 recipients were given 26 embryos recovered
between Day 12 and Day 13. These embryos were
elongating and had a diameter between 0.5 and
20 mm.

After transfer, non-return to heat and progesterone assay
were used for early determination of pregancy. The recipients
presumed pregnant were slaughtered between days 45 and 120 of
pregnancy, and the number of foetuses in each uterine horn was
determined. The position of the corpora lutea was noted for
37 recipients transferred between Day 9 and Day 13.

RESULTS

The results are shown   in Table 1.

Group 1   9 recipients out of 17 were pregnant (52.9%) and
12 embryos out of 34 gave foetuses (35.3%).

Group 2   10 recipients out of 15 transplanted at the Day 10
and Day 11 stage were pregnant (66.6%) and
implantation rate reached 56.6% (17 foetuses for
30 embryos).

Group 3   7 recipients out of 13 were pregnant (53.8%) and
10 embryos out of 26 were implanted (38.4%).

A comparison of these results shows no significant
difference at the 5% level between groups with regard to
pregnancy and implantation rates. At the 5% level, there was
no significant difference in blastocyst development in the
ipsilateral horn as compared to that in the horn contralateral
to the corpus luteum (51% implantation vs 38%).

TABLE 1

RESULTS OF CERVICAL TRANSFER AT DIFFERENT STAGES

	Group 1 D7 to D9	Group 2 D10 to D11	Group 3 D12 to D13	Total D7 D13
Number of recipients	17	15	13	45
Number of embryos transplanted	34	30	26	90
Number pregnant (%)	9 (52.9)	10 (66.6)	7 (53.8)	26 (57.7)
Number of foetuses	12	17	10	39
Implantation rate	35.3%	56.6%	38.4%	43.3%

TABLE 2

EMBRYONIC SURVIVAL RELATED TO UTERINE HORN AFTER BILATERAL TRANSFER

Transfer site	Horn ipsilateral to corpus luteum	Horn contralateral to corpus luteum
Number of embryos transplanted (1)	37	37
Number of live foetuses (2)	19	14
Survival percentage	51.3%	37.8%

(1)   Transfer between Day 9 and Day 13

(2)   Slaughter of recipients after 45 days of pregnancy or later.

High twinning rates may be reached after bilateral cervical transfer: in group 2, 7 pregnant recipients out of 10 carried twins. The 2 foetuses in 6 out of 7 pairs of twins were well distributed, and in one case the 2 were in the same horn ipsilateral to the corpus luteum.

DISCUSSION

Cervical transfer may be successfully used to transplant blastocysts between the Day 7 and Day 13 stages, when the percentage of eggs recovered and of normal embryos, remains constant after superovulation (Renard and Heyman, in press).

Whether the blastocyst is unhatched or not, spherical or elongated, its implantation potential seems to be the same. Perhaps at a certain stage (Day 10 to Day 11), embryonic development may not require such a close synchronisation between donor and recipient. The better results after transfer at Day 10 and Day 11 may be related to observations of the embryo at this stage, when blastocyst quality is easier to evaluate than at other stages; the inner cell mass is clearly visible, the blastocysts are usually composed of 300 to 500 cells and their aspect is more homogeneous than at later stages (Renard et al., 1977). In our experiments, we found no significant difference in the rate of development as related to the corpus luteum side. However, we cannot exclude the difference reported by Newcomb and Rowson (1976) and Sreenan (1976) after unilateral transfer, although after bilateral transfer there is very little difference.

The twinning rates we obtained by bilateral cervical transfer may be compared to those reported by Boland et al. (1976) (53.3%) and by Sreenan (1976) (42%) after one embryo was added by the cervical route to recipients which were already inseminated. The twin birth rate would be very similar to the twinning rate observed after slaughter at 45 days because embryonic survival after bilateral transfer is high between Day 30 and parturition (Sreenan and Beehan, 1976)

REFERENCES

Betteridge, K.J., Mitchell,D., Eaglesome, M.P., Randall, G.C.G., 1976.
Embryo transfer in cattle 10-17 days after estrus. VIIIth International
Congress on Animal Reproduction and Artificial Insemination, Krakow,
237-240.

Boland, M.P., Grosby, T.F., Gordon, I., 1976. Induction of twin pregnancy in
heifers using a simple non surgical technique. VIIIth International
Congress on Animal Reproduction and Artificial Insemination, Krakow,
241-244.

Lawson, R.A.S., Rowson, L.E.A., Moor, R.P., Tervit, H.R., 1975. Experiments
on egg transfer in the cow and ewe: dependence of conception rate on the
transfer procedure and stage of the oestrous cycle. J. Reprod. Fert., 45,
101-107.

Newcomb, R., Rowson, L.E.A., 1976. Aspects of the non surgical transfer of
bovine eggs. VIIIth International Congress on Animal Reproduction and
Artificial Insemination- Krakow, 262-265.

Phillippo, P., Rowson, L.E.A., 1975. Prostaglandins and superovulation in the
bovine. Ann, Biol. anim. Bioch. Biophys., 15, 233-240.

Renard, J.-P., Heyman, Y., du Mesnil du Buisson, F., 1977. Unilateral and
bilateral cervical transfer of bovine embryos at the blastocyst stage.
Theriogenology, 7(4), 189.

Renard, J.-P., Heyman, Y., 1977. Quantitative and qualitative aspects of
superovulation for embryo transfer in the bovine. In press.

Renard, J.P., Menezo, Y., Saumande, J., Heyman, Y., 1977. Quality
determination of bovine embryos produced by superovulation. In: 'Control
of Reproduction in the Cow', Galway, Sept. 27-28-29th, 1977.

Sreenan, J.M., 1975. Successful non surgical transfer of fertilised cow
eggs. Vet. Rec., 96. 490-491.

Sreenan, J.M., 1976. Egg transfer in the cow: effect of side of transfer.
VIIIth International Congress on Animal Reproduction and Artificial
Insemination, Krakow, 269-279.

Sreenan, J.M., Beehan, D., 1976. Embryonic survival and development at
various stages of gestation after bilateral egg transfer in the cow.
J. Reprod. Fert., 47, 127-128.

CATTLE TWINNING BY NON-SURGICAL EGG TRANSFER

I. Gordon and M.P. Boland

Faculty of Agriculture, University College, Dublin.

ABSTRACT

Work towards twinning in cattle by simple egg transfer is described. Twenty-eight mated beef heifers each received an extra fertilised egg a week after breeding by Cassou 'gun' transfer; egg survival rate was markedly higher (37.5 vs 7.7%) when temperature control ($30^{o}C$) was applied at time of transfer than when transfer was at ambient temperature (5-$15^{o}C$). In a second experiment, 12 mated cattle received one egg by transfer at one week; 9 animals became pregnant to first service, 5 transferred eggs survived and 4 cattle produced twin calves. A third experiment attempted to increase yield of fertilised eggs from beef heifers prior to dispatch at the abattoir; twelve cattle underwent two superovulation treatments in the final month before slaughter, eggs being recovered by non-surgical technique after the first treatment and by flushing the tract of the animal after slaughter on the second occasion.

INTRODUCTION

The present report deals with progress in cattle twinning
by egg transfer, in the light of data from our work and that
conducted elsewhere. An earlier paper (Gordon, 1976) dealt
with the particular approach which we have adopted towards
twinning; surplus beef heifers used as egg donors, non-surgical
transcervical transfer by Cassou 'gun' and bovine eggs previously 'stored' in the rabbit oviduct. Twinning in cattle may be
achieved with varying degrees of success by methods other than
egg transfer. McCaughey and Dow (1977) have reported on the
use of two spaced doses of PMSG (800 iu); five of fifteen cattle so treated produced twins without the occurrence of higher
litters. Mulvehill and Sreenan (1977), using a low dose of
PMSG (750 - 800 iu) at the end of short-term progestagen treatment in oestrus synchronisation studies among beef suckler
cows, report 15% twins in cows calving to first service. French
workers also reported on employing a low dose (800 iu) of PMSG
in conjunction with progestagen treatments in suckler cattle
but without recording the incidence of multiple births (Chupin
et al., 1975; Courot, 1976).

EGGS FOR TWINNING TRANSFERS

Superovulation

One hundred and ninety three post-pubertal beef heifers
(Hereford crossbreds) have received the standard PMSG-HCG
treatment for superovulation as previously described (Gordon,
1976). The average yield of fertilised eggs remains above
eight per donor treated (Table 1); the current availibility of
low-cost prostaglandin ($F_2\alpha$) and its analogues now permits a
superovulation treatment incorporating prostaglandin for use
in beef heifers to be considered as an alternative to this
procedure. A trial with 36 animals examined ovulatory response
to PMSG-prostaglandin treatment in comparison with the standard
PMSG-HCG superovulation technique (data in Table 1). Eighteen
beef heifers underwent superovulation by way of the standard

procedure (2 500 iu. PMSG $^{\pm}$ oestradiol-17β -2 000 iu. HCG) and an equal number by a prostaglandin procedure (2 500 iu. PMSG at mid-cycle + 500 mcg Cloprostenol after 48 h). Yield of embryos per donor was depressed by the fact that two (11.1%) of the prostaglandin-treated cattle failed to exhibit oestrus and under this particular hormonal regime were not bred. With the standard PMSG-HCG technique, donors are bred at a pre-determined time (relative to HCG administration and estimated time of ovulation) so that fertilised eggs can be provided by those cattle that fail to show oestrus. Total ovarian response (follicles $\geqslant$ 10 mm + CL), the percentage of eggs recovered and eggs fertilised proved to be comparable following the two contrasting treatment regimes. One possible advantage of employing low cost prostaglandin could be in permitting beef heifer cattle to be placed on gonadotrophin treatment without reference to a specific stage of the oestrous cycle (eg Day 16 in the currently used standard PMSG-HCG regime). However,

TABLE 1

SUPEROVULATORY RESPONSE TO STANDARD (PMSG-HCG) AND PROSTAGLANDIN (PG-PMSG)
TREATMENT

	Standard PMSG-HCG Treatment	Superovulation Treatment	
		Standard	Prostaglandin
Donor Cattle	193	18	18
Bred by AI at Oestrus	183	18	16
Mean Follicular Response	21.6	20.1	18.8
Mean Ovulation Rate	15.9	16.3	15.6
Mean No. Eggs Recovered	9.6	9.2	10.1
Mean No. Eggs Fertilised	8.3	7.9	9.6
Fertilised Eggs per Donor Treated	8.3	7.9	8.6

there is evidence that gonadotrophin treatment initiated during the mid-luteal phase, as in the regime reported above, gives a higher ovulation rate and yield of embryos than treatment started earlier (Phillipo and Rowson, 1975; Betteridge, 1977; Sreenan and Gosling, 1977); for that reason, it is desirable that donors should be at a known stage of the cycle when commencing gonadotrophin treatment.

Double Superovulation Treatment

In examining ways and means of increasing the average yield per donor beef heifer of eggs for twinning transfers, preliminary work with twelve cattle examined the application of two consecutive superovulation treatments in the month prior to the dispatch of the heifer at the abattoir.

The heifer is first treated at mid-cycle with 1 500 iu PMSG (2 days prior to 500 mcg Cloprostenol) and non-surgical recovery of the eggs attempted 6 - 7 days after breeding at the controlled oestrus. At the time of recovery, heifers receive two daily doses of 500 mcg Cloprostenol to induce regression of the multiple corpora lutea. After the second controlled oestrus, and in the mid-luteal period of the ensuing cycle, the PMSG-Cloprostenol regime is applied for a second time, using our usual dose level of 2 500 iu on this occasion. Results of treatment are in Table 2; superovulation was induced and some number of fertilised eggs recovered on each of the two occasions.

The non-surgical recovery method employed was based on that described by Rowe et al. (1976) and further elaborated on by Rowe et al. (1977); these workers reported recovering 50 to 60% of eggs in certain groups of superovulated cattle. Assuming that continued experience with this non-surgical recovery technique eventually provides a 50% yield of eggs, an average of 12 fertilised eggs per beef heifer donor may be achieved. In the present programme, donors are slaughtered at the local abattoir 3 - 4 days after breeding and the reproductive organs returned to the laboratory for the egg recovery operations. Of 31 fertilised eggs transferred for storage to the rabbit oviduct from such slaughtered heifers and subsequently recovered and examined, 24 (77.4%) were deemed to show evidence of normal development.

TABLE 2

SUPEROVULATORY RESPONSES IN BEEF HEIFERS UNDERGOING TWO PMSG TREATMENTS IN THE FINAL MONTH BEFORE SLAUGHTER

	PMSG Dose Employed	
	Treatment I	Treatment II
	1 500 iu	2 500 iu
Heifers	12	12
In oestrus and bred by AI	12	11
Mean ovulation rate $\pm$ SD	8.5 ± 5.02	12.9 ± 8.97
Large unovulated follicles	2.5 ± 1.89	7.0 ± 8.97
Eggs recovered	13(14.3%)	103(66.5%)
Eggs fertilised	13(100.0%)	101(98.0%)
Mean fertilised eggs/donor	1.1 ± 0.96	8.5 ± 5.88

There is no clear conclusion in the literature regarding the outcome of repeated superovulation attempts in cattle. Early reports indicated a diminution of superovulatory responses with successive gonadotrophin applications (Willett et al. 1953; Hafez, et al. 1965); more recent reports have been less clearcut (Scanlon, 1972; Miller et al., 1975; Struthers and Marshall, 1976; Saumande and Chupin, 1977). The present findings agree with those of Scanlon (1972) and Miller et al. (1975) in showing a substantial superovulatory response after the second of two consecutive gonadotrophin treatments; results also

show that mid-cycle multiple corpora lutea of the superovulated
bovine can be induced to regress by a normal dose (500 mcg) of
prostaglandin analogue.

The ability to recover cattle eggs by non-surgical techniques opens up several possibilities; recovery can be attempted on several occasions and techniques should be cheaper to
employ than surgical procedures. There has been a proliferation of reports dealing with recovery techniques; egg recovery
rates comparable to those achieved by surgery can be expected
after appropriate manipulative experience has been acquired
(Alexander et al 1976; Elsden et al 1976; Rowe et al., 1976;
Rowe et al., 1977; Greve et al., 1977).

There has been less information on the subsequent viability
of embryos recovered by these non-surgical procedures, but they
should not prove to be less viable than eggs recovered by surgical procedures. However, in employing non-surgical techniques
recovery is necessarily attempted when the eggs have been exposed to the donor's uterine environment for several days.
Non-surgical recoveries may also be attempted at ambient temperature which could expose embryos to hazard under some conditions. Among the advantages of using non-surgical recovery
techniques, on the other hand, is the possibility that consecutive superovulation attempts can be applied at shorter time
intervals than can be achieved when intervention is the chosen
method of obtaining eggs.

Yield of Transferable Eggs

There is evidence supporting the concept that a higher
incidence of morphological abnormalities occurs in eggs from
superovulated as compared to untreated donor cattle (Church
and Shea, 1976; Betteridge, 1977; Elsden et al., 1977); it is,
however, important to establish factors responsible for such
abnormalities. In our own work, there is clear evidence that
the incidence of abnormalities increases substantially among
four-day and older eggs (Table 3). A marked increase in abnormal eggs has also been reported by Newcomb, et al (1976)

with increasing intervals from oestrus to recovery, more than
50% of eggs proving to be degenerate when left in the donor's
tract for 8 days. Present evidence shows that egg abnormalities occur in cattle superovulated by way of PMSG-prostaglandin
treatment as readily as in those superovulated with the standard PMSG-HCG regime (Table 4).

TABLE 3

INCIDENCE OF ABNORMAL EGGS IN RELATION TO DAY (POST-OESTRUM) OF RECOVERY
(STANDARD PMSG-HCG) TECHNIQUE

	Day of Recovery					
	2	3	4	5	6	All Days
Heifers	12	79	40	47	15	193
Fertilised eggs recovered	169	589	314	393	131	1596
Eggs showing major abnormalities	–	24	45	178	49	296
% Eggs with major abnormalities	–	(4.1)	(14.3)	(45.3)	(37.4)	(18.6)

TABLE 4

INCIDENCE OF EGG ABNORMALITIES IN SUPEROVULATED CATTLE

	Superovulated Treatment (2 500 iu PMSG)	
	Standard	Prostaglandin
% Eggs 'normal'	69.9	59.1
% Major abnormals	14.9	11.0
% Minor abnormals	15.2	29.9
All eggs	100.0	100.0

The major abnormality, as reported previously (Gordon, 1976), is characterised by a breakdown in the blastomere membrane evident first in eggs at the morula stage. There are strong indications that it is the uterine environment of the superovulated bovine which is responsible for the abnormality rather than an inherent defect in the egg itself. Data in Table 5 show that the removal of superovulated eggs (mainly Day 3) from the bovine uterus and their subsequent transfer to the rabbit oviduct results in most showing normal development when scrutinised at Day 6. Such findings agree with those reported by the Cambridge group (Trounson et al., 1976).

TABLE 5

EFFECT OF SUPEROVULATION TREATMENT ON EGG DEVELOPMENT IN THE RABBIT OVIDUCT

	Superovulation Treatment	
	Standard	Prostaglandin
Eggs incubated in oviduct	162	67
Eggs developing %	125(77.2%)	53(79.1%)

The incidence of egg abnormalities is related to the extent of the superovulatory response in the donor; there is a significant increase in percentage abnormal eggs as total ovarian response to PMSG increases (Boland et al., unpublished data). The incidence of abnormalities associated with a dose level of 1 500 iu PMSG may be substantially less than that with doses of 2 500 iu.

In studies reported above, in which non-surgical recovery of Day 6 eggs was attempted in cattle superovulated with 1 500 iu, no instances of abnormalities were recorded but data are limited in number. Church and Shea (1976) note that high levels of ovarian stimulation in their donor cattle usually produced 'low quality' eggs. Although much remains to be done

in characterising the different types of egg abnormality that
may occur, the problem would seem to constitute a major difficulty when seeking a high yield of eggs towards the mid-cycle
stage by non-surgical recovery procedures.

For such reasons, the present approach towards twinning
involves moderate superovulation of the beef heifer donors with
1 500 iu   PMSG initially to permit an acceptable yield of
transferable eggs at Day 6, with the higher 2 500 iu PMSG dose
level being reserved for the terminal treatment at which time
eggs are recovered at Day 3 for short-term (3 days) storage in
the rabbit oviduct.

Alternatives to Superovulation

For cattle twinning by egg transfer to be commercially
viable, a cheap supply of eggs is essential. In the present
programme, eggs are recovered as a 'by-product' when slaught-
ering the beef heifer; furthermore, a single-egg rather than
a two-egg transfer technique is employed to conserve the supply
of transferable embryos. Nevertheless, superovulation as
presently employed can only be regarded as a crude and temporary procedure for the provision of eggs and it is hoped that
future supplies of eggs for twinning transfers may be based on
the utilisation of ovarian oocytes. This could enable the
oocytes of beef heifers to be utilised after the dispatch of
the animal at the abattoir, thus obviating the present requirement for gonadotrophin stimulation and the temporary holding of
cattle prior to slaughter.

Work by Scanlon (1969) showed that the vesicular follicle
population in 258 beef cattle dispatched at the local abattoir
varied from 6 to 132, with an average of 45 per animal (both
ovaries included). It is evident that heifer ovaries contain
an ample number of vesicular follicles; efforts to utilise the
bovine ovarian oocyte have as yet not been successful.

Although nuclear maturation can be achieved by culture of
the bovine follicular oocyte in various media and in vivo

conditions (Jesch et al., 1975; Shea et al., 1976; Iritani and
Niwa, 1977; Leibfried and First, 1977), such maturation is not
necessarily accompanied by cytoplasmic maturation as this occurs
within the vesicular follicle.

French workers produced evidence showing that normal cytoplasmic maturation may be achieved by culturing the intact
follicle in an appropriate culture medium and gas phase (Thibault et al., 1975); intrafollicular oocyte culture required
high oxygen tension, either by increased pressure or by means
of a gas phase providing a high (56%) oxygen content. Experiments with sheep oocytes matured within intact pre-ovulatory
follicles by Moor and Trounson (1977) in the presence of trace
amounts of FSH, LH and oestradiol-17/4 has resulted in both
normal nuclear and cytoplasmic maturation, the normality of
the developmental capacity of oocytes being confirmed subsequently by their fertilisation in the ewe oviduct and in the
production of viable lambs. Further studies at Cambridge suggest that certain essential cytoplasmic proteins may not be
synthesised in the culture of the extra-follicular oocyte
(Warnes et al., 1977).

Work in cattle has yet to show that a similar approach can
be successfully applied to the bovine oocyte, but it would not
seem unreasonable to assume that it can. Although the sheep
oocytes matured by Moor and Trounson (1977) were fertilised
in the sheep oviduct, the possibility exists that this can be
achieved under appropriate in vitro conditions. Japanese
workers have recently reported in vitro fertilisation of artificial matured oocytes, using bull spermatozoa which had been
incubated in the uterus of the rabbit (Iritani and Niwa, 1977).
Other workers have presented evidence suggesting that capacitation of bull spermatozoa may not be species specific, but may
be accomplished in species such as the sheep (Sreenan, 1968)
and pig (Bedirian et al, 1975; Shea et al. 1976). The possibility of basing the supply of low cost eggs for twinning
transfers on the artificially matured and in vitro fertilised

bovine egg is one of considerable interest.

TRANSFER OF EGGS

The successful development of a transfer twinning technique involves consideration of several factors; some of those which have been examined in this laboratory are noted in Figure 1.

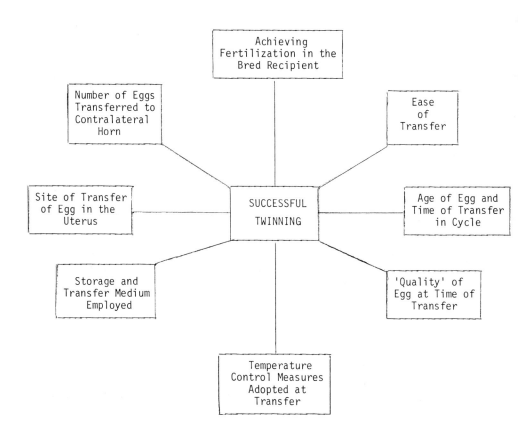

Fig. 1: Factors in successful twinning by egg transfer.

One factor of importance when working under farm, rather than laboratory, conditions is the control of temperature when manipulating the egg around the time of transfer, especially when this is being done in cold conditions. There is good evidence from laboratory studies that cooling Day 5/Day 6 cattle eggs below 10°C can substantially reduce their ability to survive (Trounson et al., 1976). There is also a report by Testart, et al.(1975) in which twinning was attempted by the transvaginal introduction of an extra egg; they suggest that certain of their failures in embryo survival may have been the result of carrying out transfer of Day 6 eggs at an ambient temperature of 5 - 6°C rather than at higher temperatures, on the other hand, Alexander and Markus (1977) conclude that cattle embryos were able to withstand successfully the temperature changes which they encountered with their particular on-farm conditions (ambient temperatures in the range 16 to 20°C).

Temperature Control Studies

Among steps involved in taking an egg transfer twinning technique on to the farm, is the short-term storage of the eggs after leaving the central laboratory until they reach the farm at which the recipients are located. One preliminary study here examined the storage of cattle eggs in 0.5 ml capacity Cassou straws for periods of 2 to 5 hours, the straws being held at 30°C in a Thermos flask. The study used Day 3 or Day 4 eggs, freshly recovered by surgery from superovulated donors and stored in a medium of phosphate-buffered saline supplemented with 15% foetal calf serum. The subsequent viability of the eggs was determined by transferring them to the rabbit oviduct and examining them for evidence of continued development two or three days later. As shown in Table 6, the majority of eggs gave evidence of continued development.

In the main part of the study, twenty-eight mated beef heifers were used as recipients a week after breeding by AI, 15 of them receiving an extra egg under temperature controlled conditions and 13 receiving eggs by transfer at ambient temperature (5 - 15°C). All eggs used in these transfers had been

stored in the rabbit oviduct for 20 to 76 hours previously.
The eggs were first loaded into 0.5 ml capacity Cassou straws
and stored for up to 3 hours at 30°C (this to simulate the
time taken for eggs to reach the farm at which the twinning
transfers were to be attempted). The temperature control measures consisted of loading the Cassou straw into the inseminating 'gun' at 30°C (in practice, this would be carried out in
the operator's car prior to moving into the cow shed) and placing the complete instrument into an insulated container for the
journey to the recipient animal.

TABLE 6

EFFECT ON TEMPORARY STORAGE IN 0.5 ml   CASSOU STRAWS (30°C) ON SUBSEQUENT
DEVELOPMENT OF EGGS IN THE RABBIT

	Storage Period (h)		Totals
	2 - 3	4 - 5	
Eggs transferred	27	11	38
Eggs recovered (%)	20 (74.1%)	10 (90.9%)	30 (78.9%)
Eggs developing (%)	16 (80.0%)	7 (70.0%)	23 (76.7%)

At the instant of commencing the transfer, when the inseminating instrument was inserted into the recipient's vagina,
a plastic insulating collar was slid back to expose the tip of
the 'gun'. The object was to introduce the egg into the cervix without exposing it to temperatures below 25 to 30°C. In
cattle receiving the extra egg in the absence of temperature
control, the inseminating 'gun' was loaded in the cow-shed at
ambient temperature and transfer carried out immediately.

Results in Table 7 support the view that temperature
control, under the conditions of this trial, facilitated embryo
survival. Trounson et al. (1976) have noted that rapid cooling
can reduce survival of the bovine embryo to a greater degree

than a slow rate of cooling. In the absence of temperature control, cattle eggs in the present work would be subject to rapid cooling (30°C to ambient temperature) when being loaded into the inseminating instrument. Devising a simple temperature control procedure may permit more successful transfers under farm conditions.

TABLE 7

PREGNANCY AND TWINNING IN RELATION TO TEMPERATURE CONTROL MEASURES APPLIED

Transfer Technique	Recipient cattle	Became pregnant	With Twin embryos	% Embryo survival
Without ($5-15^{\circ}$C) temperature control	13	7 (53.8%)	1 (14.3%)	1 (7.7%)
With (30°C) temperature control	15	9 (60.0%)	4 (44.4%)	6 (37.5%)

Twinning by One-Egg Transfer

Several reports have dealt with the establishment of twin-pregnancy by way of the introduction of an extra egg into the contralateral horn of the mated recipient bovine. French workers employed a transvaginal transfer technique and reported twin pregnancies in 3 of 8 recipient cows (Testart et al., 1975). In the United States, Anderson et al, (1977) have achieved similar conception and twinning rates in recipient cows when twinning has been attempted surgically by single-egg transfers (unilateral) to mated recipients or by two-egg transfer (bilateral) to unbred recipients. In our own work, emphasis continues to be on single-egg transfers using the non-surgical Cassou 'gun' technique. Previous data had been confined to recipient heifer cattle sacrificed in the first or second months of pregnancy and it was desirable to obtain some information for animals allowed to progress to full-term.

Twelve non-lactating cattle (nine cows and 3 well-grown heifers) were used as bred recipients for one-egg transfers. In six cattle, transferred eggs had previously been stored in the rabbit oviduct for an average of 55 hours (41 to 71 hours); in the other six, eggs were recovered surgically from untreated donors (single-ovulating) and transferred directly to recipients. Transfers were conducted one week after breeding by AI, with the recipients restrained in a crush in a room held at 24°C. Results of the work are in Table 8. There was no evidence of abortion or foetal loss during pregnancy, although it should be noted that no attempt was made to diagnose twin-pregnancy by way of palpation per rectum.

TABLE 8

CALVING OUTCOME OF TWINNING TRANSFERS IN CATTLE

	Source of Eggs Employed		All eggs
	Stored in rabbit oviduct	From unsuperovulated donors	
Recipient cattle	6	6	12
Calved (1st service)	5 (83.3%)	4 (66.7%)	9 (75.0%)
Twin sets born	2 (40.0%)	2 (50.0%)	4 (44.4%)
Survival of transferred eggs	3 (50.0%)	2 (33.3%)	5 (41.7%)

The survival rate among transferred eggs was comparable to that recorded in previous work among heifer cattle sacrificed in early gestation. Blood typing was used to determine the parentage of the six single-born calves. One of these, both from the results of the blood typing and coat markings, was found to be the offspring of a donor heifer. The duration of gestation in the four twin-bearing cattle was within the same range as that recorded in cattle producing singles.

The transfers were carried out at either 5 or 6 days after the date of breeding. The present technique is designed to use cattle eggs which have been recovered from heifer donors at Day 3 and stored in the ligated rabbit oviduct for a further three days.

There is no conclusive evidence in the literature indicating that embryo survival is enhanced by delaying the transfer beyond about Day 6. On the other hand, there is ample evidence that embryo survival rates are substantially lower with Day 3/4 eggs than with those which are 5 to 6 days old (Newcomb and Rowson, 1975; Nelson et al, 1975; Hahn and Hahn, 1976).

One reason for delaying the day of transfer when using non-surgical techniques such as the Cassou 'gun' would be that the positioning of the egg in the uterine horn could be more compatible with the usual location of the egg at this time in the untreated animal. French workers have achieved a pregnancy rate of 60% and a twinning rate of 50% by two-egg bilateral cervical transfer at days 9 and 10, the eggs being placed 5 cm into each horn beyond the uterine common body (Renard et al., 1977).

352

REFERENCES

Alexander, A.M., Markus, A.N. and Hooton, J.K. 1976. Non-surgical bovine
    embryo recovery. Vet. Rec. 99, 221.

Anderson, G.B., Baldwin, J.N., Cupps, P.T., Drost, M., Hooton, M.B. and
    Wright, R.W. 1976. Induced twinning in beef heifers by embryo
    transfer. J. Anim. Sci., 43, 272 (abs).

Anderson, L.L. and Parker, R.O. 1976. Calves produced by surgical transfer
    of embryos. J. Anim. Sci., 42, 1359 (abs).

Anderson, G.B., Cupps, P.T. and Drost, M. 1977. Twinning in beef cattle
    following unilateral and bilateral embryo transfer. J. Anim. Sci.,
    45 (1), (abs).

Baker, R.D., and Polge, C. 1976. Fertilisation in swine and cattle. Can.
    J. Anim. Sci., 56, 537-538.

Bilton, R.J. and Moore, N.W. 1976. Storage of cattle embryos, J. Reprod.
    Fert. 42, 537-538.

Bilton, R.J. and Moore, N.W. 1977. Successful transport of frozen cattle
    embryos from New Zealand to Australia. J. Reprod. Fert., 50, 363-
    364.

Bedirian, K.N. Shea, B.F. and Baker, R.D. 1975. Fertilisation of bovine
    follicular oocytes in bovine and porcine oviducts. Can. J. Anim.
    Sci., 55, 251-256.

Betteridge, K.J. 1977. Superovulation. In Embryo transfer in farm animals.
    A review of techniques and applications, Canada Dept. of Agriculture,
    Monograph 16, ed K.J. Betteridge, Ottawa, pp. 1-9.

Chupin, D., Pelot, J. and Thimonier, J. 1975. The control of reproduction
    in the nursing cow with a progestagen short-term treatment. Ann.
    Biol. anim. Bioch. Biophys., 15 (2), 263-271.

Church, R.B., and Shea, B.F. 1976. Some aspects of bovine embryo transfer.
    In Egg Transfer in Cattle, EEC Seminar, Cambridge, 73-86.

Church, R.B. and Shea, B.F. 1977. The role of embryo transfer in cattle
    improvement programmes. Can. J. Anim. Sci., 57, 33-45.

Courot, M. 1976. Studies of cattle fertility in France. Theriogenology,
    6 (5), 651-652.

Del Campo, M.R., Rowe, R.F., French, L.R. and Ginther, O.J. 1977. Unilater-
    al relationship of embryos and the corpus luteum in cattle. Biol.
    Reprod., 16, 580-585.

Drost, M., Brand, A. and Aarts, M.H. 1976. A device for non-surgical recovery of bovine embryos. Theriogenology, 6 (5), 503-507.

Elsden, R.P., Hasler, J.F. and Seidel, G.E. 1976. Non-surgical recovery of bovine eggs. Theriogenology, 6 (5), 523-532.

Elsden, R.P., Nelson, L.D., Nash, J. and Seidel, G.E. 1977. Transfer of embryos from unsuperovulated and superovulated cows. J. Anim. Sci., 45 (1), (abs).

Gordon, I. 1976. Cattle twinning by the egg transfer approach. In Egg Transfer in Cattle, EEC Seminar, Cambridge, 305-319.

Greve, T., Lehn-Jensen, H. and Rasbech, N.O. 1977. Non-surgical recovery of bovine embryos. Theriogenology, 7 (4), 239-250.

Hafez, E.S.E., Jainudeen, M.R. and Lindsay, D.R. 1965. Gonadotrophin-induced twinning and related phenomena in beef cattle. Acta Endocrinol. 50 (Suppl. 102), 5-43.

Hahn, J and Hahn, R. 1976. Experiences with ova transfer in cattle. Wld. Rev. Anim. Prod., 12 (4), 51-58.

Iritani, A. and Niwa, K. 1977. Capacitation of bull spermatozoa and fertilisation in vitro of cattle follicular oocytes matured in culture. J. Reprod. Fert., 50, 119-121.

Jesch, J., Foote, W.D. Phelps, D.A. and Tibbitts, F.D. 1975. Maturation of bovine oocytes in rabbit oviducts. J. Anim. Sci., 41, 360 (abs).

Leibfried, L. and First, N.L. 1977. Characterisation of bovine follicular oocytes. J. Anim. Sci., 45 (1) (abs).

McCaughey, W.J. and Dow, G. 1977. Hormonal induction of twinning in cattle. Vet. Rec., 100, 29-30.

Miller, W., Larsen, L.H., Nancarrow, C.D. and Cox, R.I. 1975. Superovulation and surgical collection of ova from cows on each of two successive cycles. J. Reprod. Fert., 43, 384-385.

Moor, R.M. and Trounson, A.O. 1977. Hormonal and follicular factors affecting maturation of sheep oocytes in vitro and their subsequent developmental capacity. J. Reprod. Fert. 49, 101-109.

Mulvehill, P. and Sreenan, J.M. 1977. Improvement of fertility in post-partum beef cows by treatment with PMSG and progestagen. J. Reprod. Fert., 50, 323-325.

Nelson, L.D., Bowen, R.A. and Seidel, G.E. 1975. Factors affecting bovine embryo transfer. J. Anim. Sci., 41, 371 (abs).

Newcomb, R. and Rowson, L.E.A. 1975. Conception rate after uterine transfer of cow eggs in relation to synchronisation of oestrus and age of eggs. J. Reprod. Fert., 43, 539-541.

Newcomb, R. Rowson, L.E.A. and Trounson, A.O. 1976. The entry of super-
    ovulated eggs into the uterus. In Egg Transfer in Cattle, EEC,
    Seminar, Cambridge 1-15.

Newcomb, R. and Rowson, L.E.A. 1976. Aspects of the non-surgical transfer
    of bovine eggs. Proc. 8th. Int. Congr. Reprod. AI (Krakow), 319 (abs).

Phillipo, M. and Rowson, L.E.A. 1975. Prostaglandins and superovulation
    in the bovine. Annls. Biol. Anim. Biochim. Biophys., 15, 233-240.

Renard, J.P., Heyman, Y and du Mesnil du Buisson, F. 1977. Unilateral and
    bilateral cervical transfer of bovine embryos at the blastocyst stage.
    Theriogenology, 7 (4), 189-194.

Rowe, R.F., Del Campo, M.R., Eilts, C.L., French, L.R., Winch, R.P. and
    Ginther, O.J. 1976. A single cannula technique for non-surgical col-
    lection of ova from cattle. Theriogenology, 6 (5), 471-483.

Rowe, R.F. Del Campo, M.R. and Ginther, O.J. 1977. Non-surgical collection
    and transfer of bovine embryos. J. Anim. Sci., 45 (1) (abs).

Saumande, J and Chupin, D. 1977. Superovulation: a limit to egg transfer
    in cattle. Theriogenology, 7 (3), 141-149.

Scanlon, P.F. 1969. Studies in reproduction in the cow. Ph.D. Thesis, NUI,
    Dublin.

Scanlon, P.F. 1972. Ovarian response of cows following pregnant mare serum
    gonadotrophin treatment during two successive oestrous cycles. J.
    Dairy Sci., 55, 527-528.

Shea, B.F., Hines, D.J., Lightfoot, D.E., Ollis, G.W. and Olson, S.M. 1976.
    The transfer of bovine embryos. In Egg Transfer in Cattle, EEC,
    Seminar, Cambridge, 145-152.

Shea, B.F., Latour, J.P.A., Bedirian, K.N. and Baker, R.D. 1976. Maturation
    in vitro and subsequent penetrability of bovine follicular oocytes.
    J. Anim. Sci., 43, 809-815.

Sreenan, J.M. 1968. In vivo and in vitro culture of cattle eggs. Proc.
    6th Int. Congr. Anim. Reprod. AI (Paris), 1, 577-579.

Sreenan, J.M. and Gosling, J.P. 1977. The effect of cycle stage and plasma
    progesterone level on the induction of multiple ovulations in heifers.
    J. Reprod. Fert., 50, 367-369.

Struthers, G.A. and Marshall, D.P.J. 1976. Egg transfer in cattle. Proc.
    N.Z. Soc. Anim. Prod., 36, 58-66.

Thibault, G., Gerard, M. and Menezo, Y. 1975. In vitro acquired ability of
    rabbit and cow oocyte to ensure sperm nucleus decondensation during
    fertilisation. Ann. Biol. anim. Bioch. Biophys, 15, 705-715.

Trounson, A.O., Willadsen, S.M., Rowson, L.E.A. and Newcomb, R. 1976. The storage of cow eggs at room temperatures and at low temperatures. J. Reprod. Fert., 46, 173-178.

Trounson, A.O., Willadsen, S.M., Rowson, L.E.A. 1976. The influence of in vitro culture and cooling on the survival and development of cow embryos. J. Reprod. Fert. 47, 367-370.

Willett, E.L., Buckner, P.J. and McShan, W.H. 1953. Refractoriness of cows repeatedly superovulated with gonadotrophins. J. Dairy Sci., 36, 1083.

Warnes, G.M., Moot, R.M. and Johnson, M.H. 1977. Changes in protein synthesis during maturation of sheep oocytes in vivo and in vitro. J. Reprod. Fert., 49, 331-335.

Wright, R.W., Anderson, G.B., Cupps, P.T. and Drost, M. 1976. Blastocyst expansion and hatching of bovine ova cultured in vitro, J. Anim. Sci., 43, 170-174.

SURVIVAL OF CULTURED AND TRANSPORTED BOVINE EMBRYOS FOLLOWING SURGICAL AND NON-SURGICAL TRANSFERS

J. Hahn, L.A. Moustafa, U. Schneider[*], R. Hahn[**],
W. Romanowski and R. Roselius[***]

[*] Tierärztliche Hochschule Hannover, 3000 Hannover 1

[**] Besamungsverein Neustadt a.d. Aisch, 8530 Neustadt (Aisch)

[***] Rinderproduktion Niedersachsen GmbH, Bremen-Hannover, 2820 Bremen 77.

We have been looking for a simple and reliable method for culturing, transporting and transferring preimplantation bovine embryos; we found the straw to be a very useful tool. The application of this or similar methods for transporting and transferring preimplantation embryos has been reported by Sreenan (1975), Boland et al. (1976), Alexander and Markus (1977), Brand and Drost, (1977).

At first we used preimplantation rabbit embryos to investigate the use of straws for the culture and transfer of embryos. Straws are filled with three columns of culture medium, separated by air-spaces. The embryo floats freely in the middle column. One end of the straw remains open to the air.

TABLE 1

DEVELOPMENT OF 2-CELL RABBIT EMBRYOS IN CULTURE MEDIUM: 0.2 ml BSM-2 (HAM'S F 10 MOD. KANE AND FOOTE 1970) ATMOSPHERE IN CULTURE CHAMBER: 5 % CO_2 IN AIR

Culture system	No. of embryos	24 hrs 8-16 cell-stages	48 hrs Morulae	72 hrs Early blastocysts
			In-vitro development (%)	
Microdrops under oil	90	89 (99)	87 (97)	80 (89)
Straws	85	83 (98)	81 (95)	75 (88)

Table 1 shows that there is no difference between embryos cultured in microdrops or straws during the 72-hour period from the two cell stage to the late morula. Similary the straw method shows no detrimental effect on the transfer of embryos after culture periods of 24, 48 and 72 h.

In another experiment we cultured 30 bovine preimplantation embryos in PBS for a period of 36 hours. The embryos were collected on Day 7. The culture medium used was PBS modified by Whittingham (1971), and the straws were maintained at a temperature of 35 - 37°C in a thermos container. Twenty-four of these embryos (80%) developed to expanded and hatching blastocysts. Development of the embryos was determined partly while they were still within the straw. In view of these results we had no hesitation in using the straw for long-term culture and transfer of bovine embryos.

For more than a year we have been performing non-surgical transfer of embryos on the standing cow, and have shown that results from this method are equivalent to results from the surgical method (Hahn et al., 1977).

The straw is very suitable for non-surgical embryo transfer because it is easy to handle and to fill with the embryo. The transfer can be performed anywhere, with the help of a special instrument, which is manufactured to our specification by the Cassou Company. It is very similar to the Cassou instrument used in AI. The holder is 3 mm in diameter and is thinner than the instrument used in AI. We use three lengths of holder: 44, 49 and 54 cm. The mini-straw, available commercially, fits the holder exactly. In a few animals some difficulty may be experienced in the passage of the instrument through the cervical canal, but with experience these difficulties are readily overcome. During transport, the straw is kept in a plastic carrier which guards against environmental changes.

TABLE 2

RESULTS OF EMBRYO TRANSFER USING STRAWS MEDIUM: MOD PBS AGE OF EMBRYO AT
COLLECTION: DAY 7 SYNCHRONY OF RECIPIENTS: O to -24 HRS

Interval between collection and transfer	No. of embryos	Embryonic death or abortion (very probably)	No. pregnant (%) (2 - 3 months)
1 - 4 hrs	27	0	17 (62.9)
20 -24 hrs	11	1	5 (45.5)
32-36 hrs	30	9	1 (3.3)

Table 2 shows that there is a decline of transfer results
after 20 - 24 h of culture in PBS. After 32 - 36 h culture
the decline is seen to be significant. This high incidence of
embryonic loss or abortion is striking. All the embryos which
were cultured more than 20 hours were transported over long
distances by road, rail or air. Only developed blastocysts
were transferred. These results are contradictory to those of
Trounson et al. (1976). They achieved a success rate of 50%
after 48 h culture in PBS.

We tried to determine the reason for our poor transfer
results using PBS for culture in straws. We cannot completely
exclude the possibility that, as we were transferring embryos
to recipients in foreign countries, slight errors in synchronisation may have been responsible for some failures. However,
the experiments were repeated three times and always with the
same results.

In view of this we are now concentrating on investigations
of the culture medium and the system used. At first, we again
used rabbit pre-implantation embryos, starting at the morula
stage to give a better comparison with bovine morulae. We used
two systems, straws and microdrops, and two culture media, PBS
and BSM-2, both of which were supplemented with 20% FCS. The
differences in development and viability of embryos in different media and systems were obvious after 72 h cultivation. We

divided the changes into 3 groups:

1) Development of the embryonic disc

2) Degeneration of blastomeric cells

3) Abnormal zona breakage.

TABLE 3

IN-VITRO DEVELOPMENT OF RABBIT MORULAE

Culture system	Medium	No. of embryos	In-vitro development				
			24 hrs	48 hrs		72 hrs	
			Early blastoc. (%)	Expand. blastoc. (%)	Blastoc. with developed embryonic disc (%)	Percentage of Degenerate blastom. cells	Abnormal zona breakage
Micro-drops	BSM-2	40	36 (80)	26 (85)	18 (45)	10	35
Straws	BSM-2	30	25 (83)	20 (67)	5 (17)	7	75
	PBS	30	26 (87)	12 (40)	3 (10)	73	77

On consideration of these three changes after 72 hours culture it was observed that in microdrop culture the embryonic disc was less altered than in straws. Also it was obvious that in straw-cultivation abnormal zona breakage was common in both media. We found that the percentage of degenerated blastomeres after PBS-medium is very high.

These results with rabbit embryos showed that the culture of morulae stages in straws disturbed the blastocyst expansion, especially when PBS was used as the culture medium.

After this we tried again the culture of bovine morulae in straws, using PBS and BSM-2. The embryos derived from 3

360

donors on Days 6 -7 and were cultured for 40 h. After morphological observation of blastocyst differentiation, the embryos
were fixed as flat mounts and stained for cell viability counts.

TABLE 4

DEVELOPMENT OF BOVINE MORULAE CULTURED IN STRAWS

Medium	No. of embryos	In-vitro development 40 hrs		
		24 hrs blastocysts (%)	expanded blastocysts (%)	Average percent degenerated blastomeres (range)
BSM-2	11	9 (82)	9 (82)	18 (0-47)
PBS	11	8 (73)	7 (64)	36 (8-63)

Table 4 shows that the development of bovine blastocysts
after 40 h culture in straws is reduced and the percentage of
degenerated blastomeres is increased when PBS is used as the
culture medium, whereas BSM-2 seemed to support development and
differentiation better. This indicates that PBS supplemented
with Glucose, Pyruvate and FCS does not support bovine blastocyst development properly, when cultured for more than 24 h.
It seems that the composition of BSM is more suitable for
culture of embryos for longer periods of time.

To investigate the question whether culture in BSM leads
to higher blastocyst viability we transferred bovine embryos
after 36 h storage in BSM in straws. Results of the transfer
are not yet available.

In conclusion, we can say that developed stages of bovine
embryos can be cultured in PBS and transported in straws
without any loss of viability when transfer is performed on
the day of collection. For a longer period of culture and
transportation at body temperature, our investigations suggest

that BSM-2 is superior to PBS. So far, we have been unable to prove that, in long-term culture, the straws have a negative influence on embryonic development.

REFERENCES

Alexander, A.M. and Markus, A.N. 1977. A modified technique for the trans-
    plantation of embryos into recipient cows. Vet. Rec. 100, 73-74.

Boland, M.P., Crosby, T.E. and Gordon, I. 1976. Induction of twin pregnancy
    in heifers using a simple non-surgical technique. Proc. 8th Int.
    Congr. Anim. Reprod. AI. Krakow, Vol. III, 241-243.

Brand, A. and Drost, M. 1977. Embryo transfer by non-surgical methods.
    In: K.J. Betteridge (Ed.) Embryo transfer in farm animals, Canada
    Department of Agriculture, Monograph. 16, 31-34.

Hahn, J., Schneider, U., Romanowski, W. and Roselius, R. 1977. Ergebnisse
    der chirurgischen Eiübertragung am stehenden Tier. Dtsch. Tierärztl.
    Wochenschr. 84, 229-231.

Kane, M.T. and Foote, R.H. 1970. Culture of two- and four-cell rabbit
    embryos to the expanding blastocyst stage in synthetic media. Proc.
    Soc. Explt. Biol. Med. 133, 921-925.

Sreenan, J.M. 1975. Successful non-surgical transfer of fertilised cow
    eggs. Vet. Rec. 96, 490-491.

Trounson, A.O., Willadsen, S.M. and Rowson, L.E.A. 1976. The influence of
    in-vitro culture and cooling on the survival and development of cow
    embryos. J. Reprod. Fertil. 47, 367-370.

Whittingham, D.G. 1971. Survival of mouse embryos after freezing and thaw-
    ing. Nature (Lond.) 233, 125-126.

PRACTICAL APPLICATION OF NON-SURGICAL COLLECTION
OF BOVINE EMBRYOS IN DANISH PEDIGREE CATTLE

T. Greve, H. Lehn-Jensen and N.O. Rasbech
Institute for Animal Reproduction
Royal Veterinary and Agricultural University, Copenhagen.

ABSTRACT

Non-surgical collections of bovine embryos have been carried out under farm conditions in Danish pedigree herds during the past two years. Thirty-six (36) dairy cows of the main breeds selected by the breeders' organisations were stimulated with an average of 2250 - (1500 - 3000) iu PMSG on Days 9 - 12 of their oestrous cycle followed by injection of 500 μg Cloprostenol 48 hrs later and insemination on Days 0 and 1. Non-surgical collection of embryos occurred on Days 6 - 8 by using a one-way cannula technique based on the Foley catheter. The collector is described in detail.

The procedure proved to be very reliable, stable and easy to use. The instrument could be guided through cervix to both uterine horns in all attempts without causing damage to the genital tract.

Ninety-six percent (96%) of the flushing media was recovered. Bleeding rarely occurred. The estimated recovery rate of eggs was 69%. The best rate was observed on Days 7 (74%) and 8 (76%) whereas Days 6 and 10 yielded the poorest recovery rates (54 and 43%). Eggs were recovered in 95% of the flushings. It was possible to recover 6.3 eggs/donor with a wide variation (0-29). 0.8 of these were unfertilised; 1.9 appeared obviously degenerated and 3.7 normally developed. 0.31 of these had some degree of degeneration. The highest number of eggs was recorded in the 2500 iu group (22.2). 3.1 eggs per donor could be transferred to recipients surgically or non-surgically at the farm. Fifty-five embryos were transferred surgically to the tip of the uterine horn and thirty recipients were diagnosed pregnant by rectal palpation yielding a pregnancy rate of 54%.

Thirty-three eggs were transferred non-surgically and 10 recipients *became pregnant (31%).*

Successful freezing of bovine embryos were carried out and three instruments for non-surgical transfer of embryos are being tested.

INTRODUCTION

Non-surgical collection of bovine embryos has been carried
out under farm conditions in various parts of Denmark during
the past two years.

It is the aim of this paper to describe the procedure used
and the results achieved so far in Danish pedigree herds.

MATERIALS AND METHODS

Only valuable dairy cows belonging to the main domestic
breeds served as potential donor animals. A total of thirty-
six (36) gynaecologically normal cows were stimulated with an
average of 2250 - (1500 - 3000) iu PMSG* on Days 9 - 12 of
their oestrous cycle followed 48 h later by an injection of 500
µg Cloprostenol **). An outline of the donor schedule is given
in Table 1.

TABLE 1

DONOR SCHEDULE

Experimental day	Day of oestrous cycle	Treatments
O	9-12	1500-3000 iu PMSG/ animal im.
2	11-14	500µg Estrumate/animal im
4	O (heat)	insemination (2 straws)
5	1	" " "
11	6-8	non-surgical recovery

All animals were in heat within 48 h after the prostaglandin
treatment and were inseminated twice with a double dose of deep
frozen semen. Three animals developed several anovulatory

* Antex (R). Leo Pharmaceuticals, Copenhagen.
** Estrumate (R). Kindly supplied by Mr. Jørgen Frederiksen, ICI Pharma,
Copenhagen.

follicles (polycysts) and four animals responded poorly to the
superovulation with only one corpus luteum. The number of corpora
lutea (CL) was assessed by rectal palpation at the day of
collection thus representing the estimated number of ovulations.
The average number of CL in all animals was 7.5 and in the
group selected for collection 9.2. The individual variation
greatly exceeded any conceivable breed variation or dose
dependence of PMSG. The highest number of ovulations was
achieved, when the animals were treated with 2500 iu PMSG (11.9
for all animals and 14.2 for the selected group).

The collection of embryos or ova occurred on days 6 - 8
after the onset of standing heat and all collections were
carried out on the farm. The donor animal was placed in a
chute or stanchion and given 10 mg Xylazine[+] in the caudal
vein. In most cases this administration was followed by slight
uterine concentration which made it easier to delineate the
uterus. Epidural anaesthesia was accomplished with 3 - 5 ml
2% Lidocaine. The tail, vulva, and perineal region were washed
thoroughly with soap[++] and finally disinfected with 70%
alcohol just prior to insertion of the collector.

The technique used for non-surgical recovery of bovine
embryos has been described by Rasbech (1976), Greve et al.
(1977) and Greve and Lehn-Jensen (1977).

The collector consists of an outer metal tube, 5 mm in
diameter and 49.5 cm long. A sterile two-way Foley balloon
catheter is pulled through the instrument by means of a string
or a hook (Figure 1C). Four additional holes (3.5 mm in
diameter) are made in the tip of the Foley catheter to permit
better recovery of fluid, and an open metal coil spring measuring
2.5 mm in diameter and 5.5-6 cm long placed in the tip of the
Foley catheter (Figure 3 A).

[+] Rompun Vet. [(R)] Bayer.
[++] Hibiscrup [(R)] ICI-Pharma, Copenhagen.

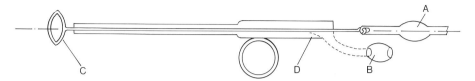

Fig. 1 A. Foley catheter, B. valve, C. hook, D. metal tube

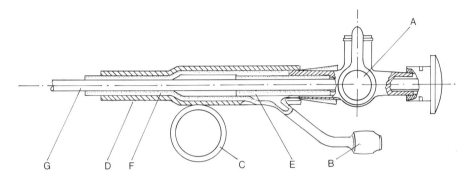

Fig. 2 A. three-way stop cock, B. valve, C. metal ring, D. metal tube,
E. connecting tube, F. Foley catheter, G. stiletto

A three-way stop cock is attached to the rear end of the
Foley catheter (Figure 2A). In order to support the flexible
tip of the Foley catheter, a thin metal stiletto (60 cm long,
Figure 3F) is placed in the Foley catheter and connected to
the stop cock by means of a Luer fitting.

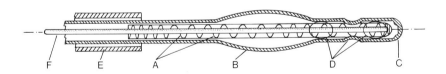

Fig. 3 A. coil, B. balloon, C. tip of catheter, D. holes, E. metal tube,
F. stiletto

The coil spring prevents the stiletto from penetrating the
Foley catheter and stabilises the tip. The whole device is 60
cm long. Before use it is sterilised at 110°C for 30 min, Under

rectal control the instrument is carefully guided through the
cervical canal and placed in the appropriate horn. When the
tip of the instrument reaches the greater curvature of the
uterus, the horn is pulled gently caudally over the instrument
and an attempt is made to extend as deep as possible into the
horn without too much manipulation and damage to the endometrium.
When the instrument reaches the desired location 20 - 35 ml of
air is inflated, and the size and position of the balloon is
carefully checked. The site of the balloon is checked
routinely during the whole flushing.

The insertion of the instrument is carried out extremely
gently and it is never placed very deep within the horn if it
is suspected that further manipulation would cause trauma to
the genital tract. The stiletto is now removed and a pre-
sterilised rubber tube attached to each outlet of the stop
cock (Luer lock fitting)(Figure 4).

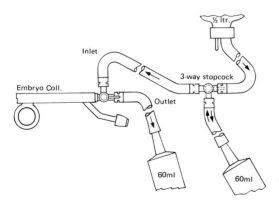

Fig. 4 Tubing system connected to instrument

A syringe containing 60 ml flushing medium*⁺ is connected
to one of the rubber tubes and an empty 60 ml syringe is
attached to the other rubber tube. Thirty to 90 ml of fluid
*⁺ Dulbecco's PBS solution

are introduced into the horn at a time, and after none or
gentle massage the fluid is evacuated by turning the valve of
the stop cock and letting the media into the syringe with aid
of slight suction. When 60 ml is collected, the fluid is
transferred to pre-warmed 250- or 500 ml separator funnels [+*]
placed in an incubator at 30°C. Each horn is flushed thoroughly
with 120 - 700 ml of the fluid.

When one horn is flushed the balloon is deflated, the
stiletto reinserted and the instrument transferred to the
contralateral horn, where the procedure is repeated. Flushing
of both horns lasts from 15 - 40 min. After 15 - 30 minutes
the fluid is examined for the presence of eggs under a stereomicroscope and, after initial examination, the fluid is
transferred to a second separator funnel and re-examined 2 - 4
h later.

Transfer of blastocysts was carried out <u>surgically</u> by
flank incision or non-surgically using the <u>Cassou </u>method or the
instrument described by <u>Rasbech</u> (1976$_1$).

RESULTS AND DISCUSSION

The estimated recovery rate of eggs was 69%. The best
rate was observed on Days 7 (74%) and 8 (76%) whereas Days 6
and 10 yielded the poorest recovery rates (54 and 43%). Eggs
were recovered in 95% of the flushings.

Because of the clear appearance of flushing medium the
identification and isolation of eggs were rapidly accomplished.

It was possible to recover 6.3 eggs/donor with a wide
variation (0 - 29). (Table 2).

+* Model: Institute for animal reproduction.

TABLE 2

MAIN RESULTS OF THE NON-SURGICAL RECOVERIES FROM 29 DONOR COWS

	Esti-mated no. of corpora lutea	Number of eggs /donor	Unfer-til. eggs	Degen. eggs	Normally developed with some degenerat-ion	Normal eggs	Frozen eggs	Lost eggs	Recove rate %
n	267	184	22	55	9	88	6	4	-
x	9.2	6.3	0.8(12%)	1.9(30%)	0.31(4.8%)	3.1(48%)	-	-	69
+SD	7.5	6.1	1.9	2.8	0.13	3.5	-	-	16
var.	1-35	0-29	0-10	0-12	-	1-19	-	-	0-10

n   : number

x   : mean

+SD   : standard deviation

var. : range

0.8 of these were unfertilised, 1.9 appeared obviously degenerated and 3.7 normally developed according to the stage of flushing in relation to occurrence of heat (day of collection). 0.31 of these had some degree of degeneration.

The proportion of degenerated eggs was high at a dose level of 3000 iu PMSG (81%). The best recovery rate was observed when donors received 1500 iu.

The highest number of eggs was recorded in the 2500 iu group (11.2). All animals used now and in the near future will receive 2500 iu.

3.1 eggs per donor could be transferred to recipients surgically or non-surgically at the farm.

Fifty-five normal appearance embryos were transferred surgically to the tip of the uterine horn ipsilateral to the ovary containing the corpus luteum. Thirty recipients were diagnosed pregnant by rectal palpation yielding a pregnancy rate of 54%.

Thirty-three eggs were transferred non-surgically and 10 recipients became pregnant (31%). This pregnancy rate is somewhat lower than described from other groups (Boland, 1976; Sreenan, 1975) and attempts to improve this method are currently being carried out.

For non-surgical transfer of bovine eggs three different types of instruments have been developed. One of these instruments is constructed for transcervical deposit of eggs into the tip of the horn and a third one is devised for transvaginal and transuterine deposit of eggs. These instruments are being used experimentally at present.

In connection with these investigations successful freezing of Day 7 embryos has been carried out and a number of practising veterinarians have been able to collect bovine eggs with success by using the procedure described after a brief training period. Further the Danish Co-operative AI association has established an experimental centre for bovine embryo transplantations as a service to breeders.

It is believed that the non-surgical method can be even more efficient and an aid in a rapid multiplication of good pedigree stock. The superovulation however must be more predictable and reliable.

CONCLUSION

The past two years of experience with non-surgical embryo-transplantations under farm conditions in Denmark has proved to be successful. The instrument used for embryo collection was very reliable and easy to use without damage to the genital tract in heifers as well as in cows. Bleeding seldom occurred and breeding records from the donor cows have shown that the superovulation and collection had no adverse effect. In practice, veterinarians have been able to use this method with success after a brief training period. Successful freezing of Day 7 embryos has been carried out in connection with these investigations and instruments for non-surgical embryo transfer of bovine embryos are being tested.

REFERENCES

Boland, M.P. 1976. Twinning in beef cattle - Surgical and non-surgical
    apparatus. 27th Ann. Meeting, EAAP, August 23rd-26th, 1976, Zürich.

Greve, T. and Lehn-Jensen H. 1977. Ikke kirugisk opsamling af bovine
    embryoner under praktiske forhold. (Practical applications of non-
    surgical embryo transplantations in valuable pedigree animals). -
    Inst. Sterilitetsforskn. Arsberetn. 20, 1-24.

Greve, T., Lehn-Jensen, H. and Rasbech, N.O. 1977. Non-surgical recovery
    of bovine embryos. Theriogenology, 7, No. 4, 239-249.

Rasbech, N.O. 1976. Non-surgical recovery and transfer of bovine embryos
    under farm conditions. 27th Ann. Meeting, EAAP, August 23rd-26th,
    1976. Zurich.

Rasbech, N.O. 1976. Nicht-chirurgische Gewinnung und Ubertragung von
    Rinderembryonen unter Praxis-Verhältnissen. (Non-surgical recovery
    and transfer of bovine embryos under farm conditions). Dtsch.
    Tierärzt. Wschr. 83, 515-586.

Sreenan, I.M. 1975. Successful non-surgical transfer of fertilised cow
    eggs. - Vet. Rec. 96, 480-491.

DISCUSSION

L.E.A. Rowson *(UK)*

Now, can we start this session with some questions on the Heyman paper please.

T. Greve *(Denmark)*

I have a procedural question. I noticed on your Table on the treatments you had only 48 h from the last Cloprostenol treatment in the recipients until you expect them to be in heat. Is that right? Usually three days elapses ....

Y. Heyman *(France)*

It was two days. The last Cloprostenol treatment was on the same day for donors and recipients. It is known that donors come in heat a little earlier than recipients.

T. Greve

How much earlier?

Y. Heyman

Half a day.

K. Betteridge *(Canada)*

Did you have any unilateral pregnancies in the contralateral horn?

Y. Heyman

Yes, we had two cases of unilateral pregnancies in the contralateral horn.

M. Boland *(Ireland)*

I may have missed a point but did you withdraw your inseminating gun to transfer in the second egg or were you able to transfer in the two eggs by just doing one entry through the cervix?

Y. Heyman

We made only one entry with the gun through the cervix; the two eggs were transferred at the same time separated by a bubble of gas and when the transfer was made in the first horn we withdrew the gun a little to pass into the other horn. The transfer usually takes two or three minutes.

A. Brand *(Netherlands)*

Was there any difference in pregnancy rate in relation to the difficulty and time involved in passing through the cervix?

Y. Heyman

We have looked at this but there was no difference in pregnancy rates in transfers which took less than three minutes compared with those which took more than three minutes. With some transfers it took ten to twelve minutes to pass through the cervix but we still obtained pregnancies.

K. Betteridge *(Canada)*

If I may comment, I think these are terrific results for non-surgical transfer. They are the same as we get with surgical transfers with similar stage embryos. We have heard some varying results today but I think these are the most encouraging.

J.P. Renard *(France)*

I would mention two points: in these experiments we don't consider that passing through the cervix is a major problem. The critical point is after we have passed through the cervix and we have placed the egg not more than 5 cm from the uterine end of the cervix. That is the first point. The second point is that we also pay particular attention to the egg quality. We are trying to find an answer to Dr. Lampeter's question - what is a good embryo? At the moment we don't know. We make very selective decisions with our embryos, we use only some 40 - 45% of the superovulated embryos that we recover.

T. Greve

I would like to make another comment regarding egg quality.

At one time we were in the position that we had to use more or
less all the embryos available but in the last six months we have
only used embryos which we believed to be of very high quality.
We have made very careful drawings of them and then we have been
retrospective, going back to see which embryos do, in fact, give
the best pregnancy rates. In most cases we have been successful
when we have said that we believe an embryo to be good because
it has given rise to a pregnancy. Where we have had rather disappointing results is with the hatched embryos, but maybe they
were poor embryos - we don't know. We also use only about 50%
of the embryos and since we have been making this rigid selection
our pregnancy rates have definitely risen.

A. Brand

Dr. Heyman, you showed that there is already a difference
in the size of the embryos at Days 11 and 12. Going back to the
time of ovulation, there is always an interval between the first
and last ovulations. If there is a twelve hour interval, then
maybe the embryo which ovulates just at the end is in a poor
situation to develop later on.

Y. Heyman

I don't think so because Days 12 and 13 embryos grow very
quickly and in a few hours their length may increase rapidly, for
embryos which have ovulated at the same time but we do not have
sufficient data to say whether long embryos develop better than
short ones.

A. Brand

Have you any indications that your pre-treatment of recipients, two $PGF_2\alpha$ injections, has any influence on the corpus
luteum and endometrium after ovulation? I am asking this because
in sheep I have had some indication, based on cytochemical analysis, that the endometrium is different from the normal endometrium
after a second injection of $PGF_2\alpha$.

Y. Heyman

I don't know.

L.E.A. Rowson

I am sorry to have to close down on that paper but I think we shall have to move on as we are a little behind time.

T. Greve

Professor Gordon, in your first Table you were showing consecutive superovulations within 30 days for at least two treatments. You said that after the first ovulation you gave them prostaglandin injections and then they came in heat. I just wanted to ask you, do the superovulated animals come into heat if you give them a prostaglandin injection, for example, on Day 8, or what is the oestrous response and have you followed with plasma progesterone in those cases?

I. Gordon (Ireland)

On the proportion that come into heat, of the 12 animals that were given the prostaglandin, it was at the time of the non-surgical collection of eggs, 11 out of the 12 were in oestrus, but instead of the normal interval, prostaglandin to oestrus, of the two days, the interval was 3 - 4 days. So it wasn't that there was a longer interval but that the oestrous response was quite surprising.

L.E.A. Rowson

Thank you very much

I. Wilmut (UK)

May I put a question to Dr. Schneider? I think there is a danger of confusion here. Did your PBS medium have foetal calf serum in it?

U. Schneider (West Germany)

Yes, it was supplemented with 20% foetal calf serum.

T. Greve

I am not quite sure that I understood you correctly. I know that some of the examinations you made were in droplets under the microscope. Did you do these just prior to transfer

and directly in the straw or did you get them out and then back
in the straw because I would have thought there was a danger of
error if you did it through the straw.

U. Schneider

You can judge the embryos through the straw when you start
with the early blastocysts and follow up the expansion of them.
This can be clearly judged through the straw. However, in some
cases, especially after a long transfer, the embryos were taken
out of the straw, judged in the microdrop under the stereo microscope and then loaded again in another straw for transfer.

F. du Mesnil du Buisson *(France)*

When you transferred embryos after the collection, are the
eggs preserved at a temperature of 20° or $37^{\circ}C$ and do they develop
or not during this period?

U. Schneider

They are kept on a heating plate at $37^{\circ}C$ for a short period
of time, up to three hours. We find that we have to change this
a little bit because if you keep a volume of 5 ml of PBS on a
heating plate without a cover, over a three hour period, you get
an increase of osmolarity. We have to guard against this by
placing it in a humid atmosphere.

J.P. Renard

In your opinion what is the clinical difference between
BSM-2 + foetal calf serum, and PBS + foetal calf serum, that may
explain the difference in preservation after culture?

U. Schneider

Well, the modified Ham's F 10 is enriched with amino acid
and the PBS is just a plain buffer with some inorganic salts
supplemented with glucose and pyrovate. At the beginning we
observed that the embryos developed in PBS so morphologically it
was not possible to detect any difference. However, after transfer when we discovered that the transfer success was so low, we
looked a little bit further to see if it is a medium effect. We

will have to wait until this other group of animals where the embryos were cultured has been evaluated for pregnancy.

W.C.D. Hare *(Canada)*

Was the same batch of foetal calf serum used in each case to supplement your medium?

U. Schneider

Yes, it was the same.

T. Greve

Have you made any comparative studies using larger volumes than the mini-straw, for example, because that is a pretty small volume? It probably has little buffer capacity. What would happen if you did a simultaneous study using a larger volume?

U. Schneider

Due to the number of eggs we can utilise we haven't done this but we were surprised that for rabbit embryos the culture conditions in these mini-straws were pretty good; in some cases the embryo survived really well.

L.E.A. Rowson

Well, now it's time to move on to the next paper - Professor Rasbech.

A. Brand

You mentioned that there was a relationship between the protein content in the feed and the response after superovulation. How do think that relationship works? Is it just the protein content? How does it affect the subsequent superovulation after PMSG treatment?

N. Rasbech *(Denmark)*

You are opening up quite a field of discussion on that. We had ovarian activity which is very dependent on energy intake, protein, the relationship between sodium and potassium, the phosphorus intake and so on. However, you asked particularly about the protein. Now, during the last two or three years, protein has been cheap for the farmer in Denmark and there has been a tendency to overfeed with protein. We have noticed that there is an increased frequency of cysts and prolonged oestrous periods in these herds which have been overfed with protein. I don't have the figures here to give you exact data. However, although we are very careful with these herds, we are not very optimistic about the results of superovulation because when we are dealing with a higher frequency of cysts in a herd and with a prolonged oestrous period, there are problems. In my opinion that is not the only thing which influences superovulation. There are many other nutritional aspects involved. On the other hand, they are also producing infertility in herds so to some extent we know them but the problem is what to do with them.

There is another big problem and that is how dairy cows are fed during the dry period. In many cases they are overfed.

L.E.A. Rowson

Well, once again, thank you for your contributions.

SESSION 3

MANIPULATION OF OVA IN VITRO

Chairman:

N. Rasbech

CULTURE OF MAMMALIAN OVA

M.T. Kane

Department of Physiology, University College, Galway, Ireland.

INTRODUCTION

The development of culture techniques for bovine ova is crucial to the development of other techniques for the manipulation of these ova. There have been a number of successful attempts to culture 8-cell and later cleavage stages of bovine embryos to the blastocyst stage (Tervit et al., 1972; Renard et al., 1976; Wright et al., 1976 a, b; and Beehan and Sreenan, 1976, unpublished observations) but culture of 1 or 2-cell bovine ova to the blastocyst has been much more difficult. Wright et al. (1976 b) reported the culture of three out of fourteen 1 - 2-cell cattle ova to blastocysts. The medium used was Ham's F10 supplemented with foetal calf serum. This is the only report to date of the culture of such early stage cattle ova beyond 8 or 12 cells and while the report is very encouraging, the presence of a very undefined fluid like foetal calf serum means that this system may be subject to great variation in response. Culture of bovine ova has been reviewed recently (Seidel, 1977).

The first part of the present review will discuss information available from work on the development of synthetic media for the culture of ova or laboratory animals and try to relate this to the culture of bovine ova. The second part of the review will point out that there are clear parallels to the problems of culturing early cleavage stages of bovine ova in work with the ova of most other mammalian species including laboratory animals and will try to suggest a possible reason for this.

COMPONENTS OF SYNTHETIC MEDIA

Table 1 lists possible components, and the physical para-

meters that must be controlled in a complete synthetic medium
for ovum culture.

TABLE 1

CONSTITUENTS OF OVUM CULTURE MEDIA

Major Inorganic Ions

    NA, K, Ca, Mg, Cl, PO_4, SO_4, HCO_3.

Osmolarity & pH

Energy Sources

    pyruvate, glucose, fatty acids.

Amino Acids

Vitamins

Trace Elements

Gas Phase

    P_{CO_2}, P_{O_2}

Miscellaneous Factors

    culture container, ambient pressure,

    density of ova, ovum movement, quality of water,

    hormonal factors.

Major Inorganic Salt Constituents

The principal function of sodium and chloride ions in the
development of cultured cells is to act as osmotically active
ions because they are responsible for most of the osmolarity
of culture media. No reports could be found in the literature
of studies in which these ions were replaced completely in ova
culture media by other ions or molecules.

The effects of all the other major inorganic salt constituents except HCO_3 have been systematically examined by Wales (1970) in the case of mouse ova. In the absence of potassium, ovum cleavage was almost completely inhibited. However, good development occurred over a range of K concentrations varying from 0.6 mM to 48 mM, indicating that the level of K may not be crucial for ovum development. It is known that the level of K in female reproductive tract fluids is considerably higher than that in plasma (Restall and Wales, 1966; Borland et al., 1977) and at least two media designed for culture of cattle ova have contained high levels of K (Tervit et al., 1972; Menezo, 1976). The only other ion found essential for ovum cleavage was Ca (Wales, 1970) but a range of concentration from 0.4 mM to 10mM allowed relatively normal development in culture. Complete absence of Ca inhibited cleavage and very low levels of Ca prevented the compaction of mouse embryos at the morula stage (Ducibella and Anderson, 1975).

The complete absence of PO_4 and Mg significantly depressed but did not abolish development of 2-cell ova to blastocysts (Wales, 1970). Omission of SO_4 had no effect on development to the blastocyst stage.

The main role of HCO_3 in culture media is to function as part of a pH buffer system but HCO_3 is also involved in metabolism, particularly in 1-carbon metabolism and in the formation of Krebs cycle metabolites. This situation is reflected in the fact that growth of early cleavage stages of mouse ova is very limited in phosphate buffered media without HCO_3 (Quinn and Wales, 1973) and that while one-cell rabbit ova will develop to morulae in a Hepes buffered medium without HCO_3, development to the blastocyst stages requires low levels of bicarbonate (Kane, 1975).

The evidence to date from work with mouse ova would suggest that, within certain relatively broad limits, levels of the major inorganic ions may not be crucial to development of preimplantation ova.

Osmolarity and pH

Work with laboratory animals indicates that ova will grow over a wide range of osmolarity and pH. Two-cell mouse ova grow to blastocysts in media varying in osmolarity from 200 to 354 mOSM (Brinster, 1965a) and 2-4 cell rabbit ova to blastocysts in media varying from 230 to 339 mOSM (Naglee et al., 1969). Two-cell mouse ova developed to blastocysts in media ranging from pH 5.87 to 7.78 (Brinster, 1965a) and 1-cell rabbit ova to blastocysts from pH 6.64 to 7.91 (Kane, 1974). There was some evidence that the optimum pH for rabbit blastocyst expansion was slightly higher (c 7.6) than for blastocyst formation (c 7.3).

These studies suggest that, while osmolarity and pH control are important in ovum culture media, they may not be causative factors in the poor results found to date with early cleavage stage ova of the cow.

Energy Sources

The early work of Whitten (1957) and Brinster (1965b) on a wide range of potential energy sources showed that only pyruvate, lactate, phosphoenol pyruvate and oxaloacetate would support growth of 2-cell mouse ova and that glucose was not effective until the 8-cell stage. More recently it has been shown that 1-cell rabbit ova will develop in a simple salt solution containing defatted albumin and either pyruvate, or any one of 5 long chain fatty acids or, to a lesser extent, either of the short chain acids acetic or propionic (Kane, 1976). The fact that short chain fatty acids may be capable of acting as energy substrates for ova is especially relevant to bovine ova since it is well known that these fatty acids and especially acetic acid form a major part of the energy substrate in bovine blood. The only synthetic medium for cattle ovum culture incorporating acetate as an energy substrate is Menezo's B2 Medium (Menezo, 1976).

Amino Acids

Early cleavage stages of the mouse ova will develop to

blastocysts in simple salt solutions containing crystallised
bovine serum albumin (BSA) or polybinylpyrrolidone but without
the presence of free amino acids (Brinster, 1965c; Cholewa and
Whitten, 1970). Amino acids are, however, essential for
development of rabbit ova to blastocysts in culture; 0% of 57
rabbit ova developed to the blastocyst stage in a complex
culture medium containing 1.5% BSA but no free amino acids as
compared with 72% of ova developing to blastocysts in the same
medium with a full complement of 20 amino acids added (Kane
and Foote, 1970a).

Daniel and Krishnan (1967) have also reported on growth
promoting effects of amino acids on rabbit blastocysts in vitro.
The difference between the two species is readily explainable
in view of the fact that the protein content of mouse ova does
not change greatly from 1-cell to blastocyst (Brinster, 1967;
Schiffner and Spielmann, 1976) whereas mass and protein content
of the rabbit embryo increases with blastocyst formation
(Lutwak-Mann, 1971). This increase in protein content
necessarily requires amino acids for protein synthesis. Since
the mass of the bovine embryo is also known to increase with
blastocyst formation it can be expected that a full complement
of amino acids will be necessary for culture of bovine ova at
the blastocyst stage.

Vitamins

Vitamins are not necessary for formation of mouse blastocysts in culture. However, it has been shown by Kane and Foote
(1970a) that omission of a group of 11 water soluble vitamins
from a complex culture medium decreased the proportion of
rabbit ova forming blastocysts from 72% to 54% and the percent
ova forming expanding blastocysts from 42% to 5. Daniel (1967)
previously found that four of these 11 vitamins were necessary
for growth of 5 day blastocysts. The species difference here
between mouse and rabbit is again probably due to the differing
growth patterns of the blastocyst of these two species. Since
the bovine blastocyst is closer in this regard to the rabbit
blastocyst than that of the mouse, it seems reasonable to expect

that vitamins may be necessary for bovine blastocyst growth. Vitamin C, which was not used in the experiments just cited, has been incorporated in one ovum culture medium in order to maintain an oxidation-reduction potential characteristic of the female tract (Menezo, 1976).

Trace Elements

Addition of trace elements to the medium is not necessary for development of mouse ova to blastocysts and Kane and Foote (1970a) found that omission of trace elements from the culture medium did not inhibit either blastocyst formation or blastocyst expansion of rabbit embryos. This does not exclude the possibility that the Analar grade salts which make up the major salt constituents of the medium may have contained sufficient trace element contamination to supply the embryo's needs. In contrast to these results, Daniel and Millward (1969) found that omission of ferrous ion from Hams F12 medium completely abolished cleavage of rabbit ova and caused collapse of rabbit blastocysts. The difference in results may be due to the fact that the medium used was apparently free of all macromolecules and these results of Daniel and Millward may reflect a macro-molecule-ferrous ion interaction.

Macromolecules

It is generally agreed that the presence of macromolecules in the culture medium is necessary for growth of mammalian ova. Crystallised BSA is very commonly used as such a macromolecule and its function has in this regard never been completely clarified. To some extent it can be replaced by polyvinylpyrrolid or Ficoll eg for culture of 2-cell mouse ova to blastocysts (Cholewa and Whitten, 1970) or 1-cell rabit ova to morulae (Kane, unpublished data). However the presence of albumin in the medium is essential for the transition from 1-cell to 2-cell in the mouse (Cholewa and Whitten, 1970) and for blastocyst form-ation in the rabbit (Kane and Foote, 1971; Kane, unpublished observations). These functions of BSA which cannot be replaced by polyvinylpyrrolidone or Ficoll may be related to the fact that commercial BSA is a very impure product both from the point of view of protein content and also because of low molecular weight material bound to it. Purification of commercial samples of BSA on SDS polyacrylamide gradient gels indicates that these samples generally contain only from 75% to 90% albumin (Kane and Headon, unpubished data). The protein impurities mostly consist of material of higher mol. wt. than the albumin. Also albumin is well known to be a carrier of small molecular weight substances such as fatty acids (Goodman, 1958), steroid hormones (Westphal, 1970) and many other compounds. Some of the fatty acid bound to the albumin may act as an energy source for rabbit ova but recent work in this laboratory indicates that the role of commercial BSA samples in promoting rabbit blastocyst growth in vitro may be connected with contaminants other than fatty acids (Kane and Headon, unpublished data). Hence claims by various workers (including the author of this review) that media containing BSA are defined media are plainly in error. Also, variations in quality of BSA are almost certainly responsible for some of the variation in results between different laboratories.

It has been claimed by Krishnan and Daniel (1967) that a specific uterine protein 'blastokinin' or uteroglobin is an inducer and regulator of blastocyst formation and expansion but subsequent research by the same and other workers has not borne out this claim. A careful exploration of the role of uterine proteins in promoting growth of rabbit blastocysts in vitro (Maurer and Beier, 1976) has shown that unfractionated uterine protein has a better growth stimulating effect than any specific uterine protein fraction including uteroglobin and that the unfractionated uterine protein appears to be only marginally more effective than BSA.

Gas Phase

The two gas phases normally used for culture of ova are either 5% CO_2 in air or 5% CO_2, 5% O_2 and 90% N_2. The 5% CO_2 is used in order to give a pH of about 7.4 with a 25 mM HCO_3 buffer although it may be used with higher levels of HCO_3 if

higher pH values are desired. There is some evidence that the
5% level of O_2 which gives a PO_2 in the culture medium more
nearly approximating that in oviducal fluid is better for the
culture of cattle ova than the 20% O_2 level of the air gas
phase (Wright et al., 1976a). An interesting point in relation
to the use of CO_2 is that many commercial samples are
contaminated with levels of carbon monoxide which could be
toxic to cells including cultured ova (McLimans, 1972).

Miscellaneous Factors

There are a number of miscellaneous factors which may
affect ovum culture and have not been studied to any great
extent eg nature of culture container, ambient pressure, density
of ova, ovum movement, water quality and hormonal factors.

(a) Nature of Culture Container

There is little information on the relative merits and
demerits of different systems of containing the ova in
culture. The different systems in use currently include
the following: microdrops of medium under paraffin liquid
in plastic tissue culture dishes (Brinster, 1963), medium
contained in the wells of plastic Disposo trays (Kane and
Foote, 1970b) or Falcon Micro Test tissue culture plates
(Seidel, personal communication) and medium contained in
stoppered test tubes (Tervit et al., 1972). The quality
of the paraffin liquid used to cover medium may be an
important factor in many cases as some batches are toxic.

(b) Ambient Pressure

Elliott et al.(1974) found that increased ambient pressure
improved the growth of rabbit ova in vitro.

(c) Density of Ova

The number of ova per ml of medium may affect ovum
development. We have found some evidence that numbers of
rabbit ova per drop of culture medium can affect response
and that single ova per drop of culture medium do

particularly badly (Kane, unpublished data).

(d)   Ovum movement

The effect of continuous movement as compared with the more usual stationary systems of ovum culture is a potentially important factor which has not been investigated to any extent.

(e)   Water Quality

The use of high quality distilled water has been shown by Whittingham (1971) to be a crucially important factor.

(f)   Hormonal Factors

To date, there appears to be no clearcut reports of a beneficial influence of hormones on the growth of ova in culture while there are a number of reports of the deleterious effects of steroid hormones (reviewed by Warner, 1977). Most work has been done on the effects of steroid hormones but it is possible that protein and peptide hormones or other hormonal factors may have important effects on embryos in vitro.

There are many other potential factors not discussed here which could be critical to in vitro embryo development and perhaps particularly to large scale blastocyst growth and expansion.

COMPARATIVE PROBLEMS IN CULTURING MAMMALIAN OVA

There are two different sets of problems involved in the culture of mammalian ova. The first involves culture through the early cleavage stages and the second involves culture of the blastocyst stages of those species in which the blastocyst enlarges greatly before implantation. Both sets of problems apply to the bovine embryo. I will examine the problem of culture through the early cleavage stages first.

The rabbit is perhaps the only species in which it has

been possible in a number of laboratories repeatedly to culture
1-cell ova to blastocysts. The same thing is possible for the
mouse in a few carefully selected strains and crosses. Otherwise there tends to be a block at the 2-cell stage and in
general it is not possible to culture the 1-cell ova of random
bred Swiss mice to the blastocyst stage. There is one report
in which this has been done (Cross and Brinster, 1973) by
changing the medium after division from 1-cell to 2-cell in
vitro to a special medium which would not have been necessary
for 2-cell ova grown from the 1-cell in vivo. Culture of 8-cell
rat ova to blastocysts is possible (Folstad et al., 1969) but
culture from the 4-cell stage or earlier is blocked (Mayer and
Fritz, 1974). One-cell hamster ova cultured in vitro remain
blocked at the 2-cell stage whereas 8-cell stages will develop
to blastocysts (Whittingham and Bavister, 1974). A limited
number of human ova have also been grown in vitro from 1-cell
to blastocyst (Edwards et al., 1970).

The rabbit, unlike all the other species discussed so far
is an induced ovulator and it is interesting to note that it
has recently been reported that 1-cell ova of two species of
induced ovulators, the ferret (Whittingham, 1975) and the cat
(Bowen, 1977) can be cultured to blastocysts. I would like to
make the very tentative suggestion that possibly spontaneous
ovulators such as the rat, hamster, sheep and cow may normally
require some hormonal trigger for the early cleavage stages of
ovum development and that this trigger is not required by the
ova of certain induced ovulators. It is also possible that
the rat and hamster might prove useful models in which to study
the problems associated with early cleavage development of ova.

The second type of problem in ovum culture is associated
with the growth of blastocysts. Cleavage of ova is usually
not associated with an increase in mass and protein content
of the embryo but blastocyst growth of a number of species
including the rabbit and domestic animals involves a large
increase in embryo mass. The rabbit is the only laboratory
species in which a very marked degree of blastocyst growth takes

place before implantation. Thus it may serve as a useful model for bovine ovum growth. Some of the recent work with rabbit blastocysts would suggest that macromolecules or small molecules bound to macromolecules exert a controlling influence here.

CONCLUSION

Progress in the area of bovine ovum culture depends to some extent on progress in the area of ovum culture generally and maximum use should be made of knowledge and insights arising from work on other species, including laboratory species whose ova can be obtained much more cheaply than cattle ova. Two particularly useful species from this point of view may be the rat and rabbit.

However the more direct growth of culturing bovine ova themselves is also necessary. Here there are perhaps two logical ways of tackling the problem. One method is to construct an artificial medium from the information obtained from analyses of the fluids of the bovine female reproductive tract (Menezo, 1976). A second approach involves the collection of oviducal and uterine fluid samples and the culture of ova in these samples. These might be purified in order to isolate the factor or factors necessary for growth. This approach is currently being used in this laboratory in co-operation with An Foras Taluntais.

ACKNOWLEDGMENT

Part of the author's research referred to in this review was supported by a grant from An Foras Taluntais.

REFERENCES

Borland, R.M., Hazra, S., Biggers, J.D. and Lechene, C.P. 1977. The
    elemental composition of the environments of the gametes and pre-
    implantation embryo during the initiation of pregnancy. Biol.
    Reprod. 16, 147-157.

Bowen, R.A. 1977. Fertilisation in vitro of feline ova by spermatozoa
    from the ductus deferens. Biol. Reprod. 17, 144-147.

Brinster, R.L., 1963. A method for the in vitro cultivation of mouse ova
    from two-cell to blastocyst. Expl. Cell Res. 32, 205-208.

Brinster, R.L., 1965a. Studies on the development of mouse embryos in
    vitro. 1. The effect of osmolarity and hydrogen ion concentration.
    J. exp. Zool. 158, 49-58.

Brinster, R.L., 1965b. Studies on the development of mouse embryos in vitro.
    11. The effect of energy source. J. exp. Zool. 158, 59-68.

Brinster, R.L., 1965c. Studies on the development of mouse embryos in
    vitro. 111. The effect of fixed nitrogen source. J. exp. Zool.
    158, 69-78.

Brinster, R.L., 1967. Protein content of the mouse embryo during the
    first five days of development. J. Reprod. Fert. 13, 413-420.

Cholewa, J.A. and Whitten, W.K., 1970. Development of two-cell mouse
    embryos in the absence of a fixed-nitrogen source. J. Reprod. Fert.
    22, 553-555.

Cross, P.C. and Brinster, R.L., 1973. Sensitivity of one-cell mouse
    embryos to pyruvate and lactate. Expl. Cell Res. 77, 57-62.

Daniel, J.C. Jr., 1967. Vitamins and growth factors in the nutrition of
    rabbit blastocysts in vitro. Growth, 31, 71-77.

Daniel, J.C. Jr. and Krishnan, R.S., 1967. Amino acid requirements for
    growth of the rabbit blastocyst in vitro. J. Cell Physiol. 70,
    155-160.

Daniel, J.C. Jr. and Millward, J.T., 1969. Ferrous ion requirement for
    cleavage of the rabbit egg. Expl. Cell Res. 54, 135-136.

Ducibella, T. and Anderson, E., 1975. Cell shape and membrane changes in
    the eight-cell mouse embryo: prerequisites for morphogenesis of
    the blastocyst. Devel. Bio. 47, 45-58.

Edwards, R.G., Steptoe, P.C. and Purdy, J.M., 1970. Fertilisation and
    cleavage in vitro of pre-ovulatory human oocytes. Nature Lond. 227,
    1307-1309.

Elliott, D.S., Maurer, R.R. and Staples, R.E., 1974. Development of
mammalian embryos in vitro with increased atmospheric pressure.
Biol. Reprod. 11, 162-167.

Folstad, L., Bennett, J.P., and Dorfman, R.I. 1969. In vitro culture of
rat ova. J. Reprod. Fert. 18, 145-146.

Goodman, D.S. 1958. The interaction of human serum albumin with long-
chain fatty acid anions. J. Am. Soc. 80, 3892-3898.

Kane, M.T. 1974. The effects of pH on culture of one-cell rabbit ova to
blastocysts in bicarbonate-buffered medium. J. Reprod. Fert. 38,
477-480.

Kane, M.T., 1975. Bicarbonate requirements for culture of one-cell rabbit
ova to blastocysts. Biol. Reprod. 12, 552-555.

Kane, M.T., 1976. Growth of fertilised one-cell rabbit ova to viable
morulae in the presence of pyruvate or fatty acids. J. Physiol.
263, 235-236P.

Kane, M.T. and Foote, R.H. 1970a. Culture of two- and four-cell rabbit
embryos to the expanding blastocyst stage in synthetic media.
Proc. Soc. exp. Biol. Med. 133, 921-925.

Kane, M.T. and Foote, R.H., 1970b. Fractionated serum dialysate and
synthetic media for culturing 2- and 4-cell rabbit embryos. Biol.
Reprod. 2, 356-362.

Kane, M.T. and Foote, R.H., 1971. Factors affecting blastocyst expansion
of rabbit zygotes and young embryos in defined media. Biol. Reprod.
4, 41-47.

Krishnan, R.S. and Daniel, J.C. Jr., 1967. 'Blastokinin': inducer and
regulator of blastocyst development in the rabbit uterus. Science,
N.Y. 158, 490-492.

Lutwak-Mann, C., 1971. Rabbit blastocyst and its environment: physiological
and biochemical aspects. In: Biology of the Blastocyst, ed. R.J.
Blandau, pp 243-260, U. Chicago Press.

Maurer, R.R. and Beier, H.M., 1976. Uterine proteins and development
in vitro of rabbit preimplantation embryos. J. Reprod. Fert. 48,
33-41.

Mayer, J.F. Jr., and Fritz, H.I., 1974. Culture of preimplantation rat
embryos and the production of allophenic rats. J. Reprod. Fert.
39, 1-9.

McLimans, W.F., 1972. The gaseous environment of the mammalian cell in culture. In: Growth, Nutrition and Metabolism of Cells in Culture pp. 137-170, ed. by G.H. Rothblatt and V.J. Gristofolo. Academic Press, N.Y.

Menezo, M.Y. 1976. Milieu synthetique pour la survie et la maturation des gametes et pour la culture de l'oeuf feconde. C.R. Acad. Sc. Paris, 282, 1967-1970.

Naglee, D.L., Maurer, R.A. and Foote, R.H., 1969. Effect of osmolarity on in vitro development of rabbit embryos in a chemically defined medium. Expl. Cell Research 58, 331-333.

Quinn, P. and Wales, R.G., 1973. Growth and metabolism of preimplantation mouse embryos cultured in phosphate-buffered medium. J. Reprod. Fert. 35, 289-300.

Renard, J.P., du Mesnil du Buisson, F., Wintenberger-Torres, S. and Menezo, Y., 1976. In vitro culture of cow embryos from Day 6 and Day 7. In: Egg transfer in cattle, ed. L.E.A. Rowson. Commission of the European Communities, Luxembourg. EUR 5491, pp 159-164.

Restall, B.J., and Wales, R.G., 1966. The Fallopian tube of the sheep. 111. The chemical composition of the fluid from the Fallopian tube. Aust. J. biol. Sci. 19, 687-693.

Schiffner, J. and Spielmann, H., 1976. Fluorometric assay of the protein content of mouse and rat embryos during preimplantation development. J. Reprod. Fert. 47, 145-147.

Seidel, G.E. Jr., 1977. Short-term maintenance and culture of embryos. In: embryo transfer in farm animals, ed. K.J. Betteridge. Canada Department of Agriculture, Monograph No. 16, pp 20-24.

Tervit, H.R., Whittingham, D.G. and Rowson, L.E.A. 1972. Successful culture in vitro of sheep and cattle ova. J. Reprod. Fert. 30, 492-497.

Wales, R.G., 1970. Effects of ions on the development of the pre-implantation mouse embryo in vitro. Aust. J. biol. Sci. 23, 421-429.

Warner, C.M., 1977. RNA polymerase activity in preimplantation mammalian embryos. In: Development in Mammals, pp. 93-136, ed. by M.H. Johnson, North Holland, Amsterdam.

Westphal, W., 1970. Corticosteroid-binding globulin and other steroid hormone carriers in the blood stream. J. Reprod. Fert. Suppl. 10, 15-38.

Whitten, W.K., 1957. Culture of tubal ova. Nature, Lond. 179, 1081-1082.

Whittingham, D.G., 1971. Culture of mouse ova. J. Reprod. Fert. Suppl. 14, 7-21.

Whittingham, D.G., 1975. Fertilisation, early development and storage of mammalian ova in vitro. In: The early development of mammals, pp 1-24, ed M. Balls and A.E. Wilde, Cambridge Univ. Press. Lond., UK.

Whittingham, D.G. and Bavister, B.D. 1974. Development of hamster eggs fertilised in vivo or in vitro. J. Reprod. Fert. 38, 489-492.

Wright, R.W. Jr., Anderson, G.B., Cupps, P.T. and Drost, M., 1976a. Successful culture in vitro of bovine embryos to the blastocyst stage. Biol. Reprod. 14, 157-162.

Wright, R.W. Jr., Anderson, G.B., Cupps, P.T. and Drost, M., 1976b. Blastocyst expansion and hatching of bovine ova cultured in vitro. J. Anim. Sci. 43, 170-174.

ATTEMPTS TO PREDICT THE VIABILITY OF
CATTLE EMBRYOS PRODUCED BY SUPEROVULATION

J.-P, Renard, Y. Menezo*, J. Saumande**, Y. Heyman

Station de Physiologie animale, INRA
78350 Jouy en Josas, France

* Laboratorie de Biologie, INSA   69621 Villeurbanne, France

** Laboratorie de Physiologie de la Reproduction,
INRA 37380 Nouzilly, France

INTRODUCTION

The establishment of pregnancy after transplantation is partly determined by embryo quality. This varies due to many factors (du Mesnil du Buisson, Renard and Levasseur, 1977). Morphological criteria or changes in vitro permitting embryonic development to be seen have been studied by Shea et al. (1976) and by Trounson, Willadsen and Rowson (1976) in embryos recovered between Day 3 and Day 7. However, we have few data on embryos recovered later, especially at the time when their cervical transfer seems to be most effective.

Parameters which could be used for better prediction of embryo quality, could be applied either to the embryo or to the donor animal. This is indicated by the wide variability between donors with regard to embryo morphology (Renard and Heyman, in press) and in vitro potential for embryonic development (Seidel, 1977) after culture.

With regard to the embryo itself, our observation that glucose was necessary in the medium for blastocyst hatching, led us to try to define the criteria related to its metabolism and to determine the glycolytic enzymes in the medium.

In superovulated donors, the relationship between PMSG, ovarian stimulation and progesterone and oestradiol-17β secretion has been shown by Saumande (1977). We believed that if the viability of the embryos produced was also involved, donor hormone profiles would constitute a test for the quality of superovulated eggs.

In this report, all of these criteria have been employed to characterise the quality of embryos collected on Day 10.

MATERIAL AND METHODS

Heat was detected twice a day by a vasectomised bull in 31 Charolais heifers which were superovulated at Day 10 of the cycle by an injection of 2 100 iu PMSG (Intervert, batch V81), followed by a 500 µg injection 48 h later of a prostaglandin analogue (Cloprostenol ICI 80996). The animals were inseminated with frozen sperm 56 and 72 h after Cloprostenol injection and the embryos recovered on Day 10 of the cycle after slaughter of the donors and perfusion of the uterine horns with a phosphate buffer saline solution.

Embryo classification

After observation with a binocular microscope (x 40), the embryos were divided into 3 classes according to their morphological aspect:

Class 1: embryos at blastocyst stage, zona pellucida shed, still spherical or elongating.

Class 2: embryos at blastocyst stage still unhatched from the zona pellucida, embryonic inner cell mass well defined and distinct from the trophoblast.

Class 3: embryos have not attained blastocyst stage previously described.

Embryo size was measured with an ocular micrometer. For each donor, one or more embryos of each group were used for histological study and to determine the number of cells (10 µ sections, haematoxylin staining).

Determination of embryonic viability

Embryonic viability was determined by:

a) The ability of embryos to develop after cervical
transfer either within 4 h after recovery or after 24 h
culture in the B2 medium. An embryo was placed in each
uterine horn of a synchronised recipient (Renard et al.,
1977). Control of foetal development after slaughter or
laparotomy of recipients between weeks 7 and 9 of
pregnancy permitted us to evaluate the viability of transplanted embryos.

b) A study of morphological changes in Class 1 and 2
embryos in culture at $37^{o}C$ for 24 h or in those of Class 3
in culture for 48 h. One, 2 or 3 embryos were cultured in
0.5 ml of B2 medium using the technique described by
Renard et al. (1976). Their size was measured after
culture and a part of them fixed for histological study.

c) A study of changes in the chemical composition of the
culture medium.

Using spectrophotometric measurements, alterations in
glucose and lactate levels were determined by difference before
and after culture (glucose assay by hexokinase method; lactate
assay using the lacticode hydrogenase method).

The presence of dehydrogenase in the culture medium was
detected by colorimetric reaction 50 µl culture medium were put
on a buffered cellulose support impregnated with both the
substrate and the cofactor of the enzyme we were looking for.
The reactions were then revealed by colorimetry using 50 µl of
a solution containing a tetrazolium salt. In the presence of
the corresponding enzyme, the support became violet (reaction
+). This reaction was compared to the control support contain--
ing only the cofactor (reaction O). We tried to determine the
presence of 7 enzymes:

D-Glucose-6-phosphate-dehydrogenase (G6PDH)
D-L-Glycerophosphate-dehydrogenase
D-L-Isocitrate--dehydrogenase (ICDH)
L-Lactate-dehydrogenase (LDH)
L-Malate--dehydrogenase (MDH)

6-Phosphogluconate-dehydrogenase
Succino-dehydrogenase.

Endocrine profile of superovulated donors

Twenty ml blood samples were taken morning and evening from the jugular vein of 11 superovulated donor heifers. Sampling was done from the day of PMSG injection until the day of embryo recovery. Plasma oestradiol-17β and progesterone levels were measured by radioimmunology according to the technique of Thibier and Saumande (1975) (Testart et al., 1977).

RESULTS

Out of 470 ovulations induced in 31 donors, 346 embryos (73.6%) were collected, observed and studied.

1 - Description of D10 embryos

a) Blastocysts hatched from the zona pellucida (Class 1)

Two hundred and four embryos (58.9%) were in the blastocyst stage and hatched from the zona pellucida. The embryonic inner cell mass was clearly visible but embryo size was variable (Figure 1). One hundred and eighty two embryos (89.2%) were spheric and had a mean diameter of 253.4 $\mu \pm$ 170 μ (plate 1). Twenty two were beginning to elongate and assume an oval shape; their mean diameter was significantly larger than that of the spheric embryos (44.6 $\mu \pm$ 409 μ). Elongation was especially noticeable above a diameter of 375 μ (plate 2). A relationship between embryo size and the number of cells ($r = 0.86$) was obtained by counting the cells in 33 embryos (Figure 1). The percentage of embryonic inner mass cells decreased when the number of cells increased, or when embryo diameter enlarged. Thus, the embryonic inner cell mass did not represent more than 25% of the total number of cells when it had more than 1 000, which corresponds to an approximate diameter of 375 μ.

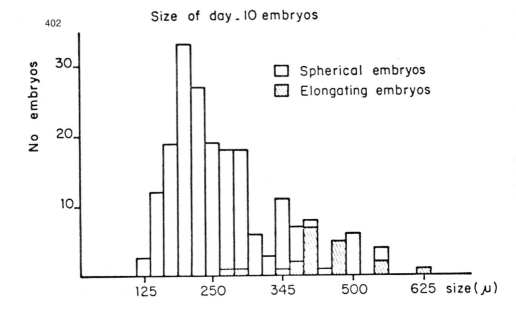

Size of day - 10 embryos

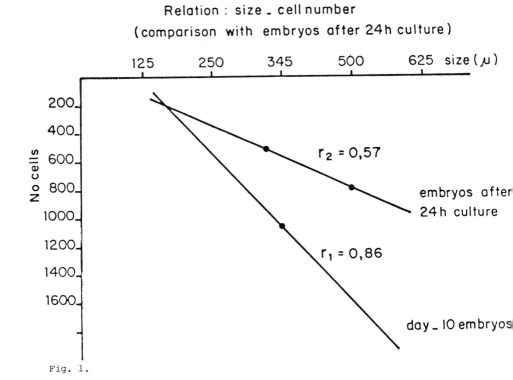

Relation : size - cell number
(comparison with embryos after 24h culture)

Fig. 1.

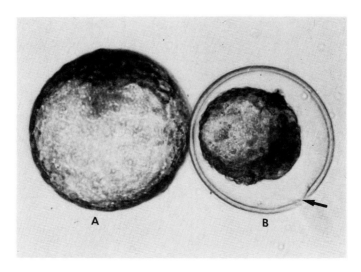

Plate 1 A: Hatched blastocyst. The inner cell mass is distinct, and
            trophoblast is spheric

        B: Blastocyst in the zona pellucida. The trophoblast is contracted
            but the inner cell mass is distinct: the zona pellucida may be
            split

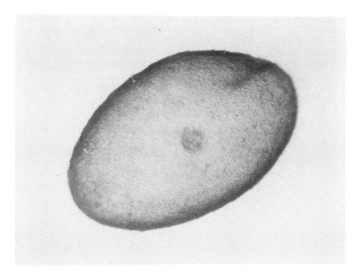

Plate 2 Elongating blastocyst. The inner cell mass is distinct and is
         small compared to the trophoblast

404

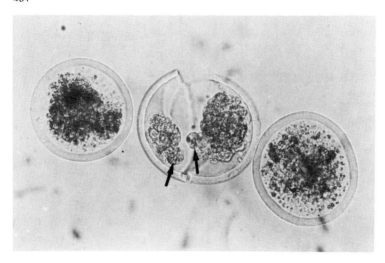

Plate 3 - Embryos have not reached blastocyst sgate.
        - They often appear disorganised, blastomeres are visible (Arrow)

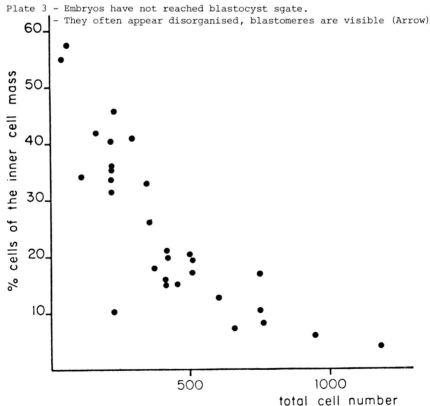

Fig. 2 Relation between total cell number of Day 10 embryos and % cells of the
      inner cell mass.

The primary endoderm was visible either at the onset of formation (plate 4) or when it completely covered the internal facet of the trophoblast. Out of 33 embryos studied histologically, 28 (85%) had few cells with pyknotic nuclei: pyknosis was found in the embryonic inner cell mass as well as in the trophoblast cells, but 5 embryos (15.1%) had more than 40% of pyknotic cells.

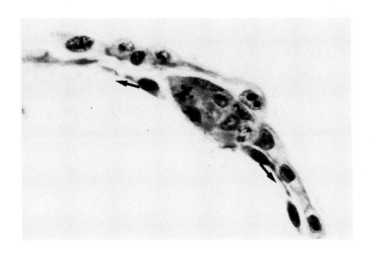

Plate 4   Hatched blastocyst
         Primitive endoderm cells form from inner cell mass (Arrow)
         When the embryo develops, they line the interior facet of the
         trophoblast

Donor distribution, related to the percentage of Class 1 embryos, is shown in Table 2. There is a wide variability between donors which does not depend on ovulation number or on collection rate. Out of the 31 animals studied, 10 had more than 75% of Class 1 embryos (91.6% of the embryos observed). These donors produced a mean of 9.9 blastocysts hatched from the zona pellucida. 12 animals had between 50% and 75% of Class 1 embryos (7.1 per animal) and 9 donors had only 20.4% of Class 1 embryos (2.1 per animal).

406

TABLE 2

CLASSIFICATION OF DONOR ANIMALS WITH REGARD TO % OF SHED BLASTOCYSTS
(CLASS 1)

	0 to 49% shed blastocysts	50 to 74% shed blastocysts	75 to 100% shed blastocysts	Total
Number of animals	9	12	10	31
Mean ovulation number	13.5	16.9	14.5	15.1
Number of examined embryos (collection rate %)	93 (76.0)	145 (71.4)	108 (74.4)	346 (73.6)
% embryos in Class 1 (Number)	20.4 (19)	59.3 (86)	91.6 (99)	58.6 (204)
% embryos in Class 2 (Number)	8.6 (8)	6.2 (9)	6.4 (7)	6.9 (24)
% embryos in Class 3 (Number)	70.9 (66)	34.4 (50)	0.1 (2)	34.1 (118)

b) Unhatched blastocysts (Class 2)

Twenty four embryos (6.9%) were in the blastocyst stage
and unhatched. The embryonic inner cell mass was clearly
visible. The trophoblast lined the zona pellucida (18 embryos)
or was contracted (6 embryos) (plate 1), the zona pellucida
often appearing split (5 out of 6); the embryo was small
(approximately 125 μ diameter) but it might reach 225 μ when
the blastocoele was large. Embryonic cell number varied
between 64 and 128. More than half of the cells composed the
embryonic inner cell mass and the percentage of pyknotic nuclei
was often high (> 40%).

c) Embryos not developing to blastocyst stage (Class 3)

One hundred and eighteen embryos (34.1%) had not reached
the blastocyst stage. They often appeared disorganised inside
the zona pellucida (plate 3). The embryonic inner cell mass
was not visible and there was a larger space betweeen the zona
pellucida and the embryo itself. Well defined or fragmented
blastomeres were often present. Histological study furnished
few data on this class of embryos because the nuclear material
was not well stained.

2 - Embryo viability

a) Hatched blastocysts
Development after transfer

TABLE 3

PREGNANCY RATE AND IMPLANTATION RATE AFTER CERVICAL TRANSFER OF SHED
BLASTOCYSTS

		Uncultured blastocysts	Blastocysts after 24 h culture
Number of embryos transferred		34	26*
(Number of donors)		(8)	(8)
Number of pregnant recipients/ total		11/17	7/13
Number of implanted foetuses/ total (%)		20/34 (58.8)	7/26 (26.9)
Site of implantation	ipsilateral horn	10/17	6/13
	contralateral horn	10/17	1/13

* Two had not shed the ZP before culture.

Thirty four blastocysts transplanted into both uterine
horns of 17 recipients gave 20 foetuses (Table 3). 58.8% of

the embryos developed after transfer; 9 of the 11 pregnant
recipients carried twins. There was no significant difference
in development between the ipsilateral horn and that contralateral
to the corpus luteum; neither was there any relation between
embryo size and ability to implant.

Morphological modifications in culture and viability

More than 70% of the embryos developed after 24 h of
culture (Table 1). Their diameter increased by a mean of 50%:
405 μ ± 173 μ after culture as against 263 μ ± 195 μ before. The
mean number of cells was significantly higher ($\alpha = 0.05$): 641
± 100 after culture (n = 20 embryos) as against 505 ± 57 before
(n ± 32 embryos). However, after culture there was a change
in the size/cell number relationship (Figure 1). For the same
number of cells, the embryos cultured for 24 h had a larger
diameter.

TABLE 1

CLASSIFICATION OF DAY 10 EMBRYOS AND DEVELOPMENT AFTER CULTURE

	Total	Class 1	Class 2	Class 3
Number of examined embryos (%)	346 (100)	204 (58.9)	24 (6.9)	118 (34.1)
Number of cultured embryos	194	129	15	50
Number of embryos developed after culture (%)	96 (49.4)	91 (70.5)	4 (26.6)	1 (2.0)

If development after the transfer of cultured or uncultured
embryos is compared (Table 3), it is shown that the percentage
of pregnant animals is comparable (53% vs 64%), but that implantation rate is clearly lower when cultured embryos are
transferred (26.9% vs 58.8%). This difference is explained by
the fact that there is practically no development in the
contralateral horn (1/13), while it is about the same as direct

TABLE 4

GLUCOSE CONSUMPTION AND LACTATE PRODUCTION BY BLASTOCYSTS CULTURED FOR 24 H

Culture Group	No embryos	Test used	Glucose consumption µg/24 h/ embryo	Lactate Production µg/24 h/ embryo
A	3	<u>Culture</u>: 48% increase in size 190 µ→ 282 µ <u>Transfer</u>: 2 embryos→ 1 foetus	18.0	9.1
B	2	<u>Culture</u>: 89% increase in size 433 µ→ 821µ <u>Transfer</u>: 2 embryos→ 1 foetus	13.1	8.1
C	2	<u>Culture</u>: 78% increase in size 412 µ→ 736µ <u>Transfer</u>: 2 embryos→ 0 foetus	0	4.1
D	3	<u>Culture</u>: 64 increase in size 175 µ→ 287µ <u>Transfer</u>: 2 embryos→ 0 foetus	2.1	0
E	3	<u>Culture</u>: No increase in size	0	2.3
F	1	<u>Culture</u>: No increase in size Histology: degenerate aspect - several pyknoses	0	8.2

transfer in the ipsilateral horn (6/13).

Biochemical changes and viability

Glucose consumption and lactate production of blastocysts cultured for 24 h is shown in Table 4. Glucose consumption rate was high (cultures A and B) when embryonic viability, confirmed by development after transfer, was maintained in culture; more than $10^{-}\mu\dot{g}$ glucose per embryo were consumed in 24 h. This was zero (cultures E and F) when the embryos did not develop in culture and presented degeneration after fixation. This consumption was zero or low (2.1 μg/embryo/24 h) when, in spite of size increase during culture, the embryos did not develop after transfer (cultures C and D).

Some lactate is produced whatever changes occur in culture and even in the absence óf glucose consumption. When this consumption is high (cultures A and B) lactate production is augmented. Dehydrogenases characterisation is given in Table 5. Dehydrogenases were not found in the culture media of un-developed embryos, when the embryos remained morphologically intact. In the culture media of growing embryos, we found several released dehydrogenases with the exception of LDH and MDH in small blastocysts and LDH and ICDH in large blastocysts.

B - Unhatched blastocysts

Out of fifteen blastocysts cultured, only four (26%) developed by shedding the zona pellucida. After transfer of two of these shed embryos, we obtained one foetus, 2 others had 122 and 247 cells, respectively, and contained only 5% of pyknotic cells. On the other hand, in embryos which had not hatched from the zona pellucida in culture, 50% were pyknotic. We found several dehydrogenases in the culture media of these unhatched, and probably degenerate, embryos (Table 5), including LDH, while this enzyme did not appear when the blastocysts developed in culture.

C - Embryos not reaching blastocyst stage

Out of fifty embryos cultured for 48 h, only one gave a

TABLE 5

DEHYDROGENASES RELEASED IN THE MEDIUM BY BLASTOCYSTS CULTURED FOR 24 H

Evolution of blastocyst size during culture		Enzymes						
		G6PDH	Glycero-Phosphate DH	ICDH	LDH	MDH	6 Phospho-gluconate DH	Succino DH
No evolution	Degenerate blastocysts into ZP (n = 4)	+	+	+	+	+	+	+
	Small blastocysts (n = 6) ϕ = 168 μ $\pm$ 16	O	O	O	O	O	O	O
size increase	Small blastocysts (n = 17) ϕ = 200μ $\pm$ 17 → 329μ $\pm$ 46	+	+	+	O	O	+	+
	Large blastocysts (n = 15) ϕ = 360μ $\pm$ 37 → 460μ $\pm$ 62	+	+	O	O	+	+	+

n = 42 blastocysts

blastocyst hatched from the zona pellucida. The others evolved
little in culture; they became dark and the whole cell mass
seemed to be disorganised. These embryos were degenerate.

Endocrine profile of donors and quality of superovulation

The results of the hormone levels obtained in 11 animals
studied are summarised in Table 6. They confirm results
previously obtained on FFPN animals (Saumande, 1977):

Maximum oestradiol-17β level was correlated with ovulation
number (r = 0.75, P < 0.02)

In some donors oestradiol-17β level increases after
superovulation, especially in animals with large follicles

at Day 10.

At oestrus, progesterone levels are abnormally high (0.7 to 1.4 ng/ml as against 0.15 $\pm$ 0.11 in the untreated animal, Yenikoye, personal communication).

After ovulation, progesterone levels are correlated with ovulation number (r = 0.87), but it must be remembered that progesterone is produced by the large follicles present on the ovary at the time of collection (correlation of number of corpora lutea + number of follicles with progesterone level reaches r = 0.91).

Animal No. 697, one of the two which did not respond to the treatment, cannot be considered as insensitive to PMSG because its oestradiol-17β level shows strong follicular stimulation and its high progesterone level at oestrus indicated that luteolysis had not occurred. There seemed to be no qualitative or quantitative parameter of plasma steroids correlated with the percentage of embryos recovered or the percentage of good quality embryos.

DISCUSSION

We used two morphological criteria to classify Day 10 embryos: the presence or absence of the zona pellucida and the presence or absence of an embryonic inner cell mass distinct from the trophoblast. These two criteria are simple and permit rapid distribution of embryos into 3 classes of very different viability. However, these morphological criteria are in-sufficient; all blastocysts hatched from the zona pellucida are not viable and implantation ability cannot be correlated with blastocyst size at the time of transfer.

The viability of embryos can be determined by culture. More than 70% of the embryos develop in vitro and 26% implant after transfer. This low implantation rate is due to poor development in the contralateral horn; local interaction between embryo and uterus plays a part. After bilateral transfer of uncultured embryos (Sreenan, 1976), there was a

TABLE 6

SUPEROVULATION AND STEROID LEVELS

Animal no	No CL	No. follicles > 5 mm	Embryos collected No.	(%)	Class 1 embryos No.	(%)	Max (E2 17β) pg/ml	Post * Ovulatory EC 17β Secretion	Pg level At oestrus ng/ml	Pg level ** on day 7
611	24	3	19	(79)	13	(68)	55	137.4	0.75	81.0
662	27	7	16	(59)	14	(87)	152	169.4	1.07	133.9
639	12	4	8	(66)	4	(50)	44	141.4	1.40	33.0
614	13	8	13	(100)	12	(92)	72	373.3	1.30	92.5
685	35	1	25	(71)	0	(0)	126	199.6	1.11	143.5
683	24	3	11	(45)	7	(63)	61	49.8	0.91	72.3
649	17	0	15	(88)	9	(60)	38	56.2	0.84	51.6
675	17	3	16	(94)	15	(93)	77	135.8	0.69	82.2
691	19	7	12	(63)	9	(75)	54	264.1	0.96	61.9
697	1	10	0	(0)	-	-	105.8	589.7	4.54	9.84
700	1	1	1	(100)	0	(0)	20.4	43.7	1.32	5.10

* Arbitrary unit of surface under E2 - 17β curve between Day 1 and Day 5

** Mean of two samples

65% survival in the ipsilateral horn in the case of single pregnancy. Without implicating embryonic viability after culture, the low implantation rate in the contralateral horn may be due to retarded development which accentuates the difference between the two horns. After 24 h of culture, the change in the relation between embryo size and cell number seems to express this retarded development. The number of cells of cultured embryos is lower than that of uncultured embryos of the same size. However, the cell number of cultured embryos increases significantly in 24 h (+ 27%). The osmotic pressure of the B2 medium is slightly lower than that of cow blood serum; using a low osmotic pressure medium permits better development of the cattle embryo at the morula stage (Bowen et al., 1975). The blastocyst may swell in culture which would also explain the modification in the embryo size, cell number relationship.

Embryo viability is related to marked glucose consumption in the medium. Measurements carried out at later stages (Day 13, unpublished data) show increased glucose utilisation of more than 200 µg/embryo/24 h. By analogy with the rabbit egg (Brinster, 1968), glucose metabolism accelerates when the blastocyst expands rapidly. When embryonic development is limited, glucose consumption is zero or low. However, a small quantity of lactate is formed and found in the medium. This metabolite could arise from alanine transamination with reduction of the pyruvate formed, or from the reduction of pyruvate present in the culture medium. Glucose consumption is thus more interesting and significant than lactate formation. At Day 10, the embryonic metabolism thus seems due to anaerobic predominance. LDH is not released into the medium when the blastocyst develops in culture, while it is found in degenerate blastocysts. It would be actively used by the embryo which synthesises it early (Wright et al., 1976). LDH loss in the medium would be a sign of degeneration. The presence of Krebs cycle dehydrogenases, revealed by their release in the medium, indicate that the embyro could use the aerobic pathway of energy metabolism. Shunt's pathway of monophosphate hexoses may also be utilised. As in the case of mouse and rabbit (Wales, 1975), the appearance of new metabolic pathways from Day 10 would permit rapid development of the cow embryo, which is characterised by elongation.

The physiological significance of dehydrogenase release into the medium and of large-sized molecules such as LCDH (PM >200,000) is unknown. It is not known either if such a release occurs in vivo; it may be related to embryonic swelling in culture. However, it does not prevent further development of the embryo after transfer.

Variations in plasma oestradiol-17β and progesterone levels of donors are a quantitative expression of superovulation but do not explain or permit prediction of variations in embryo quality. Perhaps in our experiment, samples were not taken frequently enough (twice daily) to determine the changes in hormonal balance with might occur at precise times during the evolution of follicles stimulated by PMSG or during embryonic development.

CONCLUSION

Morphological, histological and metabolic criteria were used to determine the quality of cattle embryos produced by superovulation. After collection and cervical transfer at Day 10, implantation rate of hatched embryos was high.

Metabolic activity tests may be developed for the rapid prediction of embryo quality at this stage of evolution before transfer. Measurement of glucose consumption of the embryo in vitro seems to be a more interesting criterion than lactate production.

416

REFERENCES

Brinster, R L., 1968. Carbon dioxide production from glucose by the preimplantation rabbit embryo. Exp. Cell. Pres., 51, 330-334.

Bowen, R.A., Hasler, J.F., Seidel, G.E., 1975. In vitro development of
bovine embryos in chemically defined media. Proc. 88th Ann. Res. Conf.
Colorado State Univ., Abstr. 171.

du Mesnil du Buisson, F., Renard, J.-P., Levasseur, M.-C., 1977. Embryo
transfer by surgical methods. In 'Embryo transfer in farm animals'.
K.J. Betteridge Ed., Canada, Department of Agriculture, monograph
16, 24-26.

Elsden, R.P., 1977. Embryo transfer by surgical methods. In 'Embryo
transfer in farm animals'. K.J. Betteridge Ed., Canada. Department
of Agriculture, monograph 16, 27-28.

Renard, J.-P., Heyman, Y., 1977. Quantitative and qualitative aspects of
superovulation for embryo transfer in the bovine. In press.

Renard, J.-P., Heyman, Y., du Mesnil du Buisson, F., 1977. Unilateral and
bilateral cervical transfer of bovine embryos at the blastocyst stage.
Theriogenology, 7 (4), 189.

Saumande, J., 1977. Relationship between ovarian stimulation by PMSG and
steroid secretion. In 'Control of Reproduction in the cow', Galway,
Sept. 27-28-29th.

Seidel, G.E., 1977. Short term maintenance and culture of embryos. In
'Embryo transfer in farm animals'. K.J. Betteridge Ed., Canada,
Department of Agriculture, monograph 16, 20-24.

Shea, B.F., Hines, D.J., Lightfoot, D.E., Ollis, G.W., Olson, S.M., 1976.
The transfer of bovine embryos. Seminar on egg transfer in cattle,
Cambridge 10-12 Dec. 1975. 145-158.

Sreenan, J.M., 1976. Egg transfer in the cow: effect of site of transfer.
Proc. 8th Int. Congr. Anim. Reprod. A.I. Krakow, 3, 269-272.

Testart, J., Kann, G., Saumande, J., Thibier, M., 1977. Estradiol-17β,
progesterone, FSH and LH in prepubertal calves induced to superovulate.
J. Reprod. Fert. (accepted for publication).

Thibier, M., Saumande, J., 1975. Oestradiol-17β, progesterone and 17α
hydroxy-progesterone concentrations in jugular venous plasma in cows
prior to and during oestrus. J. Steroid. Biochem., 6, 1433-1437.

Trounson, A.O., Willadsen, S., Rowson, L.E.A., 1976. The influence of
in vitro culture and cooling on the survival and development of cow
embryos. J. Reprod. Fert., 47, 367-370.

Wales, R.G. 1975. Maturation of mammalian embryos: biochemical aspects. Biol. of Reprod., 12, 66-81.

Wright, R.W., Cupps, P.T., Watson, J.G., Chaykin, S. 1976. Lactate dehydrogenase isoenzyme patterns in oocytes, unfertilised eggs and embryos from mice and cattle. J. Anim. Sci., vol. 43, 3, 614-616.

IN VITRO EXCHANGE BETWEEN THE FOLLICLE AND ITS CULTURE MEDIUM

Y. Menezo[*], M. Gerard[**], D. Szöllösi[**], C. Thibault[***]

[*] Laboratoire de Biologie, INSA 69621 Villeurbanne, France
[**] Station de Physiologie animale, INRA 78350 Jouy en Josas, France
[***] Université Pierre et Marie Curie 75005 Paris

Meiosis may resume in vitro in intrafollicular oocytes, depending upon whether or not gonadotrophins are added in the culture medium (rat : Tsafriri et al., 1972; rabbit : Thibault and Gerard, 1973; monkey and calf : Thibault, Gerard and Menezo, 1976; sheep : Moor and Trounson, 1977). Except for the mouse (Mukherjee, 1972), it is now well established that only these follicle-enclosed mature oocytes are able to support normal embryonic development after fertilisation (rabbit : Thibault, et al., 1973; sheep : Moor and Trounson, 1977). However, these results were obtained with large antral follicles, somewhat scarce in the ovary.

In order to promote the long-term culture necessary for the maturation of middle-sized follicles, we developed a continuous-flow superfusion technique simulating the physiological situation in the ovary (Menezo et al., 1976). As in vivo, the whole growing follicle is entirely encompassed in a vascular bed and metabolic exchange occurs all around the follicle.

This study presents data on metabolic activity and compares the continuous-flow technique with that of organ culture grid. Ultrastructural aspects of follicle permeability, follicular structure have been studied.

MATERIAL AND METHODS

Calf and macacque follicles were collected according to our current technique (Thibault et al., 1975[1]). They were cultured either in organ grid conditions or employing the superfusion technique. The gas phase was 57% O2 - 5% CO2 - 38% N2.

The culture medium, based upon cow follicular fluid, was
entirely synthetic (Menezo, 1976)[*]. In some experiments Ficoll
70 (Pharmacia) was substituted for BSA. Different gonadotrophins
were always added (PMS and PMS + HCG). For biochemical studies,
we analysed the medium before and after 50 h culture. Free
amino acid analyses were performed using ion-exchange chromatography on the Optica aminolyser. Glucose was determined using
the hexokinase method and lactate using LDH, after precipitation
with perchloric acid. Electrophoresis was performed for non-
enzymatic proteins using the Pharmacia PAA 4/30 system allowing
separation according to molecular weight (MW). Hydrolytic
enzymes were detected with the colorimetric method of Monget
(1975) and dehydrogenases with a modified method of Altmann (1969).

RESULTS AND DISCUSSION

After 52 h of culture, follicle diameter increased by 5
to 10% when we used superfusion.

Free amino acids (Table 1)

The uptake of free amino acids is 3 to 39 times higher
with superfusion than on grid. The high rate of glycine and
glutamic + glutamine consumption is probably related to their
possible utilisation for the synthesis of nuclear material.
This metabolic increase leads neither to degeneration (Thibault,
Gerard and Menezo, 1975b) nor to decreased differentiation as
sometimes observed.

Glucose and lactate (Tables 2 and 3)

Glucose uptake by calf follicles is about 60 to 80 µg/h.
Lactate accumulation is from 12 to 24 µg/h. However, glucose
uptake and lactate accumulation are statistically unrelated to
the oocyte maturation stage reached and to the gonadotrophin
environment in the culture medium (PMS or PMS + HCG). Similar
observations were made for the rat by Tsafriri et al. (1976[1])
and Hillensjö (1976).

* API System - 38390 Montalieu-Vercieu

TABLE 1

COMPARATIVE STUDY OF FREE AA CONSUMPTION (NMOLES/H/FOLLICLE) 6 FOLLICLES
PER TECHNIQUE

Amino-acids	Superfusion (Fl) (M)	Organ Culture (Org)	Fl/Org
Asp + AsN	29	3.5	8.3
Thre	21	1.7	12
Ser	19	0.5	38.8
Glu-GIN	248	17.7	14
Gly	597	33.8	17.7
Ala	53	4.8	11
Val	22	7.2	3.1
Met	4	1.3	2.9
Ileu	16	2.3	6.8
Leu	18	3.2	4.8
Tyr	15	0.8	18.3
Phe Ala	13	1.1	11.6
Lys	26	3.0	8.7
Hist	14	1.2	11.9
Arg	22	3.1	7.2

TABLE 2

GLUCOSE UPTAKE AND LACTATE ACCUMULATION ACCORDING TO MATURATION STAGE
REACHED

Maturation stage reached	No	Glucose uptake $\mu g/h$ M - (σ)	No	Lactate accum. $\mu g/h$ M - (σ)
D or R.M.	6	61.2 (13.3)	6	22.1 (5.2)
M_1	8	77.6 (13.8)	8	24.3 (4.2)
M_2	6	82 (13.3)	5	11.9 (3.8)

TABLE 3

SUGAR METABOLISM ACCORDING TO HORMONAL TREATMENT IN CULTURE (50 H CULTURE TIME)

Hormonal treatment in culture	No	Glucose uptake (µg/h) M - (♂)	No	Lactate accum. (µg/h) M - (♂)
PMS + HCG	12	90.3 (12)	12	19.1 (3.1)
PMS	12	73 (13)	12	21.4 (4.2)

TABLE 4

ENZYMES FOUND IN CONTROL FOLLICULAR FLUID AND IN THE CULTURE MEDIUM AFTER 50 HOURS CULTURE (CALF)

	Before culture (follicular fluid)	After culture (medium)
Phosphatase AK	+ (1)	+
Phosphatase AC	+ (1)	+
Leucyl aryl amidase	+	+
α glucosidase	+	+
Lipase (Myristate)	O	+
β N.A. glucosaminidase	+	+
β glucuronidase	+	O
Glucose 6 PH D.H.	+ (1)	+
LDH	+	+
Isocitrate DH	+ (1)	O
Malate DH	O	+

Macromolecules (Table 4)

Electrophoresis : Proteins migrate towards the medium from healthy follicles, whatever maturation stage has been reached. Main proteins transferred from calf follicle have a MW between 80 000 and 100 000, although lower MW proteins also shift. A similar transfer also occurs with macacque follicles, but proteins heavier than 150 000 are also transferred.

Culture in media using Ficoll instead of BSA showed that

follicular albumin (MW 65 000) also migrates from the follicle.
Comparison of electrophoretic mobilities and ouchterlony
immunodiffusion indicates that the proteins found in the culture
medium seem to come from the follicular fluid and are not neo-
synthesised. Our observations may be related to the work of
Tsafriri et al. (1976_2) on the inhibitory effect of follicular
fluid on meiosis resumption. Before ovulation in healthy
follicles, the inhibitor is no longer synthesised and may leave
the follicular fluid by the transfer mechanism described.

Malic dehydrogenase, not present in the follicle at the
beginning of culture, appears in the medium. Except for the
cumulus and the oocyte (Dekel et al., 1976), the presence of
this enzyme could confirm that gonadotrophins enhance the
aerobic pathway. However, a simple release from the theca
cells is not completely excluded.

Macromolecular permeability was severely impaired when
cultured calf follicles became atretic.

Enzymes
As expected some enzymes pass through the follicular wall;
this is the case of alkaline phosphatase and non-specific
esterases. However, the same MW dependence is also observed
since β glucuronidase (MW 210 000) was not detected in culture
media. Isocitrate dehydrogenase (MW 300 000) shows a similar
scheme.

CONCLUSIONS

Continuous-flow superfusion technique was an improvement
over the organ culture grid one. There was an increase of mitotic
index and of follicle volume related to augmented amino acid
consumption. Corona cell reaction was initiated.

However, the percentage of metaphase II at the end of
culture was not controlled. So, it seems for now that the
maturation process cannot be related to a simple metabolic test

or to steroid synthesis (Lieberman et al., 1976). Macromolecular
transfer from the follicle, impaired by atresia, could be a
regulation effector of ovarian follicle population in vivo
(Peters, 1973). Very little is known about normal transfer
from the follicle, either in vivo or in vitro.

A signal of complete cytoplasmic and nuclear maturation,
which could be detected by metabolite(s) analysis in the
culture medium, is yet to be found.

ULTRASTRUCTURAL ASPECTS

Employing electron-dense markers, such as horseradish
peroxidase and lanthanum nitrate on dissected follicles of pig,
macacque and calf at various stages of preovulatory development
and with various treatments we demonstrated that these markers
penetrate throughout the entire follicle. In the stages and
conditions studied, therefore, a blood-follicle barrier does
not appear to exist.

Gap junctions exist between the cells of the corona radiata
and the oocyte in antral follicles. For the clarification of
the role these junctions may play, rabbits were mated and the
oocytes were removed 2 to 7 h later from the preovulatory follicles;
calf oocytes were also sampled at various times in an in vitro
system in which they undergo maturation regularly. The junctions
can be recognised until 5 - 5½h after mating or 24 h in culture
after which they are interrupted. Even though cortical granules
are present in the cytoplasm throughout the corresponding period
of time only a few migrate peripherally to be placed at
irregular intervals along the cell membrane. Large clusters
of cortical granules are often found a few microns from the
cell membrane of the oocyte. Five and a half hours after
mating and subsequent to elimination of the gap junctions, the
cortical granules become more regularly distributed along the
cell membrane similar to their usual disposition at time of
ovulation. Only subsequent to their final migration could
they participate in the cortical reaction at the time of sperm

penetration. The maturation of the oocyte cortex, thus apparently depends on the interruption of the cell to cell contacts between corona cells and oocyte.

REFERENCES

Altmann, F.P., 1969. The use of eight different tetrazolium salts for the
    quantitative study of pentose shunt dehydrogenation. Histochemic.,
    19, 363-374.

Dekel, N., Hultborn, R., Hillensjö T., Hamberger L., Kraicer, P., 1976.
    Effect of luteinising hormone on respiration of the preovulatory
    cumulus cophorus of the rat. Endocrinology, 98, 498-504.

Hillensjö T., 1976. Oocyte maturation and glycolysis in isolated pre-
    ovulatory follicles of PMS injected immature rats. Acta endocrinol.,
    82. 809 - 830.

Lieberman, M.E., Tsafriri, A., Bauminger, S., Collins, W.P., Ahren, K., and
    Lindner, H.R. 1976. Oocytic meiosis in cultured rat follicles during
    inhibition of steroidogenesis. Acta endocrinol., 83, 151-157.

Menezo, Y., 1976. Milieu synthétique pour la survie et la maturation des
    gamètes et pour la culture de l'oeuf fécondé. C.R. Acad. Sci., 282.
    1967-1970.

Menezo, Y., Gerard, M., Thibault, C., 1976. Culture du follicle de
    Graaf bovin dans un système à courant liquide et gazeux continu. C.R.
    Acad. Sci. D., 283, 1309-1311.

Monget, D., 1975. Différences d'activités enzymatiques entre deux lignées
    cellulaires d'insectes, *Antherea eucalypti* et *Malocosoma disstria*
    (Lepidoptera). C.R. Acad. Sci. D., 281, 651-654.

Moor, R.M., Trounson, A.O., 1977. Hormonal and follicular factors affecting
    maturation of sheep oocytes in vitro and their subsequent developmental
    capacity. J. Reprod. Fert., 49, 101-109.

Mukherjee, A.B., 1972. Normal progeny from fertilisation in vitro of
    mouse oocytes matured in vitro and spermatozoa capacitated in vitro.
    Nature, 237, 397-398.

Peters, H., 1973. Development and atresia of follicles in the immature
    mouse. Ann. Biol. anim. Bioch. Biophys., 13, 167-175.

Thibault, C., Gerard, M., 1973. Cytoplasmic and nuclear maturation of
    rabbit oocytes in vitro. Ann. Biol. anim. Bioch. Biophys. 13, 145-156.

Thibault, C., Gerard, M., Menezo, Y., 1975. Acquisition par l'ovocyte de
    lapine et de veau du facteur de décondensation du noyau du spermatozoide
    fécondant (MPGF). Ann. Biol. anim. Bioch. Biophys., 15, 705-714.

Thibault, C., Gerard, M., Menezo, Y., 1976. Nuclear and cytoplasmic aspects
    of mammalian oocyte maturation in vitro in relation to follicle size

and fertilisation in sperm action. Prog. reprod. Biol., 1, 233-240.
Hubinot P.A. Ed. Karger, Basel.

Tsafriri, A., Lindner, H.R., Zor, U., Lamprecht, S.A., 1972. In vitro
induction of meiotic division in follicle enclosed rat oocytes, by
LH, cyclic AMP and prostaglandin E2. J. Reprod. Fert., 31, 39-50.

Tsafriri, A., Lieberman, M.E., Amren, K., Lindner, H.R., 1976_1. Dissociation
between LH induced aerobic glycolysis and oocyte maturation in
cultured graafian follicles of the rat. Acta endocrinol., 81, 322-326.

Tsafriri, A., Pomerantz, S.H., Schanning, C.P., 1976_2. Inhibition of
oocyte maturation by porcine follicular fluid : partial characterisation
of the inhibitor. Biol. Reprod., 14, 511-516.

IN VITRO STORAGE OF CATTLE EMBRYOS

S.M. Willadsen, C. Polge and L.E.A. Rowson

A.R.C. Institute of Animal Physiology, Animal Research Station
307 Huntingdon Road, Cambridge CB3 OJQ, UK.

ABSTRACT

The survival of late morulae and early blastocysts from the cow after storage in PBS at room temperature and after deep-freezing in PBS containing 1.5 DMSO was tested by in vitro culture. Most embryos developed normally after 3 - 12 h (39/40) or 24 h (17/17) at room temperature. After 48 h at room temperature there was a considerable decline in survival (16/26).

Slow freezing to -33° or -36°C and transfer directly to - 196°C allowed the majority of embryos (42/68) to survive rapid thawing.

INTRODUCTION

The aim of egg storage in vitro is the maintenance of the embryo in a state of suspended animation from which it may be resuscitated after a short or long period to continue normal development either in vitro or in vivo. If the potential value of embryo transplantation to the cattle industry is to be fully realised it is necessary that more emphasis is placed on developing efficient and simple methods of embryo storage or on defining, standardising and improving those already in existence.

In this paper we wish to present some experimental data concerning short term and long term storage of late morulae and early blastocysts (Days $6\frac{1}{2}$ to $7\frac{1}{2}$), ie the stages of development which are most relevant in the context of present trends in bovine embryo transplantation.

The experiments to be described were carried out in continuation of earlier work on the same topic in this laboratory (Wilmut and Rowson, 1973; Wilmut et al., 1975; Willadsen et al., 1976_1; Trounson et al., 1976_1; Trounson et al., 1976_2; Willadsen et al., 1976_2; Willadsen, 1977; Willadsen et al., 1977_1; Willadsen et al., 1977_2).

MATERIALS AND METHODS

The embryos used in these experiments were late morulae in which compaction of the blastomeres was well advanced and early blastocysts, collected surgically or non-surgically from superovulated donors of various breeds on Days $6\frac{1}{2}$ to $7\frac{1}{2}$.

PBS (Whittingham, 1971) was used as a medium for collection and short term storage. For deep-freezing of embryos, Dimethyl-sulphoxide (DMSO) was added to PBS to a final concentration of 1.5M. For culture of embryos in vitro 20% sheep serum (SS, heated to 56°C for 30 min) was added to PBS.

Immediately after collection the embryos were placed in
10 ml PBS in covered glass cups and kept at room temperature
($18 - 22^{\circ}$C) until allocated to an experiment.

After experimental treatment the embryos were placed in
ampoules prepared from 5 x 0.5 cm test tubes, containing 0.25 -
0.3 ml culture medium. The ampoules were then sealed over a
gas burner and incubated at 37.5°C. After 20 to 24 h of
culture the embryos were examined as fresh specimens, and the
majority subsequently fixed in acetoethanol (1 : 3), and
stained with 1% Lacmoid in 45% aqueous acetic acid. Gross
morphology in the fresh stage and nuclear morphology after
staining were used as criteria of survival.

RESULTS

1.   Short term storage
     Three experiments were carried out on storage of embryos
at room temperature.

     a)   Embryos were stored in $\gtrsim$ 10 ml PBS in covered glasscups
     for 3 to 12 h and subsequently cultured for 20 to 24 h.
     In this experiment about half of the embryos were re-
     incubated at 37.5°C for an additional 24 h period after
     having been checked for survival.

     b)   Embryos were stored in a $\sim$3 ml PBS in stoppered test-
     tubes (5 x 1 cm) for 24 h and subsequently cultured for
     20 to 24 h at 37.5°C.

     c)   Embryos were stored in $\sim$3 ml PBS in stoppered test-
     tubes (5 x 1 cm) for 48 h and subsequently cultured for 20
     to 24 h at 37.5°C.

The results of experiments 1a, b and c are summarised in
Table 1.

2.   Deep-freezing
     Three experiments were carried out on deep-freezing of

TABLE 1

SHORT TERM STORAGE OF COW LATE MORULAE AND EARLY BLASTOCYSTS IN VITRO

		N	N. surviving	% surviving
A.	Storage 20°C 3-12 h.			
	Culture 37.5°C 20-24 h	40	39	97.5
	44-48 h	23	18	78.3
B.	Storage 20°C 24 h			
	Culture 37.5°C 20-24 h	17	17	100.0
C.	Storage 20°C 48 h			
	Culture 37.5°C 20-24 h	26	16	64.0

embryos. In all three, the embryos were stored for 3 to 6 h in PBS at room temperature. They were then exposed to increasing concentrations of DMSO (0.5M: 10 min, 1.0M: 10 min, 1.5M: 30 to 50 min) at room temperature. While being equilibrated with 1.5M DMSO 2 to 5 embryos were transferred to ampoules containing 0.25 ml DMSO in PBS (similar to those used for culture in vitro) which were then sealed. After the embryos had been in 1.5M DMSO for 30 - 50 min the ampoules were cooled at 1°C/min to -6 to -7°C at which point the cooling rate was changed to 0.3°C/min and crystallisation of the medium was induced by local cooling of the wall of the ampoule with a pair of forceps cooled in liquid nitrogen.

a) Ampoules were cooled at 0.3°C/min between -6 to -7°C and - 30°C. From -30° to -36°C the cooling rate was 0.1°C/min. From -36°C the ampoules were plunged directly into liquid nitrogen.

b) Ampoules were cooled at 0.3°C/min to -30°C, then at 0.1°C/ min to -33°C before being plunged directly into liquid nitrogen.

c) Ampoules were cooled at 0.3°C/min to -33°C, then at

0.1°C/ min to -36°C. From -36°C they were plunged directly into liquid nitrogen.

In all three experiments, the embryos were thawed rapidly after 30 min to a few days at liquid nitrogen temperature by agitating the ampoules in water at $\sim$25°C. Immediately after thawing, the embryos were transferred to freshly prepared 1.5M DMSO in PBS. DMSO was removed from the embryos in a step-wise fashion at room temperature (1.5M 5-10 min,/1.25M, 1.0M, 0.75M, 0.5M, 0.25M: 10 min each; PBS). After 15 min to half an hour in pure PBS the embryos were transferred to ampoules containing culture medium and incubated at 37.5°C for 20 to 24 h.

The results of experiments 2a, b and c are summarised in Table 2.

TABLE 2

DEEP-FREEZING OF COW LATE MORULAE AND EARLY BLASTOCYSTS

1.5M DMSO + Rapid thawing			
Freezing rate	N	N. surviving	% surviving
A. -6° to -30°C: 0.3°C/min -30° to -36°C: 0.1°C/min From -36°C directly into LN$_2$	27	15	55.6
B. -6° to -30°C: 0.3°C/min -30° to -33°C: 0.1°C/min From -36°C directly into LN$_2$	21	16	76.2
C. -6° to -33°C: 0.3°C/min -33° to -36°C: 0.1°C/min From -36°C directly into LN$_2$	19	12	63.2

DISCUSSION

There are two main reasons why we have chosen to use very late morulae and early blastocysts for our work. Firstly, these

stages of development are relatively easy to collect non-
surgically and presumably efficient methods for their non-
surgical transfer can be developed (Brand et al., 1976;
Trounson et al., 1977).

Secondly, whereas in cattle Day 5 morulae and earlier
embryos do not generally survive deep-freezing, Day 6½ to 7½
morulae and early blastocysts survive relatively well with a
variety of combinations of DMSO concentrating freezing and
thawing rates (Willadsen et al., 1976_3;Willadsen et al., 1977_1).
This is related to an increase in tolerance to cooling per se,
which occurs between Day 5 and Day 7 (Trounson et al., 1976_2).
Furthermore, it seems that Day 6½ to 7½ embryos are generally
relatively resistant to various unphysiological conditions
which may prevail in vitro.

Finally, the ability to blastulate in a late morula and
expansion or hatching in an early blastocyst provide reasonably
good criteria of survival. Although survival thus defined
does not necessarily imply normal viability, we have found
that in cattle there is a high degree of correlation between
in vitro survival and viability upon transfer (Willadsen et
al., 1976_3). However, by delaying collection until Day 6½ to
7½ in superovulated donors it is possible that the proportion
of normally developing embryos is decreased. There seems to
be a higher proportion of retarded and morphologically abnormal
embryos when collection is made at this stage rather than on
Day 3. Although this does not invalidate the results of the
present experiments, in which only embryos which were
considered to be normal were used, it does mean that different,
probably less good results are to be expected if similar
methods are applied in a commercial context where, presumably,
many embryos which we would consider to be of inferior quality
would be used.

Summary of results:
    The results of experiment 1a show that the simple culture
system used throughout these experiments allowed virtually

all freshly collected late morulae and early blastocysts to
continue their development for a 20 - 24 h period. When the
culture period was extended to 44 - 48 h a proportion of these
embryos began to show signs of degeneration. In most instances
the deterioration of embryo morphology coincided with the fully
expanded blastocyst stage. The survival rates are marginally
higher than those reported by Trounson et al., 1976_2 when foetal
calf serum was used instead of sheep serum.

In experiment 1b the results indicate that storage for 24 h
at room temperature does not reduce embryonic survival. The
gross morphology of these embryos was on the whole better than
that of embryos cultured for 44 to 48 h without prior storage.
This suggests that with the methods used, storage at room
temperature may be preferable to culture for maintaining
embryonic viability. Also, the survival rates were higher
than those obtained when embryos were cultured after storage
for 24 h at 0°C (Trounson et al., 1976_2).

In experiment 1c a substantial proportion of embryos
showed overt signs of degeneration after 48 h storage at room
temperature followed by 20 to 24 h of culture.

The results obtained in these experiments confirm earlier
observations made in this laboratory (Trounson et al., 1976_2;
Willadsen et al., 1976_3) that late morulae and early blastocysts
are relatively hardy organisms which will generally survive for
at least a day in simple in vitro systems. On the other hand,
it seems inadvisable to hold embryos in vitro for longer if
they are intended for transplantation.

The aim of experiment 2 was to investigate embryonic
survival rates with rapid thawing, which has generally been
found to yield less good results than slow thawing. The
theoretical basis for the freezing method used in these
experiments has been discussed elsewhere (Willadsen, 1977).

Assuming that the freezing medium, including the initial

DMSO concentration of 1.5M is fully adequate for the protection of the embryos during freezing and thawing, the central problem is twofold and concerns:

   a) the temperature from which the embryos should be placed directly at liquid nitrogen temperature, and

   b) the optimal freezing rate down to the former temperature.

Although the present experiments do not solve this problem, the results give important hints as to where and how the solution should be sought. They clearly demonstrate that fast thawing is compatible with high survival rates and indicate that under these circumstances the temperature range over which slow freezing is necessary is much more limited than anticipated from previous experiments. Survival has occasionally been obtained even in embryos transferred to liquid nitrogen from $-24^{\circ}C$.

There was no difference between the results obtained in 2b and c, although the method used in b yielded the most consistent results in individual trials. The efficiency of the present approach to cow embryo freezing compares very favourably with that of more elaborate methods (Willadsen et al., 1976[2]; Willadsen et al., 1977[2]). It should be pointed out that in these experiments as in the previous ones, the majority of surviving embryos were found to contain a proportion of degenerate cells when examined after fixation and staining by the end of the culture period. The proportion of degenerate cells an embryo will tolerate before its viability is impaired, is not known, but in the present experiments no embryo was considered to have survived unless it had expanded or re-expanded into a blastocyst with a visible embryonic disc by the end of the culture period.

No transfers have so far been carried out with embryos frozen and thawed with the present methods, but there is no reason to believe that such embryos are not viable. From a practical point of view methods such as these would greatly simplify embryo deep-freezing.

Further improvements and simplifications are no doubt possible. It is already known that DMSO may be added in just one step with no apparent damage to the embryo, and the cooling rate from room temperature to -6° to $-7^{\circ}C$ seems to be immaterial. It may also be possible to circumvent the necessity of diluting DMSO after thawing. This would be particularly important in the context of non-surgical transplantation. Calves have now been produced from non-surgically transferred frozen-thawed embryos. However, non-surgical techniques are clearly not yet satisfactory for transplantation of the stages of development which survive most readily with the present deep-freezing methods.

ACKNOWLEDGMENTS

The authors wish to thank members of the staff of the ARC Institute of Animal Physiology, Animal Research Station, Cambridge for skilled assistance. Financial support from the Milk Marketing Board of England and Wales and the Commission of the European Communities is also gratefully acknowledged.

REFERENCES

Brand, A., Taverne, M.A.M., van der Weyden, G.C., Aarts, M.H., Dieleman, S.J.,
    Fontijne, P., Drost, M. and de Bois, C.H.W. 1976. Non-surgical embryo
    transfer in cattle. I. Myometrial activity as a possible cause of
    embryo expulsion. In: Egg Transfer in Cattle (Commission of the
    European Communities, Luxembourg) 41-56.

Trounson, A.O., Willadsen, S.M., Rowson, L.E.A. and Newcomb, R. 1976_1.
    The storage of cow eggs at room temperature and at low temperatures.
    J. Reprod. Fert. $\underline{46}$, 173-178.

Trounson, A.O., Willadsen, S.M. and Rowson, L.E.A. 1976_2. The influence of
    in vitro culture and cooling on the survival and development of cow
    embryos. J. Reprod. Fert. $\underline{47}$, 367-370.

Trounson, A.O., Rowson, L.E.A. and Willadsen, S.M. 1977. Non-surgical
    transfer of cattle embryos. Vet. Rec. (in press).

Whittingham, D.G. 1971. Survival of mouse embryos after freezing and
    thawing. Nature, Lond. $\underline{233}$, 125-126.

Willadsen, S.M. 1977. Factors affecting the survival of sheep embryos
    during deep-freezing and thawing. In: Freezing of Mammalian Embryos
    (Ciba Found. Symp., Elsevier Exerpta Medica, Amsterdam) (in press).

Willadsen, S.M., Polge, C. and Rowson, L.E.A. 1977_2. The viability of deep-
    frozen cow embryos. J. Reprod. Fert. (in press)

Willadsen, S.M., Polge, C., Trounson, A.O. and Rowson, L.E.A. 1977_1.
    Transplantation of sheep and cattle embryos after storage at $-196^{\circ}C$.
    In: Freezing of Mammalian Embryos (Elsevier Exerpta Medica,
    Amsterdam) (in press).

Willadsen, S.M., Polge, C., Rowson, L.E.A. and Moor, R.M. 1976_1. Deep
    freezing of sheep embryos. J. Reprod. Fert. $\underline{46}$, 151-154.

Willadsen, S.M., Trounson, A.O., Polge, C., Rowson, L.E.A. and Newcomb, R.
    1976_2. Low temperature preservation of cow embryos. In: Egg Transfer
    in Cattle (Commission of the European Communities, Luxembourg) 117-124.

Willadsen, S.M., Trounson, A.O., Rowson, L.E.A., Polge, C. and Newcomb, R.
    1976_3. Preservation of cow embryos in vitro. Proc. VIIIth Int. Congr.
    Anim. Reprod. & A.I. Krakow $\underline{III}$, 329-332.

Wilmut, I and Rowson, L.E.A. 1973. Experiments on the low temperature
    preservation of cow embryos. Vet. Rec. $\underline{92}$, 686-690.

Wilmut, I., Polge, C. and Rowson, L.E.A. 1975. The effect on cow embryos of
    cooling to 20, 0, and $-196^{\circ}C$. J. Reprod. Fert. $\underline{45}$, 409-411.

DEEP FREEZING OF BOVINE EMBRYOS - A FIELD TRIAL

H. Lehn-Jensen

Institute for Animal Reproduction
Royal Veterinary and Agricultural University,
Copenhagen, Denmark.

Low temperature preservation of cattle embryos has been carried out at the Institute for Animal Reproduction using the method earlier described by Willadsen, Cambridge.

Day 6½ - Day 7 early expanding and expanding blastocysts were transferred to glass ampoules (50 x 5 mm) containing 0.2 ml 1.5 M DMSO in PBS after being exposed to increasing DMSO concentrations. The neck of the ampoule was closed by heat (melting) and placed in the chamber of the Union Carbide CRF-1 modified biological freezer. The freezer was equipped with an extra 2 line recorder for direct transcription of both sample and chamber temperature. A digital thermometer displayed the chamber temperature directly. Figure 1 shows the biological freezer. Cooler/freezing/thawing rates were as shown in Table 1.

TABLE 1

COOLING/FREEZING AND THAWING RATES FOR CATTLE BLASTOCYSTS (Willadsen)

	Temp. area $^{\circ}C$	$^{\circ}C$/min
Cooling/freezing	20 to -7	1
	-7 to -36	0.3
	-36 to -60	0.1
Thawing	-196 to -50	360
	-50 to -10	4
	-10 to 20	360

The embryos were stored between 25 and 30 days in liquid nitrogen at $-196^{\circ}C$.

Fig. 1 Biological freezer

Thawing was accomplished by using a dewar flask containing a mixture of liquid nitrogen and 99% alcohol at -50°C. A latex rubber tube was placed in the liquid nitrogen/alcohol bath (Figure 2). By leading warm water through the tube the temperature in the bath could be raised approximately 4°C/min up to -10°C. The ampoule was then transferred to 20°C warm water.

Thirty Day 6½ - Day 7 embryos (early expanded to 3/4 expanded blastocysts) were frozen and thawed after the above mentioned method. All embryos were recovered and judged carefully under stereomicroscope while being exposed to decreasing molarities of DMSO in PBS. Twenty-five of the embryos appeared normal having equilibrated at least 20 min in pure PBS solution. Five embryos showed obvious signs of degeneration (no re-expansion, or total destruction).

Seventeen embryos were transferred to 16 synchronised recipient heifers by flank incision and placed in the horn ipsilateral to the corpus luteum approximately 5 cm from the utero-tubal junction. A tentative pregnancy diagnosis was indicated by plasma progesterone concentrations 14 days later, ie at the expected time of heat, and confirmed by rectal palpation at approximately 30 days after transfer.

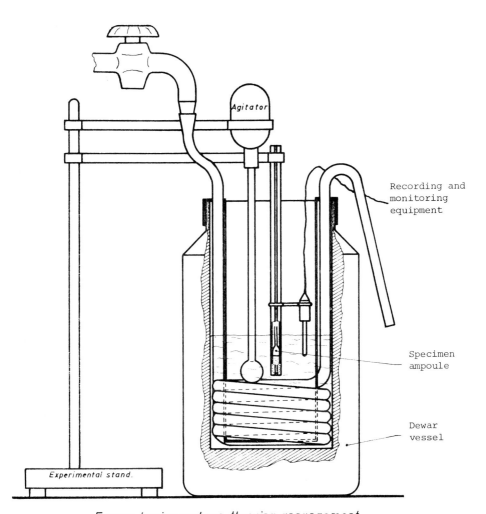

Agitator

Recording and
monitoring
equipment

Specimen
ampoule

Dewar
vessel

Experimental stand.

Frozen bovine embryo thawing arrangement.

Fig. 2 Thawing equipment
        (Drawing 1977 by Sv. Pedersen)

The overall pregnancy rate in this trial was 44% which is acceptable when compared to normal rates after surgical transfers of non frozen embryos which vary from 50 - 70% in our studies.

The whole operation was carried out on a big Danish dairy farm which at that stage had no special set-up for handling or laboratory facilities - so the description 'field trial' can very well be used.

EMBRYO SEXING WITH PARTICULAR REFERENCE TO CATTLE

W.C.D. Hare, E.L. Singh, K.J. Betteridge,
M.D. Eaglesome, G.C.B. Randall and D. Mitchell*

Animal Pathology Division, Health of Animals Branch,
Agriculture Canada, Animal Diseases Research Institute (E),
Box 11300, Station "H", Ottawa, Ontario, Canada, K2H 8P9

INTRODUCTION

There are two approaches at present to sexing embryos
during transfer; one is by the identification of sex chromatin,
the other by sex chromosomal analysis.

The sex chromatin method depends upon the identification
of a dark staining body, about 0.8 x 1.1 µ in size, lying
adjacent to the nuclear membrane. This body, known as a Barr
body, occurs when there are two X chromosomes and is seen in a
proportion of female cells. The method has been used successfully on trophoblast cells from 53/4-day-old rabbit embryos at
transfer (Gardner and Edwards, 1968).

Sexing by sex chromosomal analysis requires the preparation
of metaphase spreads from embryonic cells in mitosis. If
sufficient dividing cells are present, direct chromosome preparations can be made; if not, the cells have to be cultured
and harvested at the appropriate time for metaphase spreads.

The demonstration by Betteridge et al. (1976) that
acceptable pregnancy rates could be obtained with the transfer
of Day 14 to Day 16 bovine embryos, and the observation by
Rowson and Moor (1966) that Day 13 sheep embryos damaged before
transfer would develop into normal foetuses, led to the successful biopsy, sexing and subsequent transfer of Day 14 and Day 15
bovine embryos (Hare et al., 1976).

* Present address: Animal Pathology Division, Health of Animals Branch,
                   Agriculture Canada, Animal Diseases Research Institute,
                   Box 640, Lethbridge, Alberta, Canada. T1J 3Z4.

Since then the work has been expanded to include Days 11, 12 and 13 bovine embryos, and a micromanipulator has been used in the biopsy of smaller embryos. The methods used and the results obtained to date are described and discussed in this paper.

MATERIAL AND METHODS

One hundred embryos from 25 superovulated donors have been biopsied for sexing prior to transfer. The embryos have varied in age from Day 11 to Day 15* and in length from 0.20 mm to 85 mm.

The biopsy is done using strict aseptic procedures while maintaining the embryos as close as possible to 37°C. A piece of trophoblast, approximately 0.5 mm^2 in size, is removed taking care not to damage the inner cell mass. Initially, all biopsies were done with No. 5 extra fine forceps and $3\frac{1}{2}$" curved Vannas scissors under a dissecting microscope at X 12.5 to X 25 magnifications (Hare et al., 1976). Latterly, embryos less than approximately 9.0 mm in length have been biopsied at X 40 magnification using a Leitz micromanipulator and a Diavert inverted microscope (Wild-Leitz, Canada Ltd.). The biopsy instrument is made by glueing (Krazy Glue, Inc., Chicago, Ill) an 8 mm^2 piece of stainless steel razor blade to a 3" length of 20 gauge hardened steel wire which has been glued with epoxy (Lepages Ltd., Canada) into a 1" length of glass tubing, ID 0.05", which, in turn, has been glued with epoxy into a 2" length of glass tubing, ID 0.101" (Figure 1a). The angle that the cutting edge of the blade makes with the steel wire is critical and has to be that which allows

a) the wire to clear the glass ring on the microscope slide (described below) and

b) the entire length of the cutting edge to contact the surface of the slide evenly when the instrument is lowered to biopsy the embryo. The instrument is sterilised by

* Day 0 = first day of oestrus

gas prior to use.

The embryo is transferred in a small quantity of tissue culture medium (TCM) to a microscope slide with a glass ring (16 mm diameter x 3 mm height) fused to its surface (Fisher Scientific Company, Ltd., Canada) (Figure 1b). Looking through the microscope, the blade is positioned vertically at the desired point above the embryo and then lowered to cut the trophoblast in a guillotine-like movement (Figure 2). If this does not transect the embryo cleanly, the blade is moved forward and backward like a saw.

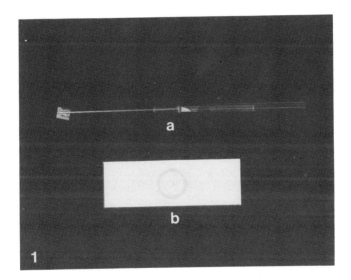

Fig.1 (a) Biopsy instrument consisting of an 8 mm^2 piece of stainless steel razor blade glued to a 3" length of 20 gauge hardened steel wire.

    (b) Microscope slide with glass ring (16 mm diameter x 3 mm height) fused to its surface.

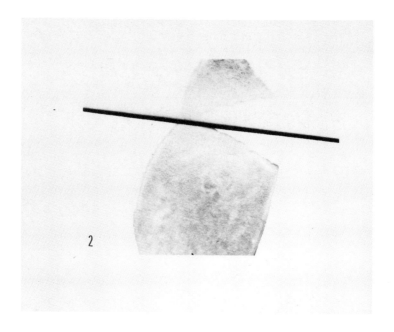

Fig.2. View through the microscope of an embryo transected by the
        biopsy instrument.

The biopsy fragment is transferred to a culture tube
containing TCM supplemented with 10% foetal bovine serum and
0.025 µ g/ml colcemid (Ciba Pharmaceutical Products, Inc.,
Montreal, P.Q.) and incubated for 1 h at 37°C.

The procedure, thereafter, is as described by Hare et al.
(1976) with the following modifications: 0.05% trypsin - 0.02%
EDTA (Grand Island Biological Company, Grand Island, NY) is
frequently used in place of 0.25% trypsin to disaggregate the
cells and, with very small fragments, a 0.5 ml or a 1.0 ml
centrifuge tube is used instead of a 2 ml tube.

Biopsied embryos are transferred to heifers, with previously
recorded oestrous cycles, in which natural or induced oestrus
has occurred synchronously with the donor ($\pm$ 1 day).

Pregnancy is diagnosed by rectal palpation at Days 38 to
42 and subsequently confirmed by rectal palpation, at surgery
or at slaughter. 'Delayed return to oestrus' is defined as
return to oestrus after Day 26, associated with elevated plasma
progesterone level at Day 21 and, in some cases, persistence
of corpus luteum determined by rectal palpation.

RESULTS

The results of sexing Days 11, 12, 13, 14 and 15 embryos
are given in Table 1 and the effect of embryo size on sexing
in Table 2.

TABLE 1

RESULTS OF SEXING DAYS 15, 14, 13, 12 AND 11 EMBRYOS

Embryo age (day)	Attempts *	Number sexed
15	40	22 (55.0%)
14	30	19 (63.3%)
13	20	13 (65.0%)
12	6	4 (66.0%)
11	4	0 (- -)
Total	100	58 (58.0%)

TABLE 2

EFFECT OF EMBRYO SIZE ON SEXING

Embryo size (length)	Attempts*	Number sexed
> 4.0 mm	32	22 (68.7%)
0.9 mm to 4.0 mm	55	35 (63.6%)
< 0.9 mm	11	1 (9.0%)

* Excluding two embryos for which there are no measurements.

In embryos 0.9 mm to 4.0 mm in length, $13/22$ (59.0%) were sexed
after being biopsied with forceps and scissors and $22/33$ (66.6%)

after biopsy using a micromanipulator.

The results of transferring 73 biopsied Days 13, 14 and
15 embryos to recipient heifers are given in Tables 3, 4 and 5.

TABLE 3

RESULTS OF SINGLE TRANSFERS OF BIOPSIED EMBRYOS

Embryo age (day)	Number of transfers	Pregnant recipients	Delayed return to oestrus >Day 26
15	4	2 (50.0%)	1 (25.0%)
14	28	10 (35.7%)	5 (17.8%)
13	7*	0 (---)	0 (---)
Total	39	12 (30.7%)	6 (15.3%)

* All non-surgical

TABLE 4

RESULTS OF TWIN TRANSFERS OF BIOPSIED EMBRYOS

Embryo age (day)	Number of transfers	Pregnant recipients	Delayed return to oestrus > Day 26
15	12+	6* (50.0%)	0 (--)
13	6+	2 (33.3%)	0 (--)
Total	18	8 (44.4%)	0 (--)

* One twin pregnancy

+ One embryo not biopsied

There is no correlation between the size of the biopsied
embryo transferred and embryo survival.

DISCUSSION

The embryo sexing success rate is virtually the same with
Days 12, 13 and 14 embryos, slightly reduced with Day 15 embryos
and very poor with Day 11 embryos. The size of the embryo does
not appear to affect the ability to sex, using the methods

TABLE 5

SURVIVAL OF BIOPSIED EMBRYOS TRANSFERRED TO RECIPIENT HEIFERS

Embryo age (day)	Number of transfers single	twin	Surgical (S) Non-surgical (NS)	Embryo survivals*
15	4	12+	S	9 (33.3%)
14	28	-	S	10 (35.7%)
	7	-	NS	0 (--)
13	-	6+	S	2 (18.1%)

* Normal development into a foetus

+ One embryo not biopsied

described, until the length is less than 0.9 mm when it is greatly reduced. All of the Day 11 embryos were less than 0.9 mm in length. Inability to sex is due to a) An absence of cells on the slide, which occurs for the most part working with fragments from embryos less than 0.9 mm in length; b) An absence of metaphase spreads in the presence of numerous cells, which occurs quite often with fragments from embryos over 15 mm in length; or (c) Very few and poor quality metaphase spreads.

A subjective observation is that success in sexing appears to be related more to a donor factor than to any other one factor: for example, from four donors, $24/_{27}$ (88.8%) Days 13, 14 and 15 embryos were sexed; whereas from another four, only $12/_{27}$ (44.4%) Days 13, 14 and 15 embryos were sexed. This is interesting in view of the observation that embryos from some donors survive better in vitro (or in vivo) than do those from others (Seidel, 1977).

The embryo survival rate following biopsy is difficult to evaluate because the transfers have been made under different experimental conditions and without strict controls. Some embryos have been flushed from the donor's uterus and held in phosphate buffered saline with bovine serum albumin, others in TCM 199; some were twin transfers, others single; some were surgical transfers, others non-surgical; and circumstances have never permitted embryos from each of a series of donors to be randomly divided into two groups, those that are biopsied and then transferred and those that are transferred without biopsy. However, there appears to be essentially no difference in the survival rate between Day 14 and Day 15 biopsied embryos (Table 5), whereas the pregnancy rate is higher with Day 15 than with Day 13 or 14 embryos (Tables 3, 4). However it should be recorded that the seven, Day 13 single transfers were first attempts with the non-surgical technique.

Although the success rate in sexing Days 12, 13, 14 and 15 embryos can reach acceptable levels, the embryo survival rate of biopsied embryos is still less than satisfactory. However, the potential value to embryo transfer, in both animal production and research, of being able to sex and eventually to combine sexing with low temperature preservation is such that the problems have to be resolved. Possible solutions are:

a) A satisfactory method of determining embryo viability prior to transfer

b) Aspirating fewer cells from late morula or early blastocyst stages, as is being done in rabbits by Rottman (1977)

c) The application of new techniques for sex determination.

ACKNOWLEDGMENTS

The skilled technical assistance of Mrs. J. Hierlihy and Messrs. R. Bériault, J. Shackleton, Y. Barbeau and G. Raby and dedicated animal care of Miss E. Cathcart and Mr. S. Shearer is greatly appreciated.

REFERENCES

Betteridge, K.J., Mitchell, D., Eaglesome, M.D. and Randall, G.C.B. 1976. Embryo transfer in cattle 10 - 17 days after oestrus. Proc. 8th Int. Congr. Anim. Reprod. A.I., Krakow. 3: 237-240.

Gardner, R.L. and Edwards, R.G. 1968. Control of the sex ratio at full term in the rabbit by transferring sexed blastocysts. Nature (Lond.) 218: 346-348.

Hare, W.C.D., Mitchell, D., Betteridge, K.J., Eaglesome, M.D., and Randall, G.C.B. 1976. Sexing two-week old bovine embryos by chromosomal analysis prior to surgical transfer: preliminary methods and results. Theriogenology 5: 243-253.

Rottman, O. 1977. Chromosomenpräparation aus einzelnen Blastomeren. 3rd. Symp. Cytogenet. Dom. Anim., Jouy-en-Josas, France.

Rowson, L.E.A. and Moor, R.M. 1966. Development of the sheep conceptus during the first fourteen days, J. Anat. 100: 777-785.

Seidel, G.E. 1977. Short-term maintenance and culture of embryos. In 'Embryo transfer in farm animals' ed. K.J. Betteridge, pp 20-24. Agriculture Canada Monograph No. 16.

MANIPULATION OF EGGS IN VITRO
ATTEMPT AT LONG TERM STORAGE OF OOCYTES AT LOW
TEMPERATURES WITHOUT FREEZING

L. Henriet, Estermans, B. Mottoul and J.H. Maison
Université catholique de Louvain, and
Institut Belge d'étude des hautes pressions,
Belgium

Prolonged preservation of living cells can be obtained by slowing down the function of the enzymes, so that cellular activity can be lowered or even suspended although the components remain fully viable.

The preservation of sperm presented this problem correctly for the first time in theory, the conditions may be schematised as follows:

in H_2O

Endogeneous factors	Energetic substances---------metabolites + energy (Motion)
	+ enzymes
Exogeneous factors	Initial pH------------------altered pH
	+ temperature
	± microbes

eg, the spermatozoon uses fructose to create pyruvic acid and hydrogen which is converted to lactic acid and energy. We can operate on the various terms of this equation, and that is what was done in artificial insemination.

1) First by supplying energy substrates, we change the data of the equation from right to left and the sperm return to the initial energy level.

2) The use of buffers lengthens the duration of the preservation by preventing the alteration of pH caused by the formation of metabolites.

3) By removing the water, an essential item in chemical reactions. However, experience has shown that lyophilisation impairs the survival of the spermatozoon.

4) We can also proceed, still from right to left,
    starting with the metabolites.

    Indeed if the pH is not constant, the spermatozoon stops
    its motion, even in the case of an excess of energy
    substrate it suffers from its own auto-antibiotic
    effect due to the lactic acid; the spermatozoon is not
    dead, but lives on without motion. We can accordingly
    make use of this auto-antibiotic property to stop the
    motion and extend the survival time. By adding lactate
    we stop the metabolism and on neutralising this acid,
    metabolism starts again (Devuyst and Henriet, 1962).
    Thus, we could maintain the living sperm, fully motion-
    less, for 7 days and investigate as we desired the
    recovery of its motility. However, after the fifth
    day, the survival was poor.

5) We can also work on the temperature. It is well known
    that enzymes need an optimal temperature for their
    function. To change that temperature is equivalent to
    blocking metabolism. As the usefulness of increasing
    the temperature is limited due to the sensitivity of
    the protein, we must inevitably go for lower temper-
    atures. For this approval to succeed, a means must be
    found of protecting the cells against the cooling
    process, ie egg yolk, which makes it possible to avoid
    the shock of cooling the sperm down from $37^{\circ}C$ to $4^{\circ}C$.

That is, in short, the way to preserve sperm, relying on the
slowing down of the enzymatic activity. Antibiotics are needed, of
course, to neutralise the exogeneous factors of the equation.
What is less well known is that energy can be saved by simple
mechanical means: we found that conservation in a cold gelatine
gel enabled sperm to regain motility after 23 days. Moreover, we
had to wait for 45 days before all the spermatozoa were dead in
the gel (Henriet, 1964). This proves that there are numerous
means of preservation. However slowing down the motility is not
enough to obtain an indefinite preservation; in the best diluents,
at 0 to $4^{\circ}C$, fertilising ability does not last longer than 3 days.

Indeed, preservation above $0^{\circ}C$ is focused upon the energy equation. But, alas, quite a lot of other chemical reactions are not inhibited in this way, so that we get damage very quickly.

It may be noted that it is the fertilisation power only that suffers, some spermatozoa are still motile after 3 weeks. However, is it not just this fertilisation power that proves the cellular integrity?

It is here that our colleague Polge has succeeded in long term storage, devising a congelation system that stops life and restores it at will.

For this we need to lower the temperatures down to below $0^{\circ}C$. This necessitates the use of an 'anti-freeze' as the living cells are essentially made up of water and crystallisation must be avoided.

This use of 'anti-freeze' does not always show advantages; the freezing impairs the fertilisation power of the sperm, and many types of 'anti-freeze' were tried out following the discovery of Smith and Polge (1949).

Are there other means of getting below zero C without the use of 'anti-freeze'?

In physics we know that water crystallises at $0^{\circ}C$ under atmospheric pressure. If we increase this pressure the crystallisation point is lowered. Therefore we planned a scheme to expose the cells to a sufficient pressure to prevent crystallisation. Difficulties arose immediately as the cohesion power of the atoms is heavy and very much increased by a fall in temperature. A diagram of pressure vs temperature of crystallisation is shown in Figure 1.

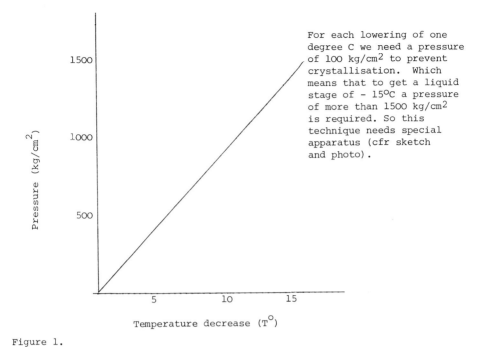

For each lowering of one
degree C we need a pressure
of 100 kg/cm^2 to prevent
crystallisation. Which
means that to get a liquid
stage of - 15°C a pressure
of more than 1500 kg/cm^2
is required. So this
technique needs special
apparatus (cfr sketch
and photo).

Figure 1.

PRESERVATION ATTEMPTS

a) Cells

Four different kinds of cells were enclosed together in
the pressure chamber - de-frosted sperm ready for insemination,
citrated blood, mature oocytes and non mature oocytes, both
aspirated from follicles in follicular fluid.

b) Containers

Cells of each type were enclosed in containers without
strong walls which allowed the pressure to be conveyed to
the fluid. As containers were used: Cassou straws, a syringe
with its plunger, a rubber bulb and metallic ointment tubes.
All are capable of withstanding high pressures. None burst
but our work did not go far enough to compare the required
qualities for cell survival.

c) Freezing stages and pressure variations

The cells must be placed under pressure before passing below the zero degree C and the apparatus has to be cooled afterwards. But as the compression chamber has a high caloric mass we had to cool it first to +1 degree C - so as to avoid a uselessly prolonged enzymatic activity. The second stage is obtained after reaching the necessary pressure (-15°C under a pressure of 1 600 kg/cm^2). When it is desired to bring the preservation process to an end the reverse procedure has to be followed: pressure maintenance up to more than 0°C.

Duration of our attempts: 5, 6 and 10 days.

RESULTS

The most convincing result is the return of the cells to a normal morphological appearance: neither the oocytes, nor the bloodcells (red + white), nor the spermatozoa looked impaired under the microscope.

All those cells showed resistance to pressures up to 1 600 kg/cm^2 without bursting. Survival is, however, something else. The spermatozoa did not move anymore - they were dead. This is the big advantage of working with this cell type: we see immediately where we are.

In the case of the oocytes we got no information on their survival because of lack of in vitro fertilisation.

The bloodcells were examined by a hematologist who after the usual staining procedure did not find any impairment of either the red or white cells.

DISCUSSION

Any interpretation of these results is premature. Improvment is essential and as we know we will have to avoid infections: If the oocytes and other cells can survive, then

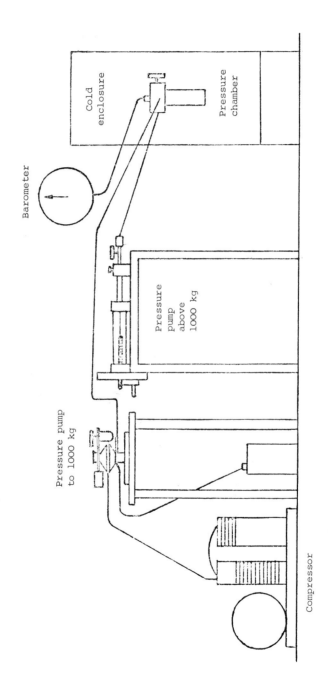

Fig. 2

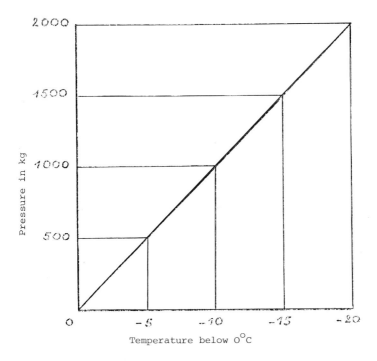

Fig. 3   Alteration of water crystallisation point with pressure

the microbes can survive also. We will need antibiotics, maybe antimycotics.

Luckily the use of filtered air flow reduces the necessity for these compounds to a minimum.

By freezing, we prevent the action of enzymes, as life is suspended under zero. But when the surroundings remain fluid, nothing is sure.

We may presume that the enzymes remain inactive, but the hypothesis of an altered activity cannot be eliminated. It would be advisable to get lower temperatures, but this requires enormous pressures and the pressure of 6 000 kg/cm^2 (needed to get to - 60°C) impairs the red bloodcells. (Dow and Matthews, 1939). There is thus, a limit to the process which must be determined.

All this describes the negative side of our work; the positive side is our hope that oocytes and embryos may be preserved by simple ways eliminating the effects of the 'anti-freeze' substances.

458

REFERENCES

Daw, R.B. and I.E. Matthews, 1939. The Disintegration of Erythrocytes
   and Denaturation of Hemoglobin by High Pressure - Philosophical
   Magazine, 7, XXVII, 637, May.

Devuyst, A. and Henriet, L. (1962). Le blocage du métabolisme du
   spermatozoïde par vie chimique. XXV° Anniversario della
   Fondazione del Istituto sperimentale italiano - 'Lazarro Spollanzam'.

Henriet, L. 1964. Rapport du Centre Expérimental d'insémination
   artificielle de l'Université de Louvain. 34-36.

Polge, C., Smith, A.U. and Parkes, A.S. 1949. Nature London 164, 666.

DISCUSSION

N. Rasbech *(Denmark)*

    I suggest we open this discussion with questions on Dr. Kane's paper.

I.Szollosi *(France)*

    I would like to consider the oocyte itself. The effects of many external factors have been investigated, as have the enzyme systems involved in oocyte metabolism. In general, in embryology books one reads that the mammalian oocyte and egg is alecithal and I hope that we have eradicated, or can eradicate, this impression for ever. This impression came from human and mouse oocytes and I think we should look at factors or chemical processes which may initiate the utilisation of products stored in the oocyte. The rabbit has, I think, yolk vesicles and we don't know their nature. Bovine oocytes have at least two distinctly different types of storage products, one is fat, (the nature of which we don't know), and the other is protein. So I think that some of the effort at studying ovum culture and metabolism should be related to the intracellular metabolism of the oocyte itself.

M.T. Kane *(Ireland)*

    I would like to make two comments on that. Firstly, I maybe did not have a chance to put this across in my talk, but I feel that generally the media we have available for culturing ova of the mammalian species are bad. It is only in a very few species, under certain circumstances, that you can get successful ovum culture. Secondly, I feel that the rabbit ovum, for instance, may be able to use endogenous lipid material because I have cultured one-cell rabbit eggs in Ficoll plus just a simple salt solution with no energy source and got something like two or three cleavages. Now, I haven't checked the cleavages by histology but they seem to me to be fairly normal which would suggest to me that maybe the rabbit ovum is capable in that situation of using endogenous sources of energy.

G.E. Seidel *(USA)*

In relating some of your work in one of the latter papers with one of your slides on the optimum osmolarity, do you think that under some circumstances we are seeing an artefact in that one gets expansion increases in size due to water uptake at these lower osmolarities and that maybe they are not as good media as maybe a higher osmolarity? Has anybody transferred embryos in those particular situations?

M.T. Kane

I don't think anyone has done studies on transferring embryos cultured in different osmolarities. I think the only way to study that kind of thing would be to take, say, a one-cell stage or two-cell stage, and strip the zona off and put the egg in media of different osmolarities to see whether one gets swelling of the vitelline membrane, i e, does the whole cell swell or shrink? That might be the best way to find out what was really the optimum osmolarity.

If I could make a comment on Dr. Willadsen's paper, when he talked about PBS + 20% foetal calf serum, I would talk about that in a different fashion; I would call it diluted foetal calf serum. Maybe if they diluted their foetal calf serum with 0.9% saline, they might do just as well.

S. Willadsen *(UK)*

Yes. I hope I didn't say that our culture medium is a perfect one or by any means defined. I think you are probably right: I think foetal calf serum + saline would do. I also think that these embryos will survive quite some time in saline alone. The other thing I would say though is that the late morula and the early blastocyst in the cow seems to be quite happy if you culture it just in PBS.

M.T. Kane

This is PBS without any protein in it?

S. Willadsen

   No, our PBS is the one which has bovine serum albumin.

M.T. Kane

   I see. Now, when you mention TC-199 in your studies, is
that TC-199 without any macromolecules in it?

S. Willadsen

   Yes, it would be.

M.T. Kane

   That is not at all a fair comparison of two media.

S. Willadsen

   Our comparison doesn't concern itself with whether the
medium is a clearly defined one or not; our comparison is really
related only to the problems that people present to us in terms
of how do we store our embryos. Now, if we could store our embryos in milk we would do that. We are not concerned with the
metabolism or anything. I agree with you that these things
should be looked at but they are not the things that we are concerned with.

M.T. Kane

   I feel if you do too many things empirically, you won't
get very far in the long run.

S. Willadsen

   I think we have proved that we have got quite a long way!

G.E. Seidel

   I would just like to make one point along those lines. Dr.
Dickmann in Kansas City wanted at one point to see how far he
could push rat embryos. He decided to collect them in distilled
water and if he got them collected and transferred within 90
seconds they were able to take it. So it depends on how long,
under what conditions, and so on, that one is doing these things.
I guess there is a place for everything and distilled water is

good enough if you are going to do it really quickly.

S. Willadsen

Could one approach the problem from a biological angle
to some extent. Our impression certainly is that if you take a
cow embryo and stick it into a rabbit oviduct, it will go on
quite nicely. If you stick a cow embryo into a sheep oviduct it
will also go on - this is up to the blastocyst stage. If you
stick a mouse embryo into a pig (which I have had the chance to
do) it will also go on. Now, it seems to me that although all
these things are oviducts, there must be a wide variation in the
sort of conditions you have there, so, would it be worth looking
at these oviducts to see what, in fact, they are providing and
approach the synthetic media from that angle? This, of course;
is what Robin Tervit did in the case of the cow and the sheep
embryo, but should this kind of approach be attempted?

M.T. Kane

I think that is reasonable and I said in my talk that that
is one of the ways to go. The problem is that there are so many
factors in the oviduct.

Y. Menezo *(France)*

I would like to make an observation on what Dr. Willadsen
was saying. He mentioned the oviducts of the sheep and rabbit,
for example, but the oviducal fluid of the rabbit and sheep are
very similar, eg, ionic composition, amino acids, pH, osmolarities
they are quite similar during the first three days following
ovulation.

However, what I would like to ask you is, when you say that
cow morulae or blastocysts are very happy in PBS, are you speaking of storage or culture?

S. Willadsen

I am speaking of the type of culture that we are doing
which means that if you stick a late cow morula which is just
about to blastulate into the PBS medium that we are using, and

you incubate at 38°C, when you come in next morning the chances
are that your late morula will be an early blastocyst. I don't
think I want to say more than that; I am not saying that it is
necessarily a normal one - I am just saying that it will maintain some sort of life and I would prefer to call that survival
and then later on we could try to assess the viability.

I. Szollosi

At this point I would like to raise another question.
Many people have used phosphate buffer as a fixative and they
have called it a biological buffer; that is why they use phosphate. We did a series of experiments on the survival of
protozoans just in a balanced saline buffer with phosphate and
at another point we took arsenate (which is in the same chemical
group) and they survived better in arsenate than in phosphate!
They were killed very rapidly in the phosphate buffer and so I
think the question of viability is one which you have to check
very closely.

M.T. Kane

Could I just ask Dr. Willadsen one very quick question.
You flushed the eggs, presumably, with PBS; do you, in fact, wash
the eggs? Could it be that your culture medium contains a significant percentage of oviducal fluid?

S. Willadsen

That is out of the question. These embryos that go for
culture go through various lots of new, fresh lots of PBS, so I
would think the oviduct or uterine contributions are very small
and, of course, in the case of the frozen embryos, it goes way
beyond that - they go through all sorts of baths on the way.

N. Rasbech

If we have time we may come back to this very interesting
topic but I propose that we move on now to the next paper which
was given by Dr. Renard.

G.E. Seidel

    I have a question that applies to many papers, including
the first one and Dr. Renard's. The bovine morulae, and probably most morulae, seem to be very flexible organisms. Has anyone found a medium in which most morulae do not go to blastocysts
if it is a reasonable medium in terms of pH, osmolarity, and you
keep the embryos warm? We tried a number of media and we got
development from morulae to blastocysts in all of them.

R.J. Ryan *(USA)*

    This is not a question but a comment which you may find of
interest here. A few weeks ago I heard a paper by Dr. Strick-
land at the Rockefeller University where he was examining early
embryos and found that there was a good deal of protease activity
elaborated by these embryos, by the trophoblast, by the inner
cell mass and, I believe by the endoderm. This could be assayed
for rather simply by just placing them on an agar plate with a
protein substrate in it and then taking them off and looking to
see whether the agar plate became clear. This might be a useful
marker for you in checking viability in eggs.

S. Willadsen

    To return to Dr. Seidel's question. I think there is one
medium which would not allow this although I haven't checked it,
and that is bovine serum, untreated bovine serum. Another medium
which I have tried, is egg yolk. Egg yolk will kill cow blastocysts - my egg yolk did!

T. Greve *(Denmark)*

    I have a brief question for Dr. Renard. It was a very
interesting paper and you mentioned checking the viability of
superovulated eggs. Have you done any similar studies in non-
superovulated animals?

J.P. Renard *(France)*

    No, we have no information on that.

M.T. Kane

    Dr. Renard, you see hatched blastocysts quite a lot in the
bovine in your experiments. Do you think that the hatching
occurs due to the embryo, as it were, digesting or chewing its
way out of the zona pellucida, or do you think the hatching occurs because there are uterine enzymes which actually lyse the
zona?

J.P. Renard

    We have done studies on this in two ways, firstly by
culturing non hatched blastocysts recovered at Day 7 or Day 8
for two or three days. The other type of study was a scanning
electron microscope study with flushed blastocysts. We think
the main process of hatching is due to activity of the blastocyst itself. There are one or several contractions. We don't
know exactly how many until we can do some cinematographic studies.
These contractions of the blastocyst occur in culture medium
that is deprived of any enzymes of the uterine fluid because the
embryos are thoroughly washed prior to culture. After the contractions the zona becomes thinner; by electron microscopic
studies we see that it becomes thinner. There is also an enzyme in the uterus - the two processes exist but the first one
is probably the most important.

K. Betteridge *(Canada)*

    May I add a comment on that. The same answer against the
uterine effect would be the fact that you can find completely
normal, unfertilised eggs in the uterus as late as Day 16 so the
zona hasn't been lysed by any uterine effect there. Admittedly
it hasn't been subjected to changes that it may undergo at fertilisation but it would argue against a uterine effect.

U. Schneider *(West Germany)*

    I have a question for the Canadian group. Did you observe
any other chromosomal abnormalities in the blastocysts you looked
at due to superovulation treatment?

W.C.D. Hare *(Canada)*

In the material which we have looked at which is admittedly fairly late on, we have not observed any chromosomal abnormalities with the exception of one case in which we were unable to detect two chromosomes, nor were we able to detect a 'y' chromosome, and there seemed to be 59 chromosomes. But we had few cells to go on so it is very speculative that this was an XO but it is a possibility. I would expect that if we were looking for chromosomal abnormality we should look very much earlier - probably as early as the 8-cell stage. It can be done; a technique has been developed for doing this in sheep. I think it would be interesting to establish what proportion of superovulated, fertilised, eggs are dying at early stages due to chromosomal abnormalities.

N. Rasbech

We are coming to the paper dealing with embryo sexing; we can continue with the discussion of chromosomes but are there any more questions on the last paper? If not we will go on to the papers by Drs. Menezo and Szollosi.

G.E. Seidel

I have a question for Dr. Szollosi. In the mature graafian follicle of the bovine, when you are talking about the stratum granulosum rather than the cumulus, is that pseudo-stratified epithelium, in other words, does every granulosa cell have a process that goes down the basement membrane, or is it, in fact, stratified? There is some debate over this in other species.

I. Szollosi

I believe it is a true stratified epithelium. They change shapes at different phases of activity, that's true, there is elongation and there is a very distinct change in the pre-ovulatory periods but I am quite convinced that it is a true stratified epithelium.

K. Betteridge

Dr. Menezo, you mentioned that glucose metabolism,

comparing the cultured follicles in the presence of PMSG and
PMSG + HCG was not different. Was the glucose metabolism of
these two sets of follicles different from follicles cultured
in the absence of either gonadotrophin?

Y. Menezo

The problem is that if we culture follicles without PMS,
FSH or LH, we get atretic follicles. If we do not add some
gonadotrophins we get atresia so it is quite impossible to test
this.

N. Rasbech

If there are no more questions we will proceed to the next
paper which was given by Dr. Willadsen.

R. Newcomb (UK)

I would like to ask Dr. Willadsen why he freezes slowly
until -30°C with his fast freezing technique, whether this is
necessary or whether you could, in fact, start freezing fast
earlier?

S. Willadsen

I have deliberately jumped over all the niceties of these
things, the complicated theories and hypotheses that people
have concerning freezing. What I was attempting to ensure was
that any variation there might be between cells in the individual
embryos and between different embryos in terms of dehydration,
would be eliminated. That's the first thing - they would all
be dehydrated to the same extent.

The second point is that I want this dehydration to be
exactly corresponding to the dehydration of the extracellular
medium. In other words, the embryos should be equilibrated with
the liquid phase of the external medium and that liquid phase
should be the equilibrium liquid phase for this particular solution at this particular temperature. If you did that, you could
then go into a phase diagram for a PMS/DMSO solution and you
could state exactly what was the composition of the liquid phase

in the cells. This means that you could start thinking in terms
of producing that sort of volume with a different liquid phase,
or, alternatively, you could try producing that composition of
the liquid phase with a different freezing rate. Now, all this
is much more interesting from a theoretical point of view than
from a practical point of view and the message which I wanted
to put across today was, that if you wanted to freeze any cell,
any embryo - say, a Day 12 embryo (although I am not sure I
could freeze that one) - we now have information which says that
probably all embryos, all cells, in fact, can survive rapid thawing. You don't know that they can survive with your particular
method but you can adjust your method so that you can achieve
maximum survival rate with rapid thawing. This makes freezing
much more simple. If I was going to freeze embryos nowadays I
would use a freezing rate of about 0.3°/min and transfer samples
to liquid nitrogen from various temperatures all the way down
to -80°C. If you go at 0.3°/min and you want them to survive
rapid thawing, you have to transfer them to liquid nitrogen from
a temperature below -45°C. To me this suggests, in the light
of the other experiments, that these embryos are not catching
up, they are being frozen too rapidly so that they are lagging
behind with respect to dehydration because apparently I am getting the same degree of dehydration at -33°C.

N. Rasbech

Are there any more questions for Dr. Willadsen's paper?

S. Willadsen

Perhaps I may make a comment on my own paper. There was
one slide which I didn't show you and that is the one which I
think may be the most pertinent in terms of the practical application of these techniques. We were involved in a project where
we were freezing embryos in England in a semi-commercial situation. All the quarantine regulations were maintained in this
unit and the embryos were frozen with a view to export to New
Zealand. These were, of course, the exotic breeds that they are
interested in in such countries. In this particular trial
we froze 33 embryos and when we thawed them out in New Zealand

we thought that some 26 of them were really good ones. When I
say 'really good ones', that means that in those 33 we did freeze,
some we thought were abnormal and those, of course were
still abnormal when they were thawed out. So the rather low
proportion of normal eggs was related to the number of the
really good ones which survived de-freezing. Those ones and
some of the more questionable ones were transferred into 32
recipients, one of which died next day. Out of those we would
have liked to see some 15+ animals pregnant. In the event, the
pregnancy rates were much less satisfactory - I think there
were only 9 pregnant and of those one or two showed signs of
early embryonic death. So we were left with seven embryos out
of the 33 which were frozen of which we considered some 28 to
be good ones. I am not saying that this should necessarily dis-
courage us, I don't think it will, but it does show that you have
to be very, very careful when you are doing these things and the
there are a lot of factors which have to be taken into consideration, such as the recipients you are using and the general set-
up. I don't think we could analyse the results beyond this, but
since I have already quoted our good figures when we have been
doing transfers back home, I should also mention that we are not
doing particularly well down under.

A. Brand *(Netherlands)*
        Do you think that the length of storage of your frozen
embryos will affect survival rates? Do you expect a better pregnancy rate from embryos which have been stored for just one
month rather from those which have been stored for, say, one
year?

S. Willadsen
        In all probability there wouldn't be any difference in
that case but there are two points about this. One is that if
it turned out to be the case then you would, of course, proceed to
a lower temperature for storage. The other point which I think
is pertinent here is that I know that embryos will survive
freezing quite well if they are frozen right after they have been
recovered from the donors, but, in our recent experiments which

I have been referring to, I have kept the embryos between three
and six hours, I think. I do know that they will survive if they
are frozen right away. On the other hand, I think it is bad to
keep them for as long as 12 hours before freezing. They should
be frozen as soon as possible after they have been recovered.
I think it might be a good idea to concentrate a bit on the short
term storage after all and it could be that they should be stored,
not at room temperature, but at 38°C. These are details.

H. Breth Hansen *(Denmark)*

These recipients which got frozen embryos, what were the
plus/minus hours of the recipient compared with the heat of the
donors?

S. Willadsen

Once we got the embryos out we let the stage of the embryo
decide what stage the recipient should be at but all of these
animals were apparently very well synchronised and we were very
careful in picking and choosing them. We thought that the oestrus detection was very good but I must say that I have to put a
bit of a question mark on that because they only caught 10 of
them returning out of the 31. We were laughing at that point
but we weren't laughing later on! There was very close super-
vision of the animals after they had been operated on but only
10 of them were picked up.

N. Rasbech

Are there any questions on the last two papers by Drs.
Hare and Henriet?

I. Wilmut *(UK)*

May I ask Dr. Hare a very simple procedural point about
the sexing. How many embryos do you think that one person could
sex in a day, given the equipment and the assistant staff that
he might need?

W.C.D. Hare

It's a good question. It depends on how many hours you

are going to work in a day for a start. I think there is a very
definite cut-off point on efficiency, human error and so on. I
know that Elizabeth Singh and I tried to do 18 in one day in a
hurry. We were trying to sex the embryos to allow colleagues to
transfer that day; it was a twinning experiment. By the end of
the day we were going round in circles. I wouldn't attempt to
have just one person do it. I would say that for two people,
which is the way we usually work, it is possible to handle up
to 12 embryos in a day.

I. Szollosi

May I return to a point that was raised earlier. Is there
any knowledge of the normally occurring abnormalities in bovines?
In humans, it has been demonstrated that the majority of spontaneous abortions caused by chromosomal abnormalities normally
occurring are not picked up because of too early abortions. Before any chromosomal abnormalities could be ascribed to superovulation the normal rate of abnormalities would have been known.

W.C.D. Hare

There has only been one study reported and that is by
McFeely and others. As I recall it, they looked at 11 or 12 day
old blastocysts in the bovine following normal ovulation and
fertilisation. They had one embryo in which they found tetra-
ploid cells. Now, the interpretation of the normality of that
is open to question. Apart from that I don't know of any study
that is being done on the chromosome spectrum in blastocysts.

I. Szollosi

Are there any important spontaneous abortions which are not
picked up? In the case of humans the majority are not picked up,
particularly in the first trimester, so there is a very important
rate of loss of normally fertilised human embryos.

W.C.D. Hare

I couldn't tell you what proportion of early embryonic
death in the bovine can be attributed to chromosome abnormalities.
I don't think that it is known.

A.J. Breeuwsma *(Netherlands)*

You have been talking about the karyotyping of your cells.
Apparently this could be done by using the Barr body as well.
How many cells would you require to base your sexing procedures
based on the presence of sex chromatin?

W.C.D. Hare

I am afraid I have misled you if I gave you the impression
that sexing by the use of sex chromatin could be used in the
bovine. Maybe it could by some other people but we have looked
at these cells and it is extremely difficult because you get a
certain amount of clumping of the chromatin. I think we saw an
example of that in some slides that Dr. Willadsen showed. I
wouldn't recommend it for the bovine. I think one is much safer
to go for the sex chromosome if it is at all possible.

N. Rasbech

We are running very short of time; are there any quick
questions to Dr. Henriet's paper?

S. Willadsen

Dr. Henriet, could you tell us what pressure you think
these embryos can stand up to?

L. Henriet *(Belgium)*

Up to 500 kg it is no problem, but above, we don't know.

N. Rasbech

Well, I think we must really break it up now but do let
me thank all this morning's speakers and the contributors from
the floor.

SESSION 4

OESTROUS CYCLE CONTROL AND FERTILITY

Chairman:

J.M. Sreenan

MANAGEMENT FACTORS IN OESTROUS CYCLE CONTROL

Bridget Drew

Agricultural Development and Advisory Service,
Ministry of Agriculture, Christchurch Road, Winchester,
Hampshire, England.

ABSTRACT

In studies involving 6 754 cattle the level of fertility following prostaglandin or short-term progestin treatment and fixed time insemination was equal to untreated controls inseminated at observed oestrus. Pregnancy rates varied widely between farms. The level of nutrition over the service period was shown to have a significant effect on the calving rates of suckler cows and dairy heifers. When compared with the usual farm ration, supplementation to provide an additional 20 M/J metabolisable energy/day for twelve weeks increased calving rates to the synchronised oestrus from 50.0% to 68.9% (P< 0.001) and 37.8% to 62.2% (P< 0.01) for heifers and cows respectively. In spring calving dairy cows with an incidence of anoestrus of 26.2% treatment with progesterone initiated cyclical activity. In autumn calving herds the calving to conception interval was nearer optimum in treated cows. The effect of inseminator efficiency, bull fertility and body condition on herd fertility are discussed. Oestrous cycle control was found to be a useful aid to herd management but not a substitute for sound husbandry.

INTRODUCTION

In recent years considerable advances have been made in the development of techniques to control the reproductive cycle of cattle. It has been shown that the pregnancy rates of cattle inseminated at fixed times following prostaglandin or short-term progestin treatment are equal to those obtained in untreated controls inseminated at observed oestrus. (Cooper, 1974; Hafs et al., 1975; Wishart et al., 1977; Roche, 1974). The proportion of treated animals calving to the fixed time insemination is one of the most important factors determining the commercial value of a controlled breeding programme. The level of fertility is known to vary widely between farms and depends not only on the incidence of anoestrus, rate of oestrus detection and conception rates of the cows but also on the efficiency with which they are inseminated and the fertility of the bull used.

Cow fertility has been shown to be influenced by the level of nutrition over the service period (Durrell, 1955; Alderman, 1970) and in changes in body weight and condition (McClure, 1970; King, 1968). In this country it is usual for suckler cows and dairy heifers to be reared on a system which maximises growth from grass. In the winter and spring when the majority are mated the amount of food is usually restricted and often poor in quality. The effect of poor nutrition over the service period on fertility can be accurately measured when controlled breeding techniques are used to standardise the other variables.

A high incidence of postpartum anoestrus and sub-oestrus has been shown to occur in both dairy and suckler cows (Kudlac, 1964; Moller, 1970). Roche (1976) has shown that progesterone and oestrogen administered as a progesterone releasing intra-vaginal device can advance the time of the first fertile oestrus. In some herds inadequate oestrus detection may be a more serious problem. Williamson et al. (1972) reported that in one study only 56% of oestrus periods were detected by herdsmen and trained observers. In such cases the use of controlled breeding followed by fixed time insemination should enable reproductive efficiency to be improved by eliminating the necessity to observe for oestrus at the synchronised service and by giving non-oestrus but synchronously ovulating animals the opportunity of becoming pregnant.

The studies reported in this paper were designed to determine the management factors involved in the successful implementation of a controlled breeding programme so as to maximise the possible benefits on commercial farms.

A total of 6 754 suckler cows, dairy cows and dairy heifers were involved. In Study I the effect of the level of nutrition over the service on the fertility of suckler cows and dairy heifers was measured. In Study II the incidence of postpartum anoestrus in two herds was determined. The effect of treatment with a progesterone releasing intravaginal device on non-cyclical cows was studied. In Study III the calving to conception interval of treated cows was compared with controls.

MATERIALS AND METHODS

A total of 364 Friesian Dairy Heifers and 165 Hereford x Friesian suckler cows on nine farms in Southern England were involved in Study I. The animals were weighed and scored for body condition on a subjective 0 (very thin) to 5 (grossly fat) scale at the start of the study. They were allocated to one of two feeding groups (control or supplemented) on a stratified random basis. The control groups were fed the ration usually provided on the farm and which the farmer considered adequate.

The foods comprising the rations fed were analysed and using estimates of food consumption predicted performance rates were calculated. For the control heifers this averaged a live weight gain of 0.3 kg/day and for control cows a milk yield of 4.5 kg/ day. The supplemented groups were given extra cereal

to provide an additional 20 MJ of metabolisable energy per day
which was calculated to increase the rate of performance to 0.7
kg LWG/day (heifers) and 9.0 kg milk yield/day (cows). The
supplemented ration was fed for a twelve week period commencing
six weeks before the planned insemination date.

All the animals were treated with norgestomet/oestradiol
valerate administered in an ear implant. These were inserted
subcutaneously using a special applicator and at the same
time an intra-muscular injection of norgestomet and oestradiol
valerate was given.

After nine days the implants were removed and the heifers
were inseminated 48 and 60 h and the cows 48 and 72 h later.
No observations for oestrus were made. Each man inseminated
an equal number of control and supplemented animals and semen
from the same bulls was also equally used. Animals in oestrus
were remated at the return to service period (18 - 28 days
after implant removal) using bulls which would colour mark
their progeny. All animals were examined for pregnancy by
palpation/rectum at 42 and again at 70 days after implant
removal. At the 48 h insemination and at the 42 day pregnancy
examination the animals were reweighed and scored again for
body condition.

In Study IIa a total of 66 Hereford x Friesian autumn
calving suckler cows at least 70 days post partum and 60
Friesian dairy heifers 13 - 24 months of age were treated with
prostaglandin to synchronise oestrus. Two intramuscular
injections of PGF2α were given at intervals of twelve days
and the animals were inseminated at 72 and 88 h after the
second injection. Observations for oestrus were made twice
daily and jugular blood was collected at the time of each
injection and 42 and 72 h later. Progesterone content of blood
serum was estimated by radio immunoassay (Louis et al., 1973).
Pregnancy was diagnosed by palpation per rectum.

In Study IIb a total of 145 spring calving Friesian dairy cows at least 42 days postpartum were balanced for milk yield and length of the postpartum interval and allocated to control or treated group. Milk samples were taken from all cows for progesterone assay (Leaming and Bulman, 1976) eight days before and at the start of the trial. Control cows were observed by the herdsmen and inseminated when detected in oestrus. A silastic coil impregnanted with progesterone with a capsule containing oestradiol benzoate attached was inserted into the vagina of all treated cows for twelve days and the animals were inseminated 56 h after removal. Pregnancy rates were determined by milk progesterone assay (Booth and Holdsworth, 1976) and by palpation per rectum.

In Study III a total of 615 autumn calving Friesian cows were balanced for milk yield and postpartum interval and allocated to control or treated groups. Control cows were inseminated when observed in oestrus by the herdsman. In treated cows oestrus was synchronised by the same method used in Study IIb but the cows were inseminated twice at 48 and 72 h after coil removal. Control cows were at least 42 days postpartum and treated cows 47 days postpartum at first insemination. Pregnancy was diagnosed by milk progesterone assay (Booth and Holsworth, 1976) and by palpation per rectum.

In the studies described and other studies involving a total of 5 339 animals herd management was monitored. Details concerning the efficiency of inseminators, fertility of bulls used, the time and intensity of oestrus and the effect of routine veterinary and other management treatments on fertility were recorded.

RESULTS

Study I

The mean body weights and condition scores at the start of the study were 348 kg at 3.2 and 432 kg at 3.0 for dairy

heifers and suckler cows respectively. There was a wide
variation in fertility between farms but on each farm a higher
proportion of supplemented animals calved to the fixed time
insemination. The calving rates were, for heifers 50.0% and
68.9% (P < 0.001) and for cows 37.8% and 62.2% (P < 0.01) for
control and supplemented groups respectively. The individual
farm results are shown in Tables 1 and 2.

TABLE 1

CALVING RATES OF TREATED (NORGESTOMET) HEIFERS ON CONTROL AND SUPPLEMENTED
RATIONS

| Farm | Calved (%) fixed time insemination | |
	Control	Supplemented (+20 MJ/d ME)
1 (62)	40	67
2 (58)	59	75
3 (56)	32	59
4 (37)	67	79
5 (78)	54	69
6 (73)	53	67
Mean	50.0	68.9

TABLE 2

CALVING RATES OF TREATED (NORGESTOMET) SUCKLER COWS ON CONTROL AND
SUPPLEMENTED RATIONS

| Farm | Calved (%) fixed time insemination | |
	Control	Supplemented (+ 20 MJ/d ME)
1 (65)	41	70
2 (58)	31	58
3 (42)	36	55
Mean	37.8	62.2

In this study there was no significant relationship between liveweight or condition score and fertility. There was a tendency for pregnancy rates to improve as the level of nutrition increased, but because of the other management factors involved it was not possible to determine the level of nutrition at which optimal fertility occurred. Acceptable pregnancy rates resulted in heifers fed a ration calculated to provide for maintenance and 0.7 kg LWG/day and in suckler cows fed for maintenance and 9.0 kg milk yield/day.

Study IIa

In this study two heifers and ten cows were not observed in oestrus during the eleven days following the second injection of PGF2 α. No animals were observed in oestrus during the first two days and the majority (68% of heifers and 62% of cows) sbowed oestrus during the third or fourth day. Oestrus was observed during Days 5 - 11 in 29% of heifers and 23% of suckler cows. The distribution is shown in Table 3.

TABLE 3

INTERVAL FROM SECOND INJECTION * OF PGF2 α TO ONSET OF STANDING OESTRUS IN DAIRY HEIFERS AND SUCKLER COWS

Time observed	Number observed in oestrus	
	Dairy heifers	Suckler cows
Not observed	2	10
Day 1	0	0
2	0	0
3	25	32
4	15	9
5	6	2
6	6	2
7	2	4
8	1	3
9	1	1
10	0	3
11	2	0
	60	66

* Two intramuscular injections of PGF2 α were given 12 days apart.

All but one of the heifers responded with reduced progesterone levels after the second injection. The ten cows not observed in oestrus showed uniformly low levels of blood serum progesterone at and following each injection. None of these cows conceived to the fixed time insemination and it was concluded that these animals were in postpartum anoestrus.

Study IIb

The initial progesterone assays showed that 38 (26.2%) of the 145 cows had low levels of milk progesterone indicating that they were probably in postpartum anoestrus at the start of the trial. Seven (36.8%) of the 19 non-cyclical treated cows conceived to the fixed time insemination and 15 (78.9%) became pregnant within 24 days. Only one (5.3%) of the non-cyclical control cows became pregnant during the same period. The period from the start of the breeding season to conception was 18.1 days for the 76 treated cows and 36.0 days for 67 of the 69 control cows (P< 0.001). Two control cows failed to become pregnant.

Study III

First service pregnancy rates were 60% (297/310) for the treated cows and 59% (280/305) for controls. The period from the start of the breeding season was, for the treated cows 20 days and for the controls 35 days (P< 0.001). The calving to conception interval was 93 and 107 days (P< 0.001) for treated and controls respectively.

DISCUSSION AND RECOMMENDATIONS

The results of these studies have shown that controlled breeding can be a useful aid to herd management but not a substitute for sound husbandry. Pregnancy rates following prostaglandin or short-term progestin treatment and fixed time insemination are equal to those obtained in untreated controls inseminated and observed oestrus but the level of fertility obtained depends on the standard of management on the farm.

The level of nutrition provided over the service period has been shown to affect fertility. A ration to provide for maintenance and 0.7 kg LWG/day for heifers and 9.0 kg milk yield/day for suckler cows has resulted in acceptable fertility. It is not usually possible to provide sufficient energy to support liveweight and milk yield in dairy cows during early lactation but if possible the cows should have returned to a positive energy balance before service. Major changes in diet over the service period should be avoided.

Poor pregnancy rates have been obtained in groups which were dosed for internal parasites or received dressing against warble fly during treatment or at insemination. The results of trials suggest that 2 - 3% of a random sample of heifers are likely to have genital tract abnormalities making remote their chances of becoming pregnant. The cost of treating most of these animals would be avoided if all twins to bulls and bought-in heifers were examined for freemartinism before treatment.

Controlled breeding techniques are only likely to be of commercial value if they are safe, easy to administer, effective on the majority of animals and economically beneficial. Prostaglandin synchronises oestrus by causing premature regression of the corpus luteum and will only be effective on animals in the luteal phase of the oestrous cycle.

Progesterone administered in the form of a progesterone releasing intravaginal coil has been shown to be effective in inducing luteal activity in some cows and is likely to be more effective than prostaglandin in herds in which there is a high incidence of anoestrus.

Fatigue has been shown to affect the efficiency of the inseminator. In studies where large groups (104, 68 and 84) of cows were inseminated continuously by one man the fertility of animals inseminated in the second half of the group was significantly lower than those inseminated in the first half.

A ratio of one man per 30 cows in one group is recommended. The possibility of fatigue is reduced when the handling facilities are designed to permit easy movement and enable the inseminator to stand at the same level as the animal in the crush.

The fertility of bulls standing at insemination centres is likely to vary by about 8 - 10%. The results of trials suggest that it is possible that these differences may be accentuated when the semen is used at a fixed time insemination. At the return to service period suckler cows and dairy heifers should be mated by natural service because accurate oestrus detection is not usually possible. Bulls should be turned in with the herd as soon as the intense oestrus behaviour has subsided usually on day 5 following injection of prostaglandin and day 4 following treatment with progesterone. This is necessary to cover the animals not completely synchronised which may be a significant proportion of the herd.

In a number of studies the pregnancy rates obtained following oestrous cycle control have fallen below the farmers' expectations. In most of these cases it has been possible to identify the cause of the poor fertility.

If controlled breeding techniques are to be used with maximum benefit a high level of management over the service period is required to obtain pregnancy rate targets of 60% for dairy cows and heifers and 55% for suckler cows.

REFERENCES

Alderman, G. 1970. Nutrition as a possible cause of infertility in cattle. Vet. Rec. 87 (II) 35.

Booth, J.M. and Holdsworth, R.J. 1976. The establishment and operation of a Central Laboratory for pregnancy testing in cows. Brit. Vet. J. 132, 518.

Cooper, M.J. 1974. Control of oestrous cycle of heifers with a synthetic prostaglandin analogue. Vet. Rec. 95, 200.

Durrell, W.B. 1955. Anoestrus in heifers associated with plane of nutrition. Canad. J. Comp. Med. 19, 144.

Hafs, H.D., Manns, J.G. and Drew, B. 1975. Onset of oestrus and fertility of dairy heifers and suckled beef cows treated with prostaglandin F2 alpha J.Br. Soc. Anim. Prod. 21, 13.

King, J.O.L. 1968. The relationship between the conception rate and changes in body weight, yield and SNF content of milk in dairy cows Vet. Rec. 83, 492.

Kudlac, E. 1964. Vorkommen und ursachen des anoestrus bei kuehen mach der geburt. Proc. V. Int. Con. Anim. reprod. and AI (Trento) 5, 188.

Lamming, G.E. and Bulman, D.C. 1976. The use of milk progesterone radio-immunoassay in the diagnosis and treatment of subfertility in dairy cows Br. Vet. J. 132, 507.

Louis, T.M., Hafs, H.D. and Seguin, B.E. 1973. Progesterone, L.H. oestrus and ovulation after prostaglandin F2 alpha in heifers. Proc. Soc. exp. Biol. Med. 143, 152.

McClure, T.J., 1970. A review of developments in nutrition as it is related to fertility in cattle 1964-1969. NZ. Vet. J. 18, 61.

Moller, K. 1970. Uterine involvement and ovarian activity after calving NZ. Vet. J. 18 140.

Roche, J.F. 1974. Effect of short-term progesterone treatment on oestrous response and fertility in heifers. J. Reprod. Fert. 40, 433.

Roche, J.F. 1976. Recent developments in reproductive physiology in cows VI Richards - Orpen Memorial Lecture.

Williamson, N.B., Morris, R.S., Blood, D.C. and Cannon, C.M. 1972. A study of oestrous behaviour and oestrus detection methods in a large commercial dairy herd. Vet. Rec. 91, 50.

Wishart, D.F., Young, I.M. and Drew, B. 1977. Fertility of norgestomet treated dairy heifers Vet. Rec. 100, 417.

OESTROUS CYCLE CONTROL AND FERTILITY IN BEEF CATTLE FOLLOWING SHORT-TERM PROGESTAGEN TREATMENTS

P. Mulvehill and J.M. Sreenan

The Agricultural Institute, Belclare, Tuam, Galway, Ireland.

ABSTRACT

A total of 2 108 heifers and postpartum beef cows were used in a series of experiments to determine the fertility level following short-term progestagen synchronisation treatments and fixed time artificial insemination and also to determine some of the management factors affecting this fertility. The administration of HCG, GnRH and gonadotrophins was also studied in conjunction with synchronisation treatments.

Intravaginal administration of progesterone or cronolone by sponge pessaries or subcutaneous administration of norgestomet all with intramuscular progestagen and oestrogen on Day 1, resulted in similar fertility levels both at the synchronised and first repeat oestrus. Calving rates of 51 - 56% in heifers and 46 - 48% in postpartum cows were obtained following a double fixed-time insemination at 48 and 72 hours after progestagen removal.

Administration of HCG (2 000 iu) or GnRH (100 µg) at 20 hours after progestagen removal had a significant effect on time of ovulation, resulting in earlier and more synchronised ovulation timing than control animals. Following the use of HCG and GnRH, fertility level was similar to control animals with a double fixed-time insemination. A single insemination at 50 hours resulted in a lower fertility level because the timing was too late relative to the time of ovulation especially in the GnRH treated group.

Body condition of the animals at the time of treatment had a significant effect on fertility. An increase in animal body condition from a score of three to a score of one resulted in an increase in calving rate from 26.9 to 57.9%. Heifers and postpartum cows synchronised during the winter months (January - March) had a significantly lower calving rate to fixed time insemination than at other times of the year. In a study on the *effect of nutrition on fertility in synchronised heifers, increasing average daily gain from .23 to .40 kg/day resulted in an increase in pregnancy rate from 47 to 67%.*

Administration of PMSG at the end of the synchronisation treatment resulted in increased ovulation rates and increased fertility in both heifers and postpartum cows.

INTRODUCTION

Following the report of Wiltbank and Kasson (1968) showing that administration of progestagens over a short period (9 - 10 days), when combined with oestrogen at the beginning of treatment to induce luteolysis, was an effective method of oestrous cycle control, many reports have appeared indicating similar results. Various methods of administering the progestagens over short periods are now available. Intravaginal administration either by silastic coils (Abbott : Prid) (Mauer et al., 1975; Roche, 1975) or polyurethane sponge pessaries (Sreenan, 1975), has been shown to be effective. The use of subcutaneous implants (G.D. Searle & Co.) has also been shown to be an effective method (Wishart and Young, 1974; Wiltbank and Gonzalez-Padilla, 1975).

In all cases, the treatments are designed to control the time of oestrus and ovulation in order to allow artificial insemination on a fixed-time basis relative to progestagen withdrawal. In order to ensure the maximum synchronised oestrous response, all treatments normally contain the administration of oestrogen at the start, to regress the current corpus luteum (CL). However, the action of oestrogen in causing regression of the CL tissue would seem to depend on cycle stage at administration. Oestrogen given early in the cycle, not only does not cause CL regression or inhibition in the ewe (Ginther, 1970; Hawk and Bolt, 1970) or the cow (Smith and Vincent, 1973) but has been shown to be luteotrophic in the cow at this early stage (Days 0 - 3) by Lemon (1975). Wishart and Young (1974) however, reported a very high oestrous response and a high degree of synchronisation following the incorporation of intramuscular progesterone with the oestrogen on the first day of a 9 day progestagen treatment.

Subsequently, Sreenan (1975), also administering progesterone intramuscularly with oestrogen at the start of a 10 day progesterone treatment, reported a high oestrous response (96%) with no difference resulting from treatment at early (Days 0 - 4), luteal (Days 5 - 16) or late (Days 17 - 23) cycle stages.

In this same communication it was reported that the level of intramuscular progesterone given with the oestrogen at the start of treatment did not affect the overall oestrous response but did affect the timing of oestrus onset and the degree of synchronisation. The administration of 900 mg progesterone intramuscularly resulted in a late onset of oestrus (Mean, 73.1 $\pm$ 6.7 hr) and low degree of synchronisation while 250 mg progesterone resulted in an early onset of oestrus (Mean, 44.8 $\pm$ 3.4 hr) and a high degree of synchronisation with 75.8% of heifers coming in oestrus on Day 2 from progressive removal.

Subsequent data, (Sreenan et al., 1977) show that following the use of intravaginal sponges there is a quick release of progesterone, it is possible that the administration of intramuscular progesterone may not be required at the start of treatment because of this. Roche and Gosling (1977) have reported no increase in oestrous response following the incorporation of intramuscular progesterone at the start of an intravaginal silastic coil treatment.

Further study is needed to understand the effect of oestrogen alone or combined with varying doses of progesterone, oestrous response and degree of synchronisation. The incorporation of intramuscular progesterone at a level of 200 - 250 mg with the oestrogen is nevertheless an effective method of controlling the cycle and the effect of such a treatment on CL function, as measured by plasma progesterone level has been shown by Sreenan and Gosling (1977).

Intravaginal progesterone impregnated sponges were inserted at two cycle stages with or without the administration of a combined injection of 250 mg progesterone + 7.5 mg oestradiol benzoate and the oestrous response and daily progesterone levels were measured.

In heifers treated at Day 2 of the cycle, effective

inhibition of CL development was clear both from the subsequent induction of oestrus and also from the lower progesterone levels following the use of the intramuscular progesterone and oestrogen. Similar results were obtained following treatment at Day 12 of the cycle, Mauer et al. (1975) showed that failure of some animals to come into oestrus early after intravaginal Prid treatment was associated with high progesterone levels in those animals. Roche and Gosling (1977) have reported similar data suggesting incomplete luteolysis in those animals.

However, perhaps the major practical advantage to be gained from oestrous cycle control techniques would be the possibility of a fixed-time insemination procedure in post-partum beef cows. In this category, the question is not so much the response at particular cycle stages, it is more the induction of oestrus in ovarian inactive cows. In the post-partum beef cow, the effect of suckling has been shown to delay cyclic ovarian activity for extended periods (Wiltbank and Cook, 1958; Oxenreider, 1968).

More recently, Chupin et al., (1976) have shown that a high proportion of post-partum beef cows (20% for Charolais and 31% for Salers) are ovarian inactive up to 60 days postpartum. It is possible to induce and synchronise oestrus in a high proportion of postpartum beef cows as indicated from the data of Sreenan and Mulvehill (1974).

Fertility levels following various short-term progestagen treatments have generally been shown to be similar to that in control animals (Wiltbank and Kasson, 1968; Chupin et al., 1972; Roche, 1974; Wishart and Young, 1974; Sreenan and Mulvehill, 1975; Mulvehill and Sreenan, 1977). More recently however, attention has been drawn to the modifying effect of management (particularly nutrition level) on fertility in both heifers and cows (Drew et al. 1976). Because the advantages to be gained from oestrous cycle control will depend on the fertility level obtained at the synchronised oestrus, it is essential to understand the factors affecting fertility at this time.

The experiments reported in this communication were carried out to:

1) Compare three methods of short-term progestagen treatments on fertility following fixed-time insemination in both beef heifers and postpartum nursing beef cows.

2) To determine some of the factors affecting fertility at the synchronised oestrus.

3) To attempt to increase the fertility rate at the synchronised oestrus through the administration of gonadotrophin (PMSG) in conjunction with a short-term synchronisation routine.

MATERIALS AND METHODS

A total of 2 108 animals were used in this series of experiments. Heifers used in all of the studies were beef heifers of Hereford cross breed type with the exception of the nutrition study in Experiment 4, where Friesian heifers were used. The cows used were all postpartum nursing beef cows of Hereford breed type.

Data on oestrous response, degree of synchronisation and effect on CL function as measured by plasma progesterone level has been previously published from these studies (Sreenan, 1975; Sreenan and Mulvehill, 1975; Sreenan et al., 1977 and Mulvehill and Sreenan, 1977). These studies showed a high oestrous response and a high degree of oestrous synchronisation following the short-term (9 - 10 day) progestagen intravaginal treatments involving intramuscular administration of progestagen (250 mg progesterone or 20 mg cronolone) and oestradiol benzoate (7.5 mg). The present communication presents data on fertility levels using similar treatments followed by fixed-time insemination.

EXPERIMENT 1

The aim of this experiment was to compare three short-term progestagen synchronisation methods in terms of fertility level at the synchronisation oestrus following a fixed-time insemination procedure. A total of 696 mature maiden heifers and 521 multi-parous nursing beef cows were used. In the case of the cows a minimum interval from parturition to treatment of 35 days was allowed.

Synchronisation treatments

Treatments were as follows:

1) Progesterone (3.0 g) sponge pessaries were inserted intravaginally for 10 days. On the day of insertion each animal received a combined intramuscular injection of 250 mg progesterone + 7.5 mg oestradiol valerate.

2) Cronolone (0.200 g) sponge pessaries were inserted intravaginally for 10 days. On the day of insertion, each animal received a combined intramuscular injection of 20 mg Cronolone + 7.5 mg oestradiol benzoate.

3) Norgestomet implants (6 g) (G.D. Searle & Co) were inserted subcutaneously for 9 days. On the day of insertion each animal received a combined intramuscular injection of 3 mg norgestomet + 5.0 mg oestradiol valerate.

Breeding

All animals were artificially inseminated with frozen-thawed semen (approx. 30×10^6 spermatazoa per straw) at 48 and 72 h from progestagen withdrawal. Single inseminations at the first repeat oestrus were carried out on detection of oestrus. The treatments were carried out on a total of 98 farms, with all treatments represented on each farm where possible.

EXPERIMENT 2

The aim of this experiment was to determine the effect of administering HCG (Intervet) or GnRH (Abbott) in conjunction with short-term progestagen treatment on timing of ovulation and on fertility following one or two fixed-time inseminations. A total of 318 heifers of Hereford breed type were used in this study.

Synchronisation treatment

Progesterone (3.0 g) sponges were used as described for Experiment 1. Following sponge removal 90 of the animals were randomly allotted to three treatment groups. The first group received 2000 iu HCG at 20 h after sponge removal while the second group received 100 μg GnRH at the same time. Group 3 did not receive any treatment after sponge removal and acted as controls. The animals in each of these treatment groups were randomly allotted to three subgroups for laparoscopy or laparotomy observations at 40, 50 or 60 h post sponge removal. The remaining 228 heifers were treated similarly in three groups but laparoscopy or laparotomy observations were not carried out. Instead, two insemination routines were applied to each group as follows:-

i)   A double insemination at 48 and 72 h, or,

ii) A single insemination at 50 h after progesterone removal. Fertility in these animals was determined by recovery of viable embryos at slaughter between 40 and 50 days from first insemination.

EXPERIMENT 3

The aim of this experiment was to determine some of the management factors affecting fertility rate at the synchronised oestrus following a double fixed-time (48 + 72 h) insemination procedure. The factors studied in this experiment were the effect of body condition and season of the year at the time of treatment on subsequent calving rates. The animals used in this study were some of those described in Experiment 1.

Body condition score

A total of 557 heifers and cows were condition scored during two seasons (April to June; July to September) at the time of short term progestagen treatment administration. Condition scoring was carried out as reported by Mulvehill (1976). Animals were allotted to three main condition scores. 1 or good, 2 or medium and 3 or poor. All animals were

inseminated at 48 and 72 h from progestagen removal as described
in Experiment 1.

Season of year at treatment

A total of 1080 heifers and cows were treated with short-
term progestagen treatments as previously described and
inseminated at 48 and 72 h after progestagen removal. Treatments were carried out in four main time groups, viz:

1)   January - March.

2)   April   - June.

3)   July   - September.

4)   October - December.

EXPERIMENT 4

The aim of this experiment was to examine the relationship
between body weight changes prior to synchronisation treatment
and subsequent fertility following a fixed-time (48 + 72 h)
insemination routine. A total of 173 mature Friesian breed
type heifers were used for this study. Animals were assigned
to one of the following three groups:

1)   Silage ad libitum (low)

2)   Silage ad libitum plus 1.4 kg barley meal per head
per day (medium)

3)   Silage ad libitum plus 2.3 kg barley meal per head
per day (high).

All animals were weighed 60 days prior to synchronisation
treatment (at the start of the feeding treatments) and again
at the time of synchronisation treatment.

Synchronisation treatment

Progesterone (3.0 g) sponges were used as described in
Experiment 1.

Breeding
    All heifers were inseminated on a fixed-time basis (48 +
72 h) from progesterone removal, as described for Experiment 1.

EXPERIMENT 5

    The aim of this experiment was to determine the effect of
administering exogenous gonadotrophin (PMSG) at the end of a
short-term synchronisation treatment on ovulation rate and
fertility in heifers and on fertility in postpartum nursing
cows. A total of 400 heifers and postpartum cows were used
in this study.

Synchronisation treatment
    Progesterone (3.0 g) sponges were inserted intravaginally
as described in Experiment 1.

Ovulation rate
    One hundred and twelve Hereford breed type heifers were
used in this study. The effect of four dose rates (0, 500,
750, 1000 iu) of PMSG (Batch No 2334; Intervet) administered
intramuscularly at progesterone removal on ovulation rate was
determined. Ovulation rate and follicular development was
determined at laparotomy three to six days from PMSG.

Fertility
    Following a selected dose of PMSG (500 iu) a total of
48 heifers were artificially inseminated on a fixed-time basis
(48 + 72 h). Half of these received 500 iu PMSG at progesterone
removal and the remaining half received no treatment at
progesterone removal. Fertility rate was based on recovery
of viable embryo following slaughter between days 40 and 60
from insemination.

    A total of 240 postpartum cows were used in a similar
study. Approximately half of the cows (121) received 750 iu
PMSG at progesterone removal while the remainder (119) received
no PMSG. Insemination was again at 48 and 72 h from progesterone

removal. Fertility in these cows was based on calving data.

RESULTS

EXPERIMENT 1

Calving rates in heifers
    Calving rates of 53.5, 55.8 and 50.7% were obtained
(following the double fixed-time insemination) with progesterone
sponge pessaries, cronolone sponge pessaries and norgestomet
implant treatments respectively (Table 1).

TABLE 1

CALVING RATE IN HEIFERS FOLLOWING PROGESTERONE CRONOLONE OR NORGESTOMET
TREATMENTS

Treatment	No. Heifers	% calved to fixed AI	% calved to 1st repeat	% calved to 1st + 2nd AI
Progesterone sponge	505	53.5	58.8	82.6
Cronolone sponge	120	55.8	60.7	83.0
Norgestomet implant	71	50.7	45.0	80.4

    No difference in calving rates were evident between
treatments to either the fixed insemination ($P > 0.05$), the
first repeat insemination ($P > 0.05$) or the combined first and
repeat inseminations ($P > 0.05$).

Calving rate in postpartum cows.
    Calving rates of 47.5, 47.4 and 46.4% were obtained
(following the double fixed-time insemination) with progesterone,
cronolone and norgestomet treatments respectively (Table 2).

    No differences were again evident between treatments to
either the fixed-time insemination ($P > 0.05$), the second

service (P > 0.05) or the combined first and second services (P > 0.05). Calving rates between the fixed-time insemination and the first repeat were not different.

TABLE 2

CALVING RATE IN POSTPARTUM NURSING COWS FOLLOWING PROGESTERONE, CRONOLONE OR NORGESTOMET TREATMENTS

Treatment	No. cows	% calved to fixed AI	% calved to 1st repeat	% calved to 1st + 2nd AI
Progesterone sponge	223	47.5	58.6	77.9
Cronolone sponge	76	47.4	51.4	83.0
Norgestomet implant	222	46.4	45.7	73.8

EXPERIMENT 2

Time of ovulation

The effect of HCG and GnRH on time of ovulation is presented in Table 3.

TABLE 3

EFFECT OF HCG OR GnRH ON TIME OF OVULATION FOLLOWING A 10 DAY PROGESTERONE SYNCHRONISATION TREATMENT

Treatment	% heifers ovulating post progesterone removal (h)		
	40	50	60
Control	0	10	20
GnRH	0	40	80
HCG	10	80	60

10 heifers per subgroup

Administration of HCG had no effect on the percentage of
animals ovulated at 40 and 50 h (P > 0.05) but significantly
increased the ovulation rate at 60 h from progesterone removal
(P < 0.01). The administration of GnRH increased the percentage
of animals ovulated at 50 and 60 h (P < 0.01) from progesterone
removal while only 20% of control heifers had ovulated.

Fertility following ovulation control

The administration of HCG or GnRH to control the time of
ovulation more precisely had no effect on fertility following
a double fixed-time insemination routine (P > 0.05). When one
fixed-time (50 h) insemination was carried out, fertility
was reduced in the control and GnRH treated groups (P < 0.01)
but not in the HCG group (P > 0.05) when compared to the double
insemination routine (Table 4).

TABLE 4

FERTILITY FOLLOWING THE USE OF HCG OR GnRH WITH EITHER SINGLE OR DOUBLE
FIXED-TIME INSEMINATION PROCEDURE

Treatment	AI Time (h)	No. Heifers	% Preg.
Control	48 + 72	68	58
HCG	48 + 72	54	61
GnRH	48 + 72	55	60
Control	50	21	24
HCG	50	16	50
GnRH	50	14	28

EXPERIMENT 3

Effect of body condition score on fertility

Data on the relationship between body condition score and
fertility are shown in Table 5.

TABLE 5

CALVING RATE (%) IN RELATION TO BODY CONDITION SCORE AT THE FIXED-TIME
INSEMINATION

	Body condition score		
	1	2	3
April - June	57.9 (264)	46.3 (80)	26.7 (15)
July - September	64.3 (126)	54.8 (42)	50.0 (30)

The data are presented here within two seasons to avoid
interaction between animal condition and season effect. A
significant effect of body condition score was observed in
animals treated in the April to June season ($P < 0.01$). This
effect of body score is due to the lower calving rate ($P < 0.05$)
in animals in the poorest body condition score 3, when
compared to score 1. A similar trend was evident in the period
July to September but the differences here were not significant.

Effect of season on fertility

Calving rates for the different seasons are presented in
Table 6.

TABLE 6

CALVING RATE TO FIXED AI (48 + 72 h) IN DIFFERENT SEASONS

Animal type	Body score	Jan/ March	April/ June	July/ September	October/ December
Heifers	1	29.3(41)	58.0(264)	64.3(126)	56.5(23)
Heifers	2	35.3(34)	46.3(80)	54.8(42)	60.5(38)
Cows	2	28.6(28)	48.1(214)	49.2(122)	54.4(68)

These data are presented for animals in a definite body
condition score category to avoid interaction with season. A
significantly lower calving rate was recorded for heifers
(body condition score 1) in the January - March season than in

the other three (P< 0.01). A similar trend was observed in
heifers of body score 2 though these differences were not
significant (P>0.05). Similarly, in the postpartum beef cows
a lower calving rate was obtained in the January - March
season than in any of the other three seasons.

EXPERIMENT 4

Effect of body weight change on fertility
        Increasing body weight gain (daily gain from 0.23 to
0.40 kg/day had a significant effect on fertility following a
double (48 + 72 h) fixed-time insemination routine (Table 7).

TABLE 7

EFFECT OF BODYWEIGHT CHANGES ON FERTILITY IN PROGESTERONE SYNCHRONISED
HEIFERS

Nutrition level	No. heifers	Average daily gain (kg/day)	Preg. rate (%)
Low	53	0.23	47
Medium	60	0.32	58
High	60	0.40	67

        This increase was both linear (P< 0.01) and quadratic
(P<0.05).

EXPERIMENT 5

Effect of PMSG on ovulation rate
        The effect of varying the dose level of PMSG on the
subsequent ovulation rate is shown in Table 8.

        The mean ovulation rate increased linearly with the dose
rates employed here. There were no unovulated large follicles
($\not> 10$ mm) evident at 0 or 500 iu but they increased from 750
to 1000 iu.

TABLE 8

EFFECT OF PMSG DOSE ON OVULATION RATE IN HEIFERS

	PMSG (iu)			
	0	500	750	1000
No. heifers	16	38	42	16
Oestrus (%)	88	87	79	81
Mean ovulations	.9$^+$.06	2.1$^+$.34	3.1$^+$.70	5.3$^+$1.4
Mean follicles ($\geqslant$ 10 mm)	-	-	0.5$^+$2.0	1.8$^+$0.6

Effect of PMSG on fertility

Fertility rates following the administration of PMSG at the end of progestagen synchronisation treatment are shown in Table 9 for heifers and in Table 10 for cows.

TABLE 9

FERTILITY RATE FOLLOWING PROGESTERONE SYNCHRONISATION $^+$ PMSG (500 iu) IN BEEF HEIFERS

	Progesterone treatment	
	Control	+ PMSG (750 iu)
No. heifers	24	24
Preg. rate (%)	63	71
Twinning rate (%) (Of pregnant heifers)	-	23

There was no difference in the pregnancy rate of heifers (P$>$0.05), however because of the increased ovulation rate there was a higher fertility rate following PMSG administration. Of the PMSG treated heifers that became pregnant 23% had viable twin foetuses.

TABLE 10

CALVING RATE IN POSTPARTUM NURSING COWS FOLLOWING PROGESTAGEN $\pm$ SYNCHRONISATION - PMSG (750 iu).

	Progestagen treatment	
	Control	+ PMSG (750 iu)
No. cows	119	121
Calving rate (%)	53	72
Twinning rate (%) (of cows calving)	–	15

There was a significant difference in calving rate between the control and PMSG treated group ($P < 0.01$). The PMSG treated group had a calving rate approximately 20% higher than the control group.

The effect of the parturition to treatment interval for 89 of these cows is shown in Table 11.

TABLE 11

CALVING RATE FOLLOWING PROGESTAGEN $\pm$ PMSG ACCORDING TO THE POSTPARTUM TO TREATMENT INTERVAL

	Progestagen treatment			
	Control		+ PMSG (750 iu)	
	60*	60	60	60
Calving rate (%)	33 (9) **	57 (28)	73 (11)	73 (41)
No. twin births	–	–	1	8

* Days postpartum

** ( ) = No. cows

Cows treated prior to 60 days postpartum had a lower calving rate than those treated over 60 days. The administration

of PMSG increased the calving rate both prior to and post
60 days postpartum but the incidence of twin births was highest
in those cows treated over 60 days postpartum.

DISCUSSION

The results presented for Experiment 1 show that similar
fertility levels were obtained for three different short-term
progestagen treatments. The calving rates obtained here
following all treatments are similar to those reported following
insemination at a naturally occurring oestrus. Roche (1974_1)
reported a conception rate of 58% (slaughter data) while
Sreenan (1975) reported a calving rate of 54% in untreated
heifers, these calving rates are similar to those reported
here following a fixed-time AI procedure.

The calving rates recorded here for nursing postpartum
cows are about 5% lower than that recorded for heifers.
Control or untreated calving rates for postpartum cows following
artificial insemination are not readily available because of
the practical difficulties of inseminating postpartum beef
cows. The calving rates here agree with pregnancy rates
following synchronisation treatments reported in the literature
generally. Calving rates to the second or repeat service were
similar to those at the synchronised oestrus and can be taken
as an indication of control or untreated calving rates.
Perhaps in the postpartum cows, the significant figure is the
combined first and second service calving rates of approximately
80%.

The data from Experiment 2 show that it is possible to
induce early and synchronised ovulation in progestagen treated
animals following the use of either HCG or GnRH at the end of
the progestagen synchronisation treatment. However, the
timing of a single insemination must be accurate to take
advantage of this precise ovulation control. Both HCG and GnRH
significantly increased the percentage of animals ovulated at
60 hours from progesterone removal ($P < 0.01$). The use of a

single fixed-time insemination at 50 h from progesterone removal reduced the conception rate when compared to a double (48 + 72 h) insemination except for the HCG treated group. In the GnRH treated group, ovulation preceded the time of insemination in the majority of animals treated. It is probable that prolonging the time of insemination to a later time would yield a higher pregnancy rate. The normal pregnancy rate recorded for the HCG treated group may be due to the fact that while the insemination was at 50 h, ovulation occurred in 80% of these animals between 50 and 60 h from progesterone removal, thus allowing normal fertilisation to occur.

While the results from Experiments 1 and 2 show that normal fertility may be expected in both heifers and postpartum cows following short-term progestagen synchronisation treatments, the results from Experiment 3 indicate that this fertility level may be increased by manipulating some of the management factors involved at this time. An increase in animal body condition from a score of three to a score of one, was accompanied by an increase in calving rate from 26.9 to 57.9%.

The increase was greater in the early part of the breeding season (April - June) than later in the year (July - September) possibly reflecting the poor body condition of cows after the winter feeding season. This effect of body condition is in agreement with the work of McClure (1970) who reported that a 10% fall in body weight at the time of breeding was associated with a lower fertility rate. Likewise, the data of Drew et al. (1976) showed that increasing the level of nutrition resulted in an increase in calving rate.

From this experiment it is also clear that the season of treatment affects the outcome, with low pregnancy rates in both heifers and postpartum cows in the winter (January - March). This again reflects the poor level of nutrition pertaining at this time of year, traditionally, beef cow herds are maintained on a low plane of nutrition over the winter period.

In Experiment 4, the effect of nutrition level on fertility following the fixed-time insemination was examined under controlled conditions. The results of this study clearly indicate that bodyweight changes prior to and during the time of treatment have a significant effect on fertility. Increasing bodyweight gains were associated with increasing fertility ($P < 0.01$) in the synchronised heifers and the effect was both linear and quadratic. Various studies have shown that reproductive performance is affected by energy intake. Moller and Shannon (1972) reported that a 2% increase in bodyweight during the three weeks prior to service improved the conception rate. The increase in pregnancy rate found in these studies is consistent with the increase reported by Drew et al. (1976) in Friesian heifers synchronised with a 9 day norgestomet treatment. In their study an increase in average daily gain from 0.3 to 0.7 kg resulted in an increase in pregnancy rate of 18%.

In previous studies with postpartum nursing cows, Wiltbank et al. (1964) reported that increasing the TDN requirements (NRC recommended) to 150% of the normal level resulted in increase in fertility ranging from 13 to 22%. In a later study Wiltbank (1972) showed that decreasing the recommended (NRC) energy level by 50% resulted in a 50% reduction in pregnancy rate.

Thus, it is clear from these studies on body condition, season and nutrition level that manipulation of these factors to increase bodyweight changes for a period prior to synchronisation treatment will enhance the pregnancy or calving rates obtained following the fixed-time insemination procedure.

The data from Experiment 5 indicate a beneficial effect of gonadotrophic stimulation at the end of a progestagen synchronisation treatment in postpartum nursing beef cows.

It has been reported that in the first three weeks following parturition, there is a decline in pituitary FSH level (Labhsetwar et al., 1964; Saiduddin et al., 1968). More recent work has shown ovarian activity, based on plasma progesterone levels, in less than 20% of Charolais cows and only 31% of Salers cows by 60 days postpartum.

The data presented here, indicate that the progestagen - PMSG treatment used, stimulated follicular growth and an increased ovulation rate as evidenced by the high incidence of twin births. The incidence of twin births was higher after 60 days postpartum suggesting that the endocrinological state of these animals is similar to that of cyclic cows and thus the PMSG augmented the endogenous FSH level to increase the ovulation rates, Mulvehill and Sreenan (1977). Thimonier et al. (1976) have reported a higher oestrous response and improved degree of synchronisation following the use of PMSG in heifers.

The results here indicate that gonadotrophic stimulation may be used to stimulate follicular growth and ovulation resulting in an increased level of fertility. However, the data on ovulation response to varying the dose level of PMSG indicates that dose level may be a critical factor because of its effect on ovulation rate.

Overall, this series of experiments shows that short-term progestagen treatments based on either progesterone, cronolone or norgestomet can be used successfully to induce and synchronise oestrus and ovulation in both heifers and postpartum cows and that normal fertility levels can be attained following a double fixed-time insemination procedure. A more precisely controlled ovulation response can be achieved through the use of HCG or GnRH but the advantage in terms of changing to a single fixed time insemination depends on the timing of the insemination in relation to the timing of ovulation.

Fertility levels at the synchronised oestrus, however, can be modified according to management practice. Increased levels of nutrition resulting in increased bodyweight changes at the time of synchronised breeding will increase the pregnancy or calving rate obtained. The use of gonadotrophic stimulation in conjunction with synchronisation may increase the fertility rate in the postpartum nursing cow.

ACKNOWLEDGMENTS

The authors wish to thank Mr. A. McDonagh, Mr. G. Morris, Mr. D. Morris and Mr. P. Creaven for excellent technical assistance; Mr. F. McGee and the North West Cattle Breeding Society for help in the farm trial series.

These studies have been supported by the Commission of the European Communities.

508
REFERENCES

Chupin, D., Pelot, J., Alonso De Miquel, M. and Thimonier, J., 1976.
Progesterone assay for study of ovarian activity during postpartum
anoestrous in the cow. Proc. 8th Congr. Anim. Reprod. and AI
Cracow, 1 : 54.

Chupin, D., Le Provost, F., Mauleon, P., Ortovant, R., Parez, M., and
Petit, M. 1972. Utilisation d'implants sais-cutanes contenant un
Progestagene (SC 21009) pour synchroniser la reproduction de vaches
alliatantes. Resultats preliminaires. Proc. 7th Int. Congr. Anim.
Reprod and AI Munich, 2 : 851.

Drew, B., Wishart, D.F. and Thompson, D. 1976. The effect of nutrition
on fertility in dairy heifers. Proc. BSAP. Winter meeting.

Ginther, O.J. 1970. Length of oestrous cycle in sheep treated with
oestradiol. Am. J. Vet. Res., 31; 973.

Hawk, H.W. and Bolt, D.J. 1970. Luteolytic effect of oestradiol 17-B when
administered after midcycle in the ewe. Biol. Reprod. 2, 275.

Labhsetwar, A.P., Collins, W.E., Tyler, W.J. and Casida, L.E., 1964. Some
pituitary ovarian relationships in the periparturient cow. J. Reprod.
Fert., 8 : 85.

Mauer, R.E., Webel, S.K. and Brown, M.D. 1975. Ovulation control in cattle
with a progesterone intravaginal device (Prid) and gonadotrophin
releasing hormone (GnRH). Ann. Biol. Anim. Bioch. Biophys. 15 : 291.

McClure, T.J. 1970. An experimental study of the causes of nutritional
and lactational stress in fertility of pasture fed cows associated
with loss of body weight at about the time of mating. Res. Vet.
Sci., 11 : 247.

Moller, K. and Shannon, P., 1972. Bodyweight changes and fertility of
dairy cows. N.Z. Vet. J., 20 : 47.

Mulvehill, P. and Sreenan, J.M. 1977. Improvement of fertility in postpartum beef cows by treatment with PMSG and progestagen. J. Reprod.
Fert., 50 : 323.

Oxenreider, S.L. 1968. Effect of suckling and ovarian functions on postpartum reproductive activity in beef cows. Am. J. Vet. Res., 29 : 2099.

Roche, J.F. 1974. Effect of short-term progesterone treatment on oestrous
response and fertility in heifers. J. Reprod. Fert., 40 : 433.

Roche, J.F. 1974$_1$. Synchronisation of oestrus in heifers with implants of progesterone. J. Reprod. Fert., 44 : 337.

Roche, J.F. 1975. Synchronisation of oestrus in heifers and cows using a 12 day treatment with progesterone coils with or without GnRH. Proc. Seminar, 'Egg transfer in cattle', Cambridge, pp 231 - 242.

Roche, J.F. and Gosling, J.P. 1977. Control of oestrus and progesterone levels in heifers given intravaginal progesterone coils and injections of progesterone and oestrogen. J. Anim. Sci. 44, 1026.

Saiduddin, S., Riesen, J.W., Tyler, W.J. and Casida, L.E., 1968. Relation of postpartum interval to pituitary gonadotrophins, ovarian follicular development and fertility in dairy cows (effect of suckling, feeding level, breeding management and genetic level of milk production). Univ. Wis. Res. Bull., 270 : 15.

Smith, L.E., Vincent, C.K. 1973. Stage of cycle effect on bovine oestrous control. J. Anim. Sci. 36, 216.

Sreenan, J.M. and Mulvehill, P. 1974. Short-term (9 - 10 day) progestagen treatments for oestrous cycle control in cattle. Anim. Prod. Res. Reports, Agricultural Institute, Dublin, p 199.

Sreenan, J.M. 1975. Effect of long - and short-term intravaginal progestagen treatments on synchronisation of oestrus and fertility in heifers. J. Reprod. Fert. 45, 479.

Sreenan, J.M., Mulvehill, P. and Gosling, J.P. 1977. The effects of progesterone and oestrogen treatment in heifers on oestrous cycle control and plasma progesterone levels.

Thimonier, J., Pelot, J., Chupin, D. 1976. Synchronisation of oestrus in the cow with progestagens and prostaglandins. Proc. EEC Seminar 'Egg transfer in cattle'. Cambridge, pp 279 - 288.

Wiltbank, J.N., Rowden, W.W., Ingalls, J.E., and Zimmerman, D.D. 1964. Influence of postpartum energy level on reproductive performance of Hereford cows restricted in energy intake prior to calving. J. Anim. Sci., 23 : 1049.

Wiltbank, J.N., 1972. Reproductive patterns in beef cattle. Proc. of short course for Vets. 'Beef cattle reproduction'. Colorado State Univ. pp 1 - 7.

Wiltbank, J.N. and Cook, A.C. 1958. The comparative reproductive performance of nursed and milked cows. J. Anim. Sci., 17 : 640.

Wiltbank, J.N. and Goncalez-Padilla, E. 1975. Synchronisation and
    induction of oestrus in heifers with a progestagen and oestrogen.
    Proc. Colloquim. Control of sexual cycles in domestic animals.
    Ann. Biol. Anim. Bioch. Biophys. 15 : 255.

Wiltbank, J.N. and Kasson, C.W. 1968. Synchronisation of oestrus in
    cattle with an oral progestational agent and an injection of an
    oestrogen. J. Anim. Sci. 27, 113.

Wishart, D.F. and Young, I.M. 1974. Artificial insemination of cattle at
    a predetermined time following treatment with a potent progestin
    (SC-21009). Vet. Rec. 95, 503.

FACTORS INVOLVED IN OESTROUS CYCLE CONTROL IN THE BOVINE

S. Mawhinney and J.F. Roche

An Foras Taluntais, Grange, Dunsany, Co. Meath

ABSTRACT

Experiments were carried out on 3 173 cattle to determine factors influencing the onset of oestrus and fertility following a progesterone releasing intravaginal device (PRID) treatment.

Injection of 5 mg oestradiol benzoate (ODB) with or without 200 mg progesterone or 10 mg ODB by gelatin capsule adhered to the coil did not influence oestrous response after a 9 or 12 day PRID treatment. An interaction between pattern of onset of oestrus and length of treatment was found. When oestrogen is given, the pattern is more precise after a 12 than after a 9 day treatment. With a 14 day treatment period, a precise onset of oestrus is obtained without oestrogen.

Calving rate following 9 or 12 day treatments using injection or capsule with ODB was similar to that in controls in all experiments except one. A single insemination at 56 h gave similar fertility to two inseminations at 56 and 74 h and to control cows. The timing of a single insemination is critical with 52 h insemination giving lower fertility than single 56 or 60 h inseminations. In this experiment, the calving rate in controls was higher than insemination at 56 or 60 h. Following a 14 day treatment without oestrogen, fertility was significantly lower than that after a 12 day treatment with oestrogen or in control cows.

INTRODUCTION

The oestrous response and the pattern of onset of oestrus following any synchronisation treatment is critical when artificial insemination is to be carried out at a previously determined time (fixed-time AI). In order to achieve normal fertility to fixed-time AI, the onset of oestrus following treatment should be concise, ideally with all treated animals coming into oestrus in a 24 h period.

Administration of exogenous progesterone or synthetic progestagens for 18 to 21 days results in a concise onset of oestrus in treated animals but fertility at this oestrus is lower than normal (Hansel, 1967; Mauleon and Chupin, 1971; Jochle, 1972; Roche and Crowley, 1973). Short-term administration of progesterone or progestagens for 9 to 12 days results in normal fertility but requires the administration of oestrogen at the start of treatment to cause premature luteal regression (Wiltbank and Kasson, 1968; Wiltbank et al., 1971; Chupin et al., 1972; Roche, 1974_2; Wishart and Young, 1974; Sreenan and Mulvehill, 1975). The luteolytic action of oestrogen is dependent on the stage of the cycle at which it is administered and is ineffective in animals on Days 0 to 3 and Days 17 to 20 of the cycle (Denamur, 1972; Lemon, 1975). When progesterone is administered by subcutaneous implant, 5 mg oestradiol benzoate (OB) causes ovulation in animals between Days 17 and 20 of the cycle (Roche, 1974_1). Incorporating 50 mg progesterone into this injection inhibits the ovulatory effect of oestrogen and improves synchronisation in these animals (Roche, 1974_1; Wiltbank and Gonzalez-Padilla, 1975).

When a progesterone releasing intravaginal device (PRID, Abbott Laboratories, USA) is used instead of a subcutaneous progesterone implant, plasma progesterone levels rise rapidly, attaining luteal levels within 90 minutes of insertion (Mauer et al., 1975; Roche and Gosling, 1977).

In the light of this information, experiments were carried out to determine whether extra progesterone is required at time of PRID insertion. The effect of the duration of PRID treatment on oestrous response was studied and trials were carried out to determine if the treatment could be simplified by an alternative method of oestrogen administration. Field trials were also carried out to determine fertility following these treatments and following various fixed-time insemination regimes.

MATERIALS AND METHODS

In experiments 1 and 2 cyclic Hereford cross heifers at pasture on the research station and weighing over 275 kg body-weight were used. They were run with vasectomised bulls fitted with chin-ball marking devices and observed for signs of oestrus twice daily.

Field trials were carried out on commercial dairy farms using Friesian cows which had calved at least five weeks prior to treatment. In trials 2 and 5 Friesian replacement heifers were also used. In all trials control animals were allocated at random on each farm to be inseminated at a detected oestrus during a 24-day period. All animals were inseminated with frozen semen stored in straws by inseminators from commercial AI stations, and the dates of calving were recorded by farmers and subsequently obtained.

PRIDs where used consisted of progesterone impregnated silastic coils and were inserted into the anterior vagina by means of a plastic speculum and removed by pulling on a nylon string attached to the coil (Roche, 1976).

EXPERIMENT 1

This experiment was conducted to determine the necessity to inject extra progesterone with the OB at the start of

treatment following 9 or 12 day treatment periods. Sixty-four
heifers on Days 0 to 3 of the oestrous cycle, as determined
from previous oestrous dates, received PRID treatment for
either 9 or 12 days. A further 20 heifers on Days 17 to 20 of
the oestrous cycle received PRID treatment for 9 days. At the
time of PRID insertion, half of the animals in each group
received an intramuscular injection of 5 mg ODB and half
received an intramuscular injection of 5 mg ODB plus 200 mg
progesterone. Following PRID removal, all heifers were
individually checked for oestrus with the aid of a vasectomised
bull three times daily at 8.00 am, 4.00 pm and 9.00 pm.
Heifers not observed in oestrus by 9.00 pm were run overnight
with a vasectomised bull fitted with a chin-ball marking device.

EXPERIMENT 2

This experiment was carried out to determine whether the
oestrous response following 9, 12 or 14 day treatments would
be improved by either OB or extra progesterone at the start
of treatment. Fifty-seven heifers between Days 0 and 3 of the
oestrous cycle, as determined from previous oestrous dates,
received PRID treatments for either 9, 12 or 14 days. Animals
in each of these groups were further randomised into three sub-
groups at the time of PRID insertion. One sub-group received
no further treatment and the other two sub-groups received
either 10 mg ODB or 200 mg progesterone contained in a gelatin
capsule adhered to the inside of the PRID with a medical
adhesive (Dow Corning, Silicone Type A). These capsules
dissolve and release their contents into the vagina within two
hours (Webel, 1977) thus avoiding the need for injection.
Animals were checked for oestrus as in Experiment 1.

TRIAL 1

This trial was carried out to determine the fertility
following oestrogen administration by injection or by a gelatin
capsule adhered to the PRID at the start of a 9 or 12 day
treatment period. Three hundred and thirty six cows were

randomised into three groups. Animals in groups 1 and 2
received PRID treatment for 9 and 12 days respectively. Half
the animals in each group received an intramuscular injection
of 5 mg ODB plus 200 mg progesterone at PRID insertion and
the other half received 10 mg ODB by means of a gelatin capsule
adhered to the PRID. Animals in group 3 received no treatment
and acted as controls. The cows were checked for oestrus at
least 3 times daily and inseminated at a detected oestrus
following PRID removal or during the 24 day detection period
for the controls.

TRIAL 2

This trial was also carried out to determine the fertility
following oestrogen administration by injection or by capsule
at the start of a 12 day treatment. One hundred and fourteen
heifers received PRID treatment for 12 days and were randomised
into two groups. At PRID insertion, animals in group 1
received an intramuscular injection of 5 mg ODB plus 200 mg
progesterone and animals in group 2 received 10 mg ODB by means
of a gelatin capsule adhered to the PRID. The heifers were
inseminated at a detected oestrus following PRID removal.

TRIAL 3

This trial was carried out to determine the fertility
following either one or two fixed-time inseminations after
treatment where oestrogen was administered either by injection
or by capsule at the start of a 12 day treatment. Three
hundred and eighty five cows were randomised into three groups.
Animals in group 1 received no treatment and acted as controls.
Animals in the other two groups received a 12 day PRID
treatment with oestrogen administered at PRID insertion either
by intramuscular injection or by a gelatin capsule adhered to
the PRID. The treated animals were inseminated at a fixed time
with either one insemination at 56 h after PRID removal or
two inseminations at 56 and 74 h after PRID removal. The
controls were inseminated at a detected oestrus occurring during
a 24-day observation period.

TRIAL 4

This trial was carried out to determine the fertility
following either one or two fixed-time inseminations after a
12 or 14 day treatment period. Three hundred and thirty seven
cows were randomised into three groups, animals in one of which
received no treatment and acted as controls. The other two
groups received either a 12 day PRID treatment with oestrogen
administered intramuscularly at PRID insertion or a 14 day
PRID treatment without oestrogen administration at PRID
insertion. The treated animals were inseminated at a fixed
time with either one insemination at 56 h after PRID removal
or two inseminations at 56 and 74 h after PRID removal. The
controls were inseminated at a detected oestrus during a 24
day observation period.

TRIAL 5

This trial was carried out to determine how the time of a
single fixed-time insemination would affect fertility following
a 12 day treatment. One thousand and ninety cows and 107
heifers were randomised into four groups, animals in one of
which received no treatment and acted as controls. Animals
in the remaining three groups received a 12 day PRID treatment
with oestrogen administered at PRID insertion by gelatin
capsule and were inseminated once at 52, 56 or 60 h
following PRID removal. Control animals were inseminated at
a detected oestrus during a 24 day observation period.

RESULTS

EXPERIMENT 1

Following either a 9 or 12 day PRID treatment, 93% of
animals came into oestrus on the first four days after PRID
removal, with 82% coming into oestrus on Days 2 and 3 (Table 1).
The administration of extra progesterone with the oestrogen at

the start of treatment did not affect the oestrous response or pattern of onset following PRID treatment. Following a 9 day treatment, the oestrous response after treatment was similar in animals on Days 0 to 3 of the cycle and in animals on Days 17 to 20 at the start of treatment. The pattern of onset was more concise in animals which were on Days 17 to 20 at the start of treatment with 65% of animals coming into oestrus on Day 2 after PRID removal compared to 38% of animals which were on Days 0 to 3 of the cycle at the time of PRID insertion, but this difference is not statistically significant. In animals on Days 0 to 3 of the oestrous cycle at PRID insertion, the length of treatment does not affect oestrous response, following coil removal but signicantly ($P < 0.05$) affects the pattern of onset. Following a 9 day treatment 38% of animals came into oestrus on Day 2 after PRID removal and following a 12 day treatment 66% of animals came into oestrus on Day 2.

TABLE 1

THE EFFECT OF STAGE OF CYCLE AND LENGTH OF TREATMENT ON SUBSEQUENT OESTROUS RESPONSE AFTER PRID TREATMENT

Stage of cycle at start of treatment	Treatment	No. Heifers	Onset of Oestrus			
			1	2	3	4
0-3	9-day + OB	16	–	6	6	2
	9-day + OB + P	16	–	6	8	1
0-3	12-day + OB	16	–	12	2	2
	12-day + OB + P	16	–	9	4	2
17-20	9-day + OB	10	–	7	2	–
	9-day + OB + P	10	1	6	1	1

EXPERIMENT 2

Significantly more heifers were in oestrus on Days 1 - 4
(P < 0.05) following a 9 or 12 day PRID treatment when 10 mg
ODB was administered in a gelatin capsule attached to the PRID
than when 200 mg progesterone was administered or when no
further treatment was given (Table 2). Oestrogen administration
did not enhance the oestrous response following a 14 day PRID
treatment. The response after this treatment was high whether
oestrogen, progesterone or nothing was given at the start of
treatment.

TABLE 2

THE EFFECT OF LENGTH OF TREATMENT AND ADMINISTRATION OF OESTROGEN OR
PROGESTERONE BY GELATIN CAPSULE ADHERED TO THE COIL ON SUBSEQUENT OESTROUS
RESPONSE

Contents of gelatin capsule	No. in oestrus on days 1 - 4 after PRID treatment for		
	9 days	12 days	14 days
None	$\frac{4}{7}$	$\frac{2}{6}$	$\frac{5}{6}$
10 mg OB	$\frac{6}{6}$	$\frac{6}{7}$	$\frac{6}{6}$
100 mg P	$\frac{1}{6}$	$\frac{6}{7}$	$\frac{6}{6}$

TRIAL 1

The retention rate of PRID in this trial was 97% (Table 3).
Of the treated animals which retained the PRID, 88% were
detected in oestrus following PRID removal and inseminated.
Over a 24 day observation period, 70% of the control animals
were detected in oestrus and inseminated.

The treated animals receiving oestrogen by means of a
capsule had a similar oestrous response and calving rate to
those receiving oestrogen by means of an injection with

progesterone. Ninety-six out of 108 animals injected with
oestrogen were inseminated and 54% of these calved to the
first insemination compared to 109 inseminated of 124 animals
receiving the capsule, of which 52% calved to the first
insemination. The calving rates of both treated groups were
similar to that for controls where 52% of animals inseminated
calved to the first insemination.

TABLE 3

CALVING RATE FOLLOWING OESTROGEN BY MEANS OF CAPSULE OR INJECTION AT PRID
INSERTION

	10 mg OB capsule	5 mg OB + 200 mg P injection	Controls
No. treated	127	111	98
No. PRID's removed	124	108	
No. AI'ed	109	96	69
No. repeats	42	38	31
No. calved to 1st AI	57	52	36
% calved to 1st AI	52%	54%	52%

TRIAL 2

There were no control animals in this trial. The oestrous
response and calving rates of the animals receiving oestrogen
by means of a capsule and those receiving an oestrogen injection
with progesterone were similar (Table 4). In both groups, 91%
of treated animals were inseminated and the calving rates were
43% and 42% for those receiving the capsule and those injected,
respectively.

TABLE 4

CALVING RATE IN HEIFERS FOLLOWING OESTROGEN BY MEANS OF CAPSULE OR
INJECTION AT PRID INSERTION

	10 mg OB capsule	5 mg OB + 200 mg P injection
No. treated	59	55
No. PRIDs removed	58	53
No. AI'ed	53	48
No. repeats	24	18
No. calved to 1st AI	23	20
% calved to 1st AI	43%	42%

TRIAL 3

The results of this trial show that using fixed-time AI
allows all treated animals to be inseminated whereas only 73%
of the control animals were detected in oestrus and inseminated
(Table 5). The calving rates of those animals which were
inseminated were similar in both groups being 52% for treated
animals and 57% for control animals. However, when the
animals originally allocated to each group are considered the
calving rate of 52% for the treated animals is significantly
higher $(P < 0.05)$ than that for the control animals which becomes
42%.

Calving rates following oestrogen administration by
capsule or by injection were similar and did not differ from
that of control animals (Table 6).

A single fixed-time insemination at 56 h after PRID
removal gave similar fertility to that following two fixed-time
inseminations at 56 and 74 h after PRID removal (Table 7). The
fertility following fixed-time insemination did not differ
from that following insemination at a detected oestrus in the
control animals.

TABLE 5

CALVING RATE FOLLOWING ONE OR TWO FIXED-TIME AIs AFTER PRID TREATMENT WITH AN OESTROGEN CAPSULE OR INJECTION AT PRID INSERTION

	Capsule + 56	Capsule + 56 + 74	Injection + 56	Injection+ 56 + 74	Controls
No. treated	73	62	67	59	169
No. AI'ed	73	62	67	59	124
No. repeats	32	30	24	26	45
No. calved to 1st AI	36	31	38	31	71
% calved to 1st AI	49%	50%	57%	53%	57%
% treated calved to 1st AI	49%	50%	57%	53%	42%

TABLE 6

CALVING RATE FOLLOWING OESTROGEN BY CAPSULE OR INJECTION AT PRID INSERTION

	Capsule	Injection	Controls
No. AI'ed	135	126	124
No. repeats	62	50	45
No. calved to 1st AI	67	69	71
% calved to 1st AI	50%	55%	57%

TABLE 7

CALVING RATE FOLLOWING ONE OR TWO FIXED-TIME INSEMINATIONS AFTER PRID REMOVAL

	56	56 + 74	Controls
No. AI'ed	140	121	124
No. repeats	56	56	45
No. calved to 1st AI	74	62	71
% calved to 1st AI	53%	51%	57%

TRIAL 4

The calving rate following a 12 day PRID treatment was similar following one or two fixed-time inseminations and did not differ significantly from that of controls (Table 8). However, the fertility following a 14 day PRID treatment was significantly (P< 0.01) depressed and the lower fertility was particularly noticeable after a single insemination at 56 h.

TABLE 8

CALVING RATE FOLLOWING A 12 OR 14 DAY PRID TREATMENT

	12-1	12-2	14-1	14-2	Control
No. treated	82	75	85	75	151
No. AI'ed	79	72	82	72	121
No. repeats	31	32	51	38	46
No. calved to 1st AI	42	36	25	31	68
% calved to 1st AI	53%	50%	30%	43%	56%
% treated calved to 1st AI	51%	48%	29%	41%	45%

TRIAL 5

The calving rates in dairy cows to a single fixed-time insemination at 56 or 60 h after PRID removal are similar and significantly (P <0.05) higher than that following a single insemination at 52 h (Table 9). In heifers the fertility following insemination at 56 or 60 h is the same as in controls inseminated at a detected oestrus (Table 10), but in dairy cows it is significantly (P<0.01) lower than that of the controls in this trial. When the numbers of animals originally allocated to each treatment are taken into account, fixed-time insemination at 56 or 60 h gives similar calving rates to that of controls, in both cows and heifers.

TABLE 9

CALVING RATE IN COWS FOLLOWING A SINGLE FIXED-TIME INSEMINATION AT
DIFFERENT TIMES AFTER PRID REMOVAL

	52	56	60	Controls
No. treated	252	254	241	343
No. AI'ed	252	254	241	214
No. repeats	156	131	132	83
No. calved to 1st AI	69	95	89	114
% calved to 1st AI	27%	37%	37%	53%
% treated calved to 1st AI	27%	37%	37%	33%

TABLE 10

CALVING RATE IN HEIFERS FOLLOWING A SINGLE FIXED-TIME INSEMINATION AT
DIFFERENT TIMES AFTER PRID REMOVAL

	52	56	60	Controls
No. treated	27	29	16	35
No. AI'ed	27	29	16	24
No. repeats	11	13	6	9
No. calved to 1st AI	11	15	9	12
% calved to 1st AI	41%	52%	56%	50%
% treated calved to 1st AI	41%	52%	56%	35%

DISCUSSION

From the results in Experiment 1 it can be seen that
additional progesterone is unnecessary at the start of treatment
when a PRID is used as the synchronisation treatment rather
than a subcutaneous implant of progesterone. The initial high
release of progesterone from the PRID seems to be sufficient
to overcome the ovulatory effect of oestrogen in animals on
Days 17 to 20 of the oestrous cycle at the start of treatment.
Animals on Days 0 to 3 of the cycle at the start of treatment
have a less concise onset of oestrus following treatment than

those on Days 17 to 20 and this spread is affected by the
duration of treatment.

The general trend for all treatments, as seen in Experiment
1 and in Trial 4 as well as in reports in the literature seems
to be 'the shorter the treatment, the greater the spread in
onset of oestrus after treatment'. This may be related to
high endogenous levels of progesterone at the end of treatment
due to poor luteolysis at the start of treatment. With a
longer treatment, a greater proportion of animals will undergo
natural luteolysis during treatment and thus reduce the need
for administering a luteolytic agent at the start of treatment.
The greater spread in onset in those animals on Days 0 to 3 of
the oestrous cycle at the beginning of treatment is probably
due to the fact that oestrogen may be luteolytic, anti-luteolytic,
luteotrophic or anti-luteotrophic, according to the stage of
cycle at which it is administered and is most likely to be
luteotrophic at this time (Lemon, 1975). Animals which have
low levels of plasma progesterone at the time of PRID removal
have a more precise onset of oestrus than animals with high
levels of progesterone at PRID removal which take a longer
time to come into oestrus (Roche and Gosling, 1977).

The results of Experiment 2 indicate that the treatment may
be simplified by administering the oestrogen at the start of
treatment by means of a gelatin capsule adhered to the inside
of the PRID. This avoids the need to inject the animals at
the start of treatment. Experiment 2 shows that this form of
oestrogen administration is effective following a 9 or 12 day
treatment as it gives a higher oestrous response on Days 1 to
4 following PRID removal than does progesterone alone or no
further treatment. Oestrogen administration is unnecessary at
the start of a 14 day treatment as this alone gives a concise
pattern of onset of oestrus. The effectiveness of the
oestrogen capsule is further borne out by the results of Trials
1, 2 and 3 which show normal calving rates are obtained
following this method of oestrogen administration.

Trial 3 shows that normal fertility can be achieved with fixed-time insemination following PRID removal. A single insemination at 56 h after the end of a 12 day PRID treatment gives similar fertility to that in control animals. In our trials, control animals are observed for a 24 day period which is slightly longer than the length of an average oestrous cycle. The number of animals showing oestrus during this time is an indication of the proportion of cyclic animals in the herd and the ability of the stockman to detect oestrus. The fertility in the control animals gives an indication of the general fertility in the herd as animals are allocated at random to the various treatments.

Two inseminations at 56 and 74 h after PRID removal does not improve fertility compared to a single fixed-time insemination or insemination at oestrus in control animals. However, we have found in all our trials involving fixed-time insemination about 8 to 10% of animals will exhibit oestrus one to four days after the fixed-time insemination. These animals must be re-inseminated at the detected oestrus in order to have a normal chance of getting in calf at that oestrus.

The reason for the longer interval to oestrus in these animals may be due to inadequate synchronisation resulting in high levels of progesterone during the last three days of treatment, which has been shown to lead to a longer interval to oestrus (Roche and Gosling, 1977). It may also be due to individual variation in the rate of follicular development (Stabenfeldt et al., 1969). Gonadotrophic stimulation at the end of treatment might reduce this variability.

The results of Trial 4 show that increasing the length of treatment from 12 to 14 days results in lowered fertility especially following a single fixed-time insemination at 56 h after PRID removal. This is similar to the results from earlier experiments using long-term progesterone treatments. The increased length of treatment may alter the hormonal pattern

at the synchronised oestrus and so interfere with normal
fertilisation. It has been shown that following a long-term
progestagen treatment, animals with normal hormonal patterns
conceived whereas the animals which did not conceive had
abnormal patterns, mainly premature or greatly delayed LH peaks
(Hansel et al., 1975). Abnormal hormone levels at the induced
oestrus may alter the rate of sperm or ovum transport or ovum
cleavage and thus affect fertility (Britt and Ulberg, 1972;
Chang and Harper, 1966; Hill et al., 1971; Henricks et al.,
1973; Quinlivan and Robinson, 1969; Rodetter et al., 1972).
This altered transport rate may explain why fertility following
two fixed-time inseminations is higher than that following
only one after a 14 day treatment.

The lowered fertility following a single fixed-time
insemination at 52 h after PRID removal compared to 56 or 60 h
shows that the time of insemination is vital. In this trial
the level of fertility in dairy cows following a single fixed-
time insemination at 56 or 60 h is lower than that in the
control animals whereas in all other trials, fertility following
a single fixed-time insemination at 56 h was comparable to
that of the controls. This may have been due to stress caused
in the treated animals by handling them three times on the day
on which they were inseminated. The control animals were not
inseminated on this day unless they exhibited oestrus so the
majority of control animals were inseminated under less
stressful circumstances. Similarly in heifers, where the
numbers involved were small and thus handling was easier, the
fertility following a fixed-time insemination at 56 or at 60 h
was comparable to that of the controls.

Thus the recommended treatment with the PRID is to insert
it with an attached oestrogen capsule at least 35 days after
calving and to leave it in the vagina for 12 days. The animals
should be inseminated once at 56 h following PRID removal and
observed for oestrus for four or five days following insemination.
Any animals exhibiting oestrus at this time should be re-
inseminated. If this is not done then overall calving rate

will be reduced by 3 to 5%.

The use of a short-term progesterone treatment can ease the management problems experienced by the farmer in the use of artificial insemination, especially in beef herds where oestrous detection has always been the main drawback to the use of artificial insemination.

ACKNOWLEDGMENTS

We would like to thank the European Economic Community for supporting this study, Dr. F.J. Harte for providing facilities and Abbott Laboratories for supplying the PRIDs. We would also like to thank Mr. D. Prendiville and Mr. B. Davis for excellent technical assistance and Mr. K. Bannon for invaluable practical assistance. Thanks are also due to the artificial insemination stations and the farmers who co-operated in the field trials.

528

REFERENCES

Britt, J.H. and Ulberg, L.C. 1972. Melengestrol acetate administration to
    dairy heifers and progestagen levels in the peripheral blood plasma.
    J. Reprod. Fert. 29 : 119.

Chang, M.C. and Harper, M.J. 1966. Effects of ethinyl estradiol on egg
    transport and development in the rabbit. Endocrinology 78 : 860.

Chupin, D., Le Provost, F., Mauleon, P., Ortavant, R., Parez, M. and
    Petit, M. 1972. Utilisation d'implants sous-cutanes contenant un
    progestagene (SC 21009) pour synchroniser la reproduction de vaches
    alliatantes. Resultats preliminaires. Proc. 7th Int. Congr. Anim.
    Reprod. and AI., Munich, 2 : 851.

Denamur, R. 1972. Regulation neuroendocrinienne du cycle oestrien chez
    les animaux domestiques. Proc. 7th Int. Congr. Anim. Reprod. and AI.,
    Munich, 1 : 20.

Hansel, W. 1967. Control of the ovarian cycle in cattle. In reproduction
    in the female mammal, p. 419. Eds. G.E. Lamming and E.C. Amoroso,
    Butterworths, London.

Hansel, W., Schechter, R.J., Malven, P.V., Simmons, K.R., Black, D.L.,
    Hackett, A.J. and Saatman, R.R. 1975. Plasma hormone levels in 6-methyl
    -17-acetoxy progesterone and estradiol benzoate treated heifers.
    J. Anim. Sci. 40 : 671.

Henricks. D.M., Hill, J.R. and Dickey, J.F. 1973. Plasma ovarian hormone
    levels and fertility in beef heifers treated with melengestrol acetate
    (MGA). J. Anim. Sci. 37 : 1169.

Hill, J.R., Lamond, D.R., Henricks, D.M., Dickey, J.F. and Niswender, G.D.
    1971. The effect of melengestrol acetate (MGA) on ovarian function
    and fertilisation in beef heifers. Biol. Reprod. 4 : 16.

Jochle, W. 1972. Pharmacological aspects of the control of the cycle in
    domestic animals. Proc. 7th Int. Congr. Anim. Reprod. and AI.,
    Munich, 1 : 97.

Lemon, M. 1975. The effect of oestrogens alone or in association with
    progestagens on the formation and regression of the corpus luteum of
    the cyclic cow. Ann. Biol. Anim. Bioch. Biophys. 15 (2) : 243.

Mauer, R.E., Webel, S.K. and Brown, M.D. 1975. Ovulation control in cattle
    with progesterone intravaginal device (PRID) and gonadotrophin releasing
    hormone (GnRH). Ann. Biol. Anim. Bioch. Biophys. 15 : 291.

Mauleon. P. and Chupin, D. 1971. Maitrise des cycles sexuels chez les
    bovins. Econ. et Med. Anim. 12 : 31.

Quinlivan, T.D. and Robinson, T.J. 1969. Numbers of spermatozoa in the genital tract after artificial insemination of progestagen-treated ewes. J. Reprod. Fert. 19 : 73.

Roche, J.F. 1974$_1$. Effect of short-term progesterone treatment on oestrous response and fertility in heifers. J. Reprod. Fert. 40 : 433.

Roche, J.F. 1974$_2$. Synchronisation of oestrus in heifers with implants of progesterone. J. Reprod. Fert. 41 : 337.

Roche, J.F. 1976. Calving rate of cows following insemination after a 12-day treatment with silastic coils impregnated with progesterone. J. Anim. Sci. 43 : 164.

Roche, J.F. and Crowley, J.P. 1973. The fertility of heifers inseminated at predetermined intervals following treatment with MGA and HCG to control ovulation. J. Reprod. Fert. 35 : 211.

Roche, J.F. and Gosling, J. 1977. Control of oestrus and progesterone levels in heifers given intravaginal progesterone coils and injections of progesterone and oestrogen. J. Anim. Sci. (in press).

Rodetter, G.H., Hopwood, M.L. and Wiltbank, J.N. 1972. LH and oestrogens following oestrous control. J. Anim. Sci. 34 : 901.

Sreenan, J.M. and Mulvehill, P. 1975. The application of long- and short-term progestagen treatments for oestrous cycle control in heifers. J. Reprod. Fert. 45 : 367.

Stabenfeldt, G.H., Ewing, L.L. and McDonald, L.E. 1969. Peripheral plasma progesterone levels during the bovine oestrous cycle. J. Reprod. Fert. 19 : 433.

Webel, S.K. 1977. Control of the oestrous cycle in cattle with a progesterone releasing intravaginal device. Proc. 8th Int. Congr. Anim. Reprod. and AI., Krakow, 3, 521.

Wiltbank, J.N. and Gonzalez-Padilla, E. 1975. Synchronisation and induction of oestrus in heifers with a progestagen and oestrogen. Proc. Colloq. on Control of Sexual Cycles in Domestic Animals. Held at INRA, France, October 27-30, 1974. Ann. Biol. Anim. Bioch. Biophys. 15 : 255.

Wiltbank, J.N. and Kasson, C.W. 1968. Synchronisation of oestrus in cattle with an oral progestational agent and an injection of an oestrogen. J. Anim. Sci. 27 : 113.

Wiltbank, J.N., Sturges, J.C., Wideman, D., Le Fever, D.G. and Faulkner, L.C. 1971. Control of oestrus and ovulation using subcutaneous implants and oestrogens in beef cattle. J. Anim. Sci. 33 : 600.

Wishart, D.F. and Young, L.M. 1974. Artificial insemination of progestin
   (SC 21009) treated cattle at predetermined times. Vet. Rec. 95 : 503.

MODIFYING FACTORS OF FERTILITY
AFTER DIFFERENT OESTROUS CONTROL TREATMENTS IN BEEF CATTLE

P. Mauleon, D. Chupin, J. Pelot, D. Aguer*

INRA - Station de Physiologie de la Reproduction
37380 - Nouzilly, * Searle Vétérinaire, Périsud,
Bd. Romain Rolland 92128 - Montrouge, France

ABSTRACT

Fertility after oestrous control treatment depends on three groups of factors: type of treatment, parity and ovarian cyclicity of females and environmental conditions.

1. Treatments

In a given population of cows, subcutaneous implants of norgestomet (Searle) and vaginal coils of progesterone (Abbott) give the same calving rates (54.1 vs 53.2% respectively).

With norgestomet implants, higher calving rates are obtained with higher doses (60.1% with 12 mg vs 44.4% with 6 mg) and a short duration (53.8% with 7 days vs 28.9% with 13 - 15 days).

With both treatments (implants or coils), not using oestradiol benzoate or valerate at the beginning of treatment, results in a decrease of 6 to 10% in calving rates.

Injection of PMSG (500 to 800 iu depending on the breed) increases the calving rate by 12 to 15%

2. Female characteristics

At the time of treatment (between February and May), nursing cows are characterised by a very low level of ovarian activity: 16 and 25% of Charolais and Salers cows respectively have a corpus luteum before treatment as detected by plasma progesterone levels.

Such a physiological state modifies treatment efficiency and is different between breeds in terms of depth. So, after 12 mg norgestomet implants + 600 iu PMSG, only 68% of non-cycling Charolais cows are induced to ovulate vs 100% of Salers cows with 800 iu of PMSC. PMSG (800 iu) injected into Charolais cows increased ovulation induction, but also the number of multiple ovulations, which detracts from the practical value of this method. Moreover, fertility of those induced ovulations is lower (59.5%) than that of previously cycling cows (73.0%) possibly due to an unfavourable uterine environment.

Primiparous cows of the two breeds have lower ovarian activity (10.3 vs 35.8%) and lower fertility (37 vs 53%) than multiparous cows indicating the difficulties during the first reproductive cycle in such an extensive animal husbandry.

3. Environmental factors

A very low nutritional level during winter is the main reason for the delay in returning to cyclicity. When cows are turned out on pasture, the percentage of cycling cows rises abruptly. The effect is evident upon fertility since pregnancy rates 21 days after AI (assessed by progesterone assay) vary from 77.8% in the first week of March to 60.6% in the first week of April, 25% in the last week of April and 57.6% in May (the cows are turned out around April 15th). Thus, an effect of flushing is actually evaluated.

The fertility of bulls chosen from AI must be considered since we have obtained calving rates varying from 29.1 to 61.1% according to the bull. Progestagens create an unfavourable situation in the genital tract for spermatozoa which emphasizes the importance of spermatazoa number and quality.

Type of maternal behaviour and suckling stimuli are known to delay oestrus and ovulation. Following Wiltbank, we are testing the effect of a temporary weaning before or after treatment on ovulating rate.

INTRODUCTION

Pregnancy rates obtained after oestrous control treatments
are variable and sometimes disappointing. Some improvements can
be found through modifications of treatments, but most of these
variations of fertility can be explained by factors related to
females (parity, ovarian activity...) and to environmental
conditions.

1. - FACTORS RELATED TO TREATMENTS

In beef cattle, we have shown that most of the females are
in anoestrus at the time of breeding (Chupin et al., 1975), ie
during the end of the winter period. This means that treatment
with prostaglandins or their analogues have to be discarded.
The proof can be found in an experiment made by the union of Al
centres in which 620 two year old Charolais heifers have been
treated with two injections of Estrumate (1Cl) and have given a
calving rate of 28.5% (Chupin et al., 1977_1). So most of our
experiments have been made with progestagens, mainly norgestomet
implants (Searle and Co.) (Chupin et al., 1974). Some animals
have also been treated with progesterone coils (Abbott).

1) During five years of field trials, the calving rates
after two Al at fixed times increased with the dose of norgestomet
contained in the hydron* implants (Table 1) and decreased with
the duration of treatment (Table 3).

The highest calving rate is obtained in Salers nursing cows
with the 12 mg dose. The same study in Charolais nursing cows
gave little differences between 6 mg plus 3 mg injected and 12
mg (42.7 vs 45.8% calving rates). Experiments are now in progress
to test the effect of higher available daily doses of norgestomet
progestagen by the way of implants of Silastic allowing a
constant release throughout treatment instead of a curvelinear
decrease (Table 2).

* Without further definition, the term of 'norgestomet implants' means SC
21009 included in hydron support (Searle and Co).

TABLE 1

CALVING RATE AFTER TWO AI AT FIXED TIME FOLLOWING NORGESTOMET IMPLANT IN
SALERS NURSING COWS. EFFECT OF DOSE OF NORGESTOMET

Dose of norgestomet (mg)	Number of cows	Calving rates
6	414	44.4
9	81	55.5
6 + 3 Injected (im) the first day of treatment	258	55.4
12	672	60.1

Norgestomet: SC 21009 = dose included in the hydron matrix of implant,
incompletely released during the period of insertion (30%).

TABLE 2

CALVING RATE AFTER TWO AI AT FIXED TIME FOLLOWING TWO TYPES OF NORGESTOMET
IMPLANT (CHAROLAIS NURSING COWS)

	12 mg Hydron implants	300 µg Silastic implants
% Pregnant cows at 21 days post AI (progesterone assay)	54.8 (117)	69.6 (112)

( ) Number of treated animals.

As regards to the duration of treatments, implants must be
removed after 7 to 9 days in order to obtain high calving rates
(Table 3).

2) The comparison between norgestomet implants and progesterone coils has been measured in Salers nursing cows and in
Charolais nursing cows (Table 4). When used in the same physiological conditions of treatment (duration and post-progestagen
PMSG injections), the two techniques are equivalent.

3) At the beginning of both treatments an oestrogen is
usually administered (5 mg of oestradiol valerate injected with

implants or 10 mg of oestradiol benzoate in a capsule sealed on the coil). The necessity of such an injection is not evident in a bovine population in which few cows show ovarian activity since the first reason for its administration was luteolytic or antiluteotrophic effects in cycling females (Lemon, 1975). However, oestrogen injected on the first day of both treatments slightly improved the calving rates (6 to 12%) of such cows (Table 5).

TABLE 3

CALVING RATES AFTER TWO AI AT FIXED TIME FOLLOWING NORGESTOMET IMPLANTS IN SALERS NURSING COWS. EFFECT OF DURATION OF TREATMENT

Duration of implant insertion (days)	Number of cows	% Calving rates
7	219	58.8
9	1092	54.4
11	114	45.6
13 - 16	166	28.9

TABLE 4

CALVING RATE OF NURSING COWS AFTER TWO AI AT FIXED TIME FOLLOWING NORGESTOMET IMPLANTS AND PROGESTERONE COILS (TIME OF AI FOR IMPLANTS = 48 - 72 h
COILS = 56 - 74 h)

Breed	Norgestomet implants	Progesterone coils	References
Salers	55.2 (295)	53.2 (94)	INRA, 1976
Charolais	45.5 (321)	45.9 (74)	UNCEIA, 1976

( ) Number of treated animals.

4) Most of our experiments with beef cattle involved an intramuscular injection of PMSG on the day of removal of implants. This was the result of our drastically unsuccessful preliminary experiment in 1970 (Table 6).

TABLE 5

EFFECT OF OESTRADIOL VALERATE OR OESTRADIOL BENZOATE ADMINISTERED ON THE
FIRST DAY OF NORGESTOMET IMPLANT OR PROGESTERONE COIL TREATMENT (% CALVING
RATES OF SALERS NURSING COWS)

Treatment	with oestrogen[+]	without oestrogen[+]
Norgestomet implant 9 days	54.1 (37)	48.3 (31)
Progesterone coils 12 days	53.2 (94)	41.2 (102)

( ) Number of treated animals

+ 5 mg oestradiol valerate with implants
   10 mg oestradiol benzoate with coils

TABLE 6

EFFECT OF PMSG INJECTED AT THE END OF AN im NORETHANDROLONE TREATMENT IN
CHAROLAIS HEIFERS AND COWS.

Dose of PMSG	Calving rate	
	Heifers	Nursing cows
0 iu	20.4 (108)	28.7 (237)
800 iu	45.6 (171)	41.1 (258)

In conclusion, higher calving rates are obtained in beef
cattle with a progestagen or progesterone treatment of short
duration completed by an injection of oestrogen the first day
of treatment and an injection of PMSG on the day of removal of
implants or coils. Usually two AI are made at fixed times which
vary slightly according to the progestagen compound. Experiments
in dairy cows have shown that the reduction to AI after
progestagen treatments is associated with a decrease of 10%
in calving rates (Chupin et al., 1977$_2$). No experiment was
carried out on this subject in beef cattle, but this can be
easily done since a good oestrus synchronisation was observed
at least for Salers cows (Pelot et al., 1975).

2. - FACTORS RELATED TO FEMALES

Calving rates after the same hormonal treatment are different between breeds. Such a difference is not explained by the
slight variation between breeds of the mean percentage of
cycling cows (Table 7).

TABLE 7

BREED DIFFERENCES IN OVARIAN ACTIVITY BEFORE TREATMENT AND IN CALVING RATES
AFTER NORGESTOMET IMPLANTS

Breed	% of cycling cows	% calving rates
Salers	27.8 (1081)	60.1 (672)
Charolais	18.2 (853)	45.8 (225)

( ) Number of treated animals.

In the same breed, fertility after oestrous control is
lower in the primiparous cows than in the multiparous ones.
The primiparous cows show a higher rate of ovarian inactivity
during similar postpartum intervals (Table 8).

TABLE 8

DIFFERENCES IN OVARIAN ACTIVITY AND CALVING RATES AFTER NORGESTOMET IMPLANTS
IN PRIMIPAROUS AND MULTIPAROUS CHAROLAIS COWS

	% of cycling cows	% calving rates
Primiparous	13.3 (75)	37.2 (43)
Multiparous	27.7 (155)	53.5 (43)

( ) Number of treated animals.

In the Charolais breed, the replacement rate of females is
particularly high and the high percentage of primiparous females
modifies unfavourably the fertility of the Charolais cow
population.

Moreover, in the non-cycling cows, the efficiency of the hormonal treatment in terms of induction of ovulation (checked by a progesterone assay 10 days after Al) is lower in Charolais non-cycling females than in Salers (Table 9). So, the breed differences of the depth of anoestrus is expressed not only by the variation of ovarian activity, but also by the variation of hypophyseal-ovarian refractoriness to stimulation. However, a relationship exists between these two parameters not only between breeds but also between herds. Induction of ovulation in anoestrous cows is higher in herds where more cows are cycling. This rate of ovarian activity within herds appears as a criterion of ease of ovulation induction of the non-cycling cows. Moreover, the fertility of these induced ovulations is related to the induction efficiency; when the induction is difficult to obtain, its quality is decreased (Table 8).

TABLE 9

OVULATION RATES AFTER NORGESTOMET IMPLANTS IN RELATION WITH OVARIAN ACTIVITY BEFORE TREATMENT (PROGESTERONE ASSAY)

Breed		Number of animals	Ovarian activity		% ovulation after treatment	% pregnancy +
			cycled	non cycled		
Charolais	Cows	110	31.8	68.2	85.7 68.2	73.3 58.8
	Heifers	59	38.9	61.1	100.0 72.2	73.9 69.1
Salers	Cows	40	45.0	55.0	100.0 95.4	55.5 76.1

+ For 100 ovulations

An improvement of fertility after treatment in beef cattle can be achieved in two ways, to increase the efficiency of ovulation induction or to increase the number of cows regaining ovarian activity before treatment. As regards to the first solution, this can be done by an increase in the dose of PMSG as shown in Table 10. Unfortunately, in practice, it is a double-edged sword since any increase in the PMSG dose induces a rise in multiple pregnancies, especially undesirable ones, ie 3 or 4 calves (Table 11). An increase in the percentage of cows ovulating after treatment could also be obtained by increasing the dose of progestagen available at the plasmatic level as is done with the new silastic norgestomet implants tested in Charolais cows (Table 12).

TABLE 10

EFFECT OF DOSE OF PMSG ON THE INDUCTION OF OVULATION IN CHAROLAIS HEIFERS (TOTAL PERCENTAGE OF OVULATION: CYCLING AND NON CYCLING HEIFERS)

	Dose of PMSG		
	400 iu	500 iu	600 iu
Number of heifers	378 *	59 +	38 **
% ovulation	67.5	83.1	92.1**

*   INRA 1976 - 1977
** UNCEIA 1976

TABLE 11

EFFECT OF DOSE OF PMSG ON MULTIPLE PREGNANCIES (CHAROLAIS NURSING COWS)

Dose (iu)	Number of calving	Twins	Triplets and Quadruplets
700	425	7.3	1.2
800	649	9.6	5.0
1000	80	20.2	6.3

UNCEIA, 1970 - 1974

TABLE 12

EFFECT OF TYPE OF IMPLANTS (HYDRON OR SILASTIC) ON INDUCTION OF OVULATION IN CHAROLAIS NURSING COWS

	12 mg hydron implant	300 µg silastic implant
% ovulation	70.6 (177)	76.3 (114)

( ) Number of treated animals.

For increasing the number of cycling cows before treatment, modifications of environment could finally be more efficient.

3. - FACTORS RELATED TO ENVIRONMENT

In order to discuss the effects of environment on the efficiency of oestrous control treatment, it is necessary to describe the herd management of beef cattle in France.

From November 15th to April 15th, cows are sometimes kept inside, free with their calves or, more often tethered, with the calves allowed to suckle twice a day. Calvings occur between January and March. Breeding is traditionally by bulls after April 15th on pasture. To increase income, herdsmen wish to begin breeding as soon as early March, during the winter stabling period.

1) Effects of nutrition levels

The nutrition levels are generally low, food mainly consisting of medium or bad quality hay. From beginning of winter to calving, cows maintain a constant weight. During the whole winter period, they lose weight.

The effect of these management practices is deleterious on ovarian activity (Table 13). During the whole winter period, whatever the postpartum interval, very few cows are cycling and it is only when they go out onto pasture that reproductive activity starts again although the mean interval from calving is decreased. The favourable nutrition effect of pasture is delayed in the Charolais management.

With regard to management influence, ovarian activity differs when cows are free as opposed to being kept inside (Table 14). Unfortunately, this effect is not only related to the stabling system, since free stabling is generally used by more competent farmers, and is also associated with higher winter nutrition levels.

TABLE 13

OVARIAN ACTIVITY OF CHAROLAIS AND SALERS NURSING COWS BETWEEN FEBRUARY AND MAY

Breed		February	March	April	May
Charolais	Cycling cows (%)	-	16.0 (131)	14.2 (382)	25.2 (113)
	Mean interval since calving (days)	-	65.9	63.2	59.2
Salers	Cycling cows (%)	25.6 (133)	25.5 (632)	28.5 (263)	50.0 (22)
	Mean interval since calving (days)	57.9	60.8	53.2	48.3

( ) Number of treated animals

Turn out to pasture = April 15th in Charolais
                      May 1st in Salers.

TABLE 14

OVARIAN ACTIVITY IN SALERS NURSING COWS, EFFECT OF STABLING SYSTEM

	Free	Tied
% of cycling cows	35.9 (117)	23.2 (826)

( ) Number of treated animals.

Pregnancy rates after oestrous control treatment do not follow the same pattern as ovarian activity and a decrease only appears at the end of the winter and goes on during the beginning of the pasture period (Table 15). It is difficult to distinguish between the effect of low winter nutrition levels and the stress caused by the beginning of the pasture period (changes in nutrition, changeable weather...). The decrease is less pronounced with the 12 mg dose of norgestomet and can be related to the better ovulation inducing potency of this high dose in the period of deep anoestrus.

If this decrease in pregnancy rates were really due to low

nutrition levels, it would be overcome by feeding extra food at the time of breeding. To check this hypothesis, a comparison was carried out between a controlled group and a test group, in the same herds, receiving 2 kg of concentrate during the norgestomet treatment (Table 16). Although in this experiment the mean ovulation rate of the control group was higher than usually found, the feeding concentrates during treatment resulted in an increase in the percentage of nursing cows which ovulated and a greater increase in pregnancy rates.

TABLE 15

EVOLUTION OF PREGNANCY RATES AFTER NORGESTOMET TREATMENT IN CHAROLAIS NURSING COWS AT THE END OF WINTER PERIOD, (PREGNANCY RATES 21 DAYS AFTER A1 - PROGESTERONE ASSAY)

Date of beginning of treatment	Dose of progestagen	
	6 mg + 3 mg injected	12 mg
March 15th - 31st	72.1 (43)	71.9 (41)
April 1st - 15th	50.0 (122)	57.2 (119)
April 15th - 30th	25.4 (59)	46.1 (52)
May 1st - 15th	56.7 (80)	69.2 (13)

( ) Number of treated animals.

TABLE 16

EFFECT OF FLUSHING DURING OESTROUS CONTROL TREATMENT ON OVULATION INDUCTION IN CHAROLAIS NURSING COWS

	Control	Flushing (2 kg/ day for 9 days)
% ovulations	70.0	87.2
% pregnant	42.5	69.2

40 cows per groups

2) Choosing the bull for A1

Progestagens create an unfavourable situation in the genital tract for spermatozoa which emphasises the importance of their number and quality. Differences in fertility between bulls are

greater after Al at induced oestrus than after Al at normal
oestrus. For example, Table 17 shows the calving rates obtained
with five Charolais bulls of a commercial Al centre used after
norgestomet treatment.

TABLE 17

CALVING RATE AFTER OESTROUS CONTROL TREATMENT EFFECT OF BULL CHOOSEN FOR Al
(CALVING RATE - CHAROLAIS COWS - CHAROLAIS BULLS)

Bull	Primiparous		Multiparous		Total	
I	16.7	(36)	38.0	(50)	29.0	(86)
II	12.5	(8)	42.9	(14)	31.8	(22)
III	16.7	(6)	45.4	(11)	35.3	(17)
IV	54.5	(33)	56.0	(41)	55.4	(74)
V	55.6	(18)	63.9	(36)	61.1	(54)

( ) Number of treated animals.

3) Maternal behaviour

The long postpartum anoestrus found in nursing cows is
generally explained by the relationship between young and mother
and by the stimuli of the udder. According to this idea,
Wiltbank (1977) tried to induce ovulation by a temporary weaning
(24 to 48 hours) alone or at the end of a norgestomet treatment.
He obtained an increase in cows ovulating compared to either
controls or norgestomet implants alone (without PMSG).

An experiment has been carried out to study the effect of
this practice in Charolais cows. Results are not yet known.

In conclusion, it must be possible to get a regular
satisfactory calving rate after oestrous control treatment with
minor modifications of the treatment itself. The more widely
used progestagen (norgestomet) and progesterone administered in
implants, coils or pessaries, seem equivalent. The addition to
this treatment of oestradiol on the first day and PMSG on the
last day gives increase of more than 15% in calving rates.

544

With regard to non-cycling females, the main problem is to
get 100% ovulation after treatment. This seems difficult to
obtain by modification of treatments (PMSG doses) without
superovulation. A more promising way would be to decrease the
depth of postpartum anoestrus during winter by management
practices: nutrition levels at the time of breeding, housing
systems, temporary weaning before or after treatments.

ACKNOWLEDGEMENT

The authors wish to thank Mrs. H. Cahill for assistance with
the preparation of the English manuscript.

REFERENCES

Chupin, D., Deletang, F., Petit, M., Pelot, J., Le Provost, F., Ortavant, R. and Mauleon, P., 1974. Utilisation de progestagènes en implants sous-cutanés pour la maîtrise des cycles sexuels chez les bovins. Ann. Biol. anim. Bioch. Biophys. 14, 27-39.

Chupin, D., Pelot, J., Alonso de Miguel, M., Thimonier, J., 1976. Progesterone assay for study of ovarian activity during postpartum anoestrus in the cow. VIIIth intern. Cong. anim. Reprod. artif. Insem., Krakow, 346-349.

Chupin, D., Pelot, J., Petit, M., 1977. Le point sur la maîtrise des cycles sexuels chez les bovins. Bull. Tech. Insem. Artif. 5, 2-17.

Chupin, D., Pelot, J., Mauleon, P., 1977. Control of oestrus and ovulation in dairy cows. Theriogenology 7, (6), 339-347.

Lemon, M., 1975. The effect of oestrogens alone or in association with progestagens on the formation and regression of the Corpus Luteum of the cow. Ann. Biol. anim. Bioch. Biophys. 15 (2), 243-253.

Pelot, J., Olivier, J.P., Chupin, D., 1975. Utilisation d'implants progestagènes sous-cutanés pur la maîtrise de cycles chez les vaches allaitantes de race Salers. Détermination de la dose et de la durée de séjour optima. Ann. Biol. anim. Bioch. Biophys. 15 (1), 29-36.

Wiltbank, J.N., 1977. Breeding at a predetermined time following synchro-Mate-B treatment. Proceed. 11th Conf. artif. Insem. Beef Cattle, Denver, 57-65.

IMPROVEMENT OF THE OESTROUS CONTROL IN ADULT DAIRY COWS

D. Chupin, J. Pelot and P. Mauleon

INRA - Station de Physiologie de la Reproduction
37380 Nouzilly, France

ABSTRACT

As dairy cows are supposed to be cycling 40 to 60 days post-partum, at the time of breeding two kinds of treatments can be used : progestagens and/or prostaglandins.

During a field trial, a comparison was made between the pregnancy rates obtained after progestagen implants (Searle) during 9 days with an injection of 'Estrumate' (ICI) on the day of implant removal, with those obtained after two 'Estrumate' injections 11 days apart. One or two AI were performed at predetermined times. The highest pregnancy rate was obtained with implant + prostaglandin + 2 AI (48.7%). This was significantly (P< 0.05) higher than that obtained with the same treatment but only one AI (34.9%) and with the 2 prostaglandin injections and 2 AI (37.7%). After 2 injections of 'Estrumate' treatment, the number of AI has no effect (37.7 vs 43.5% NS).

In the same experiment, a large decrease in the number of animals diagnosed pregnant has been shown between 21 days and 5 - 6 months of gestation. This fall varies according to treatment from 19.7% with the combined progestagen-prostaglandin treatment to 35.1% with 2 prostaglandin injections (P< 0.05). This difference can be explained mainly by synchronisation failures and embryonic mortality.

In a second experiment, synchronisation rates were recorded following three treatments : progestagen implants for 9 days with 'Estrumate' injected either on the day of implant removal or two days before, and 2 'Estrumate' injections 11 days apart. After this last treatment, synchronisation was poor since only 60.2% of cows were observed in oestrus in a 48 h period and 31.8% were not observed during the first 96 hours after the end of treatment. When 'Estrumate' is injected on the day of removal, the corresponding figures are 82.0 and 18.0% respectively. This can be improved by injection of 'Estrumate' two days before implant removal (86 and 14%, respectively). This second experiment shows that the main reason for low fertility after oestrus synchronisation in the dairy cow is an insufficient synchronisation to allow AI without detection of oestrus. To improve this, some experiments are now in progress :

> *- Effect of interval between the two 'Estrumate' injections.*

> *- Effect of 'Estrumate' injection two days before implant removal and/or PMSG on the day of implant removal.*

Preliminary results show that pregnancy rates can be increased by these two methods and a calving rate higher than 50% can be achieved.

INTRODUCTION

In large herds of high yielding dairy cows the first oestrus may be seen only 70 to 80 days after calving. The general opinion is that dairy cows begin cyclic ovarian activity as soon as 10 to 30 days after calving and that the delay is due to poor oestrus detection. Control of oestrus for these cows with active ovaries can be achieved with three kinds of treatments : two intramuscular injections of prostaglandin* analogues or subcutaneous progestagens implants supplemented by an injection of progestagen and oestrogen on the first day of treatment, or the same progestagen implant associated with an injection of prostaglandin analogues. The effects of these three treatments have been described previously (Thimonier et al., 1975).

This report deals with the results of several experiments : the first was conducted on a research station to study the onset of oestrus, the second was a field trial to study the conception rate after artificial insemination at a predetermined time after the end of treatments. Following these results, two other experiments were conducted in an attempt to increase oestrus synchronisation and conception rates.

I.   OESTRUS SYNCHRONISATION

Between January 1975 and September 1976, 298 adult dairy cows of the FFPN breed were treated on an experimental farm (free stabling, high nutritional level, 5 000 kg milk per cow per year, no pastures). These cows received one of the five following treatments:

Group 1A : One subcutaneous implant of SC 21009 (6 mg of Norgestomet - Searle & Co) for 7 days with an intramuscular injection of 3 mg of SC 21009 the day of implantation and 500 µg of prostaglandin analogue (Estrumate : ICI 80996) on the day of implant removal.

Group 2A : As group 1A but implant left in place during
9 days.

Group 3A : As group 2A but ICI 80996 prostaglandin analogue
injected two days before implant removal.

Group 4A : As group 1A but implant left in place during
11 days.

Group 5A : Two intramuscular injections of ICI 80996
prostaglandin analogue 11 days apart.

Occurrence of oestrus was checked twice daily with the
aid of an androgenised cow fitted with a chin ball marking
harness. Between two observation periods this 'teaser' stayed
with the cows and each female marked between two observations
was considered as coming in heat at the following period.

Distribution of the onset of oestrus after the end of
treatment was estimated by 4 parameters:

1. Interval between the end of treatment and observation
period when the maximum number of cows are in oestrus.

2. Maximum percentage of cows in oestrus during a 24 h
period around the mode.

3. Maximum percentage of cows in oestrus during a 48 h
period around the mode.

4. Percentage of cows not detected in oestrus during the
first 96 h after the end of treatment.

Synchronisation was poorer after the two prostaglandin
treatments than after simultaneous use of Norgestomet implants
and ICI 80996 (60.2% vs 78.7% for all the combined treatments :
P < 0.01). In the first 96 h post-treatment the percentage of
animals not in oestrus was higher after the treatment of two
prostaglandin injections (31.8% vs 20.7%) (Tables 1 and 2).
Between all the progestagen-prostaglandin groups, the better
synchronisation was obtained with implants for 9 days and ICI
80996 injected two days before implant removal.

Problems of oestrus detection do not explain these differences since all the treatments were made simultaneously in the same herd throughout the whole year. Moreover, 94.3% of heifers in the same herd treated with two prostaglandin injections were observed in oestrus in a 12 h period, and the remaining 5.7% were not observed in oestrus. Hence oestrus detection in this herd seems to be sufficiently accurate.

With the same synthetic analogue of prostaglandin as in our study (ICI 80996, Estrumate), McMillan and Morris (1976) observed only 39.3% of cows in oestrus during the first 96 h after the second injection. Whether this is due to a failure to ovulate or to exhibit oestrus is not clear. Leaver et al. (1976) found 25% of cows with silent ovulations after the same prostaglandin analogue treatment and Smith (1976) observed 9 ovulation failures out of 86 heifers after the second injection.

After one injection during the mid luteal phase, Baishya et al. (1977) obtained 90% of ovulations in heifers but only 60% in adult dairy cows. We have studied this response after the second injection in 74 dairy cows in relation to the response after the first (Table 3) (Chupin and Pelot, unpublished) If luteolysis is obtained after the first injection, only 75.7% of cows ovulate following the second injection of ICI 80996. If luteolysis is not obtained the first time all cows ovulate after the second injection. Some ovulations occur too late after the first injection so that the new corpus luteum cannot respond to the second injection that takes place before day 6 of the new oestrous cycle. To increase the efficiency of this two injection treatment, one must obtain a second luteolysis even after delayed ovulations. This must be obtained by increasing the interval between the two injections to 13 or 14 days.

II. CONCEPTION RATES

The reduction in the number of AI at a predetermined time is necessary for economic purposes and it also allows an indirect

TABLE 1

DISTRIBUTION OF ONSET OF OESTRUS AFTER THE END OF TREATMENTS FOR CONTROL
OF OESTRUS IN DAIRY COWS (EXPERIMENT A)

Group	No of cows	Time to Onset of Oestrus (H)								% Not detected in the first 96 h
		12	24	36	48	60	72	84	96	
1A	36		11.1	8.3	33.3		11.1		5.5	30.5 a
2A	90	10.0	23.3	14.4	21.1	6.6	6.6			18.0 c
3A	50		32.0	32.0	22.0					14.0 c
4A	36		30.5	5.5	30.5		8.3			25.0 ac
5A	86		6.9˙	6.9	16.2	16.2	15.1	5.8	1.1	31.8 a

A - C $P < 0.05$

Figures with the same subscript are not significantly different.

TABLE 2

SYNCHRONISATION OF OESTRUS AFTER PROGESTAGEN AND/OR PROSTAGLANDIN
TREATMENT IN DAIRY COWS (EXPERIMENT A)

Group	No of cows	Interval end of Treatment till maximum No in oestrus	Maximum % in oestrus in a period of :		% Not detected 96 h post treatment
			24 h	48 h	
1A	36	48 h	41.6 b	63.8 bc	30.5 a
2A	90	24 - 48 h	42.1 b	82.0 a	18.0 c
3A	50	24 - 36 h	86.0 a	86.0 a	14.0 c
4A	36	24 - 48 h	36.0 b	74.8 abc	25.0 ac
5A	86	48 - 60 h	47.5 b	60.2 b	31.8 a

a - b $P < 0.01$

a - c $P < 0.05$

Figures with the same subscript are not significantly different.

TABLE 3

RESPONSE TO THE SECOND ESTRUMATE INJECTION ACCORDING TO THE RESPONSE TO
THE FIRST INJECTION

Luteolysis after the First Injection	No of Cows	Response to the second injection		
		Luteolysis	Oestrus	Ovulation
Yes	45	84.4	86.6	75.7
No	29	100.0	93.0	100.0

estimation of synchronisation efficiency in field trial
conditions.

Three hundred and eighty four dairy cows of the FFPN
breed were treated in several herds of three types :

Free stabling : high nutritional level ; 5000 kg of milk.

Tied stabling : experimental nutritional levels ; more
than 5000 kg of milk.

Commercial herds : free or tied stabling ; various
nutritional levels; 4000 kg of milk.

The cows were divided into four groups :

Group 1B. One subcutaneous implant of SC 21009 (6 mg of
Norgestomet - Searle & Co) for 7, 8 or 9 days with an
injection of 3 mg of SC 21009 the day the implant was
administered and 500 µg of prostaglandin analogue (ICI 80996)
on the day of implant removal. Two inseminations were
performed on a fixed time basis, 48 and 72 h after implant
removal.

Group 2B. As group 1B, but with only one insemination 56 h
after implant removal.

Group 3B. Two intramuscular injections of ICI 80996 11
days apart. Two inseminations were performed on a fixed
time basis. 72 and 98 h after the second injection.

Group 4B. As group 3B, but with only one insemination
80 h after the second ICI 80996 injection.

Bulls for AI were chosen according to their genetic value by the farmer. Commercial doses of semen, 20 to 30 x 10^6 total spermatozoa were used.

Fertility was estimated either by progesterone level 21 days after AI or by the non-return rate 5 to 6 months after AI accompanied in some cases by rectal palpation at the same time.

The mean fertility level was low in this experiment (Table 4) (41.9% pregnancy rate 5 - 6 months after AI). In a previous experiment in the same herd as Experiment A we have shown that the calving rate of treated animals was not significantly different to the calving rate of a control group formed by the cows inseminated after returning to oestrus (39.9%) (Chupin and Pelot, 1976). Cooper et al. (1976) observed a calving rate of 47.3% in their control group.

TABLE 4

PREGNANCY RATES OF DAIRY COWS INSEMINATED ON A FIXED TIME BASIS AFTER END OF TREATMENT (EXPERIMENT B)

Treatment	Group	AI	% Pregnant at 21 Days	5 - 6 Months	Difference
Progestagen implants + Prostaglandins	1B	48-72h	68.4 (114)	48.7 a	19.7 a
	2B	56h	61.2 (80)	34.9 b	26.3 a b
2 Prostaglandin Injections	3B	72 and 96 h	72.8 (102)	37.7 b	35.1 b
	4B	80h	70.4 (88)	43.5 a b	26.9 a b

(   ) No of cows

a vs b = $P < 0.05$

The highest pregnancy rate was obtained in group 1B (progestagen implant plus prostaglandin injection plus two AI - 48.7%). This figure differs significantly ($P < 0.05$) from those of group 2B (same treatment but only one AI - 34.9%) and group 3B (two prostaglandin injections plus two AI - 37.7%), but not from that of group 4B (two prostaglandin injections but only one AI - 43.5%).

A reduction in the number of AI at a predetermined time after progestagen treatment resulted in a significant decrease in pregnancy rate. After prostaglandin treatment, such a reduction seems to be ineffective. This can be explained by the poor synchronisation : some females come into oestrus too soon or too late to conceive after one or two AI at a pre-determined time (Hearnshaw, 1976). This is in opposition to the results of Cooper et al. (1976) showing a significant decrease in calving rate after only one AI at a pre-determined time (36.1%) compared to two AI (44.0%) or control group (47.3%).

In heifers, several authors have published results showing no differences in pregnancy rates after one or two AI at a pre-determined time with SC 21009 implants (Wishart, pers. comm.), with progesterone coils (Roche, 1976), or with ICI 80996 (Parez and Mauleon, 1976). Our own observations show that after one AI, there is a fall in pregnancy rate of about 10% compared with two AI after SC 21009 implants. This is usually explained by the harmful side effects of progestagens on the endometrium and uterine motility. As observed in the ewe, an increase in the total number of spermatozoa overcomes these side effects. The same results have been obtained in dairy heifers by an increase in the total number of spermatozoa from 30 to 60 million (Chupin, unpublished).

The most important point in Table 4 is the great difference between pregnancy rates 21 days after AI and 5 to 6 months after AI (from 19.7 to 35.1% depending on groups). This figure is greater than that previously observed with the same treatment. After two prostaglandin injections in dairy heifers, this fall varied from 9.0 to 15.2% depending on AI schemes (Parez and Mauleon, 1976).

This could be explained by an inaccuracy in pregnancy diagnosis and or by embryonic mortality.

Pregnancy diagnosis made 21 days after AI at a predetermined time following oestrus synchronisation in dairy cows may have a poor accuracy for several reasons:

It has been shown (Shemesh, pers. comm.) that in the dairy cow results are better if samplings are taken 24 days after AI instead of 21 days.

The first return after synchronised oestrus may be abnormal (less than 18 or more than 24 days). We have checked intervals between synchronised oestrus and first return (Table 5). After the two PG regimes, 27.7% of returns are observed before 18 days: actually these oestruses are not a first return after AI but the first oestrus after the second PG injection - these cows have not been synchronised. After the progestagen + PG treatment only 7.1% of these short returns are observed. Moreover 31.8% of all returns are observed more than 50 days after AI following the two PG treatment compared to 16.3% following the combined progestagen prostaglandin treatments. These figures indicate the importance of embryonic mortality associated with synchronisation treatments. It could be in some cases a deleterious effect of the drug but more often it is due to insemination at a wrong time. In this case, it has been shown (Sokolvskaya, 1973) that non-return rates at 30 days are quite normal with other returns appearing later.

III. IMPROVEMENT OF TREATMENTS

The foregoing results lead to the conclusion that conception rate can be increased if oestrus synchronisation is closer. Moreover, preliminary results (Chupin et al., 1977) have shown that 50 to 80 days after calving 30% of cow producing more than 20 kg of milk per day seem to be in anoestrus (estimated by two progesterone assays 10 days apart). Two experiments are

in progress to improve the double prostaglandin injection
schedule, and to adapt the combined progestagen prostaglandin
treatment.

TABLE 5

DISTRIBUTION OF RETURNS IN OESTRUS AFTER AI AT A PREDETERMINED TIME
FOLLOWING OESTRUS SYNCHRONISATION IN DAIRY COWS

Treatment	No of Returns	Interval synchronised oestrus-return (days)					
		18	18-26	27-35	36-46	47-53	53
Progestagen + Prostaglandins	98	7.1	47.9	5.2	19.4	4.1	16.3
Two Prostaglandins (11 days)	22	22.7	22.7	4.5	18.2	-	31.8

From Parez and Mauleon 1976.

Improvement of the two prostaglandin treatments : we have
compared 11 or 13 days between two injections of 500 μg of ICI
80996 in 114 adult dairy cows (tied stabling, experimental
nutritional levels, more than 5 000 kg of milk). Results in
Table 6 show that we obtained an increase of 10% in pregnancy
rate after one AI 80 h after the second injection. Unfortunately,
the oestrus date of these cows is not available and we cannot
conclude if oestrus synchronisation was improved.

TABLE 6

IMPROVING THE DOUBLE PROSTAGLANDIN INJECTION SCHEDULE, EFFECT OF LENGTH
OF TREATMENT, (PREGNANCY DIAGNOSIS COMPLETED BY NON-RETURN AT 3 MONTHS)

	11 Days	13 Days
Pregnancy rate	41.1 (56)	51.7 (58)

( ) No of cows

Adaptation of the combined progestagen-prostaglandin treatment to a mixed population of cycling and non cycling cows : 187 dairy cows in commercial herds (free or tied stabling, various nutritional levels, 4 000 kg of milk) were divided into four groups :

Group 1. One subcutaneous implant of SC 21009 (6 mg of Norgestomet - Searle & Co) during 9 days with an injection of 5 mg of oestradiol valerate and 3 mg of Norgestomet on the day the implants were administered.

Group 2. As group 1, plus one injection of 500 iu of PMSG on the day of implant removal.

Group 3. As group 1, plus one injection of 500 µg of ICI 80996 two days before implant removal.

Group 4. As group 3, plus one injection of 500 iu of PMSG on the day of implant removal.

All cows were inseminated twice, 48 and 72 h after implant removal. PMSG was injected in an attempt to synchronise ovulation in cycling cows and to induce ovulation in non-cycling ones.

Conception rates were estimated by assay of blood progesterone 24 days after AI and the non-return rate after three months.

Preliminary results are shown in Table 7. With the implant alone and two inseminations, the pregnancy rate was equivalent or higher to that obtained with two injections of ICI 80996 and one AI. This confirms our first experiment.

Injection of 500 iu of PMSG decreased pregnancy rate to 48.0%. Injection of 500 µg of ICI 80996 did not modify pregnancy rates (57.8% vs 58.7%).

Injection of both PMSG and ICI 80996 increased pregnancy rates up to 69.6%.

TABLE 7

PROGESTAGEN IMPLANTS IN DAIRY COWS, EFFECT OF PROSTAGLANDIN AND PMSG
INJECTIONS

Group	Treatment	% Pregnant
1	Progestagen implant alone	58.7 (46)
2	Progestagen implant + 500 iu PMSG	48.0 (50)
3	Progestagen implant + 500 µg estrumate	57.8 (45)
4	Progestagen implant + 500 iu PMSG + 500 µg estrumate	69.6 (46)

( ) No of cows

Although these are only preliminary results, we can
conclude that a very high fertility can be achieved in dairy
cows by a treatment associating progestagen implants, PMSG and
ICI 80996. It is thought that this has been obtained through
a high synchronisation of ovulation. At the end of the implant
treatment without ICI 80996, some cows may still have an active
corpus luteum. It has been shown (Saumande, 1977) that after
PMSG injection, progesterone levels increase. The poor
fertility result of the implant - PMSG group reflects the
prolonged life span of the corpora lutea resulting in oestrus
8 or 10 days after implant removal. Analysis of intervals to
first return in this group may confirm this hypothesis if we
observe more returns less than 18 days after the fixed time AI.

These results must be confirmed by calving rates and
continued by studies of time of ovulations, but it seems that a
calving rate higher than 50% can be achieved in dairy cows. The
second conclusion of this report is that a universal treatment
does not exist. Treatment for beef heifers and suckling cows
will be different from that of dairy heifers. In dairy cows,
several treatments will be necessary depending on milk
production, nutritional levels and herd management. In some
herds, all dairy cows are cycling at the time of breeding and
in others, a percentage of 10 to 30% of anoestrous cows

necessitate use of a more sophisticated treatment. Finally, the use of prostaglandin schemes must be confined to herds where ovarian activity is certain.

ACKNOWLEDGMENTS

The authors wish to thank Mr. L. Cahill for assistance with the preparation of the English manuscript.

REFERENCES

Baishya, N., Leaver, J., Pope, G.S. 1977. Fertility of dairy cattle inseminated at fixed times after injection of luteolytic doses of cloprostenol, an analogue of prostaglandin F. Joint Meet. SSF-SNESF, Nottingham, Abstr. 17.

Chupin, D., Pelot, J., 1976. Synchronisation des cycles de reproduction chez les vaches laitiéres. In 'Maitrise des cycles sexuels chez les bovins' (Searle), Janv. 1976, 95-100.

Chupin, D., Pelot, J., Petit, M. 1977. Le point sur la maitrise des cycles sexuel chez les bovins. Bull. Tech. Insem. Artif. 5, 2-17.

Cooper, M.J., Hammond, D., Harker, D.B., Jackson, P.S., 1976. Control of the bovine oestrous cycle with ICI 80996 (Cloprostenol). Field results in 3 810 beef cattle. VIIIth intern. Cong. anim. Reprod. artif. Insem., Krakow, Abstr. p. 57.

Hearnshaw, H., 1976. Synchronisation of oestrus and subsequent fertility in cattle, using the prostaglandin $F_2\alpha$ analogue ICI 80996 (Cloprostenol) Aust. J. Exp. Agric. Anim. Husb. 16, 437-444.

Leaver, J.D., Mulvany, P.M., Baishya, N., Pope, G.S. 1976. Fertility of Friesian cattle inseminated at fixed times after inducing luteolysis with the analogue of prostaglandin $F_2\alpha$(Cloprostenol). VIIIth intern. Cong. anim. Reprod. artif. Insem., Krakow, Abstr. p. 154.

McMillan, K.L., Morris, G.R., 1976. Oestrus synchronisation in large herds with seasonally concentrated calving pattern. VIIIth intern. Cong. anim. Reprod. artif. Insem., Krakow, Abstr. p. 163.

Parez, M., Mauleon, P., 1976. Elevage programmé chez les bovins. Econ. Med. Anim, 17, (4-5), 215-226.

Roche, J.F., 1976. Calving rates of cows following insemination after a 12 day treatment with silastic coils impregnated with progesterone. J. Anim. Sci. 43 (1), 164-169.

Saumande, J., 1977. Relationships between ovarian stimulation by PMSG and steroid secretion. EEC Conference, Galway, Sept. 1977.

Smith, J.F. 1976. Use of a synthetic prostaglandin analogue for synchronisation of oestrus in heifers. N.Z. Vet. J. 24 (5), 71-73.

Sokolovskaya, I.I. 1970. Méthode électrométrique de détermination du moment optima d'insémination des animaux en période de chaleurs. Zhivotnowdstvo 32 (11), 71.

Thimonier, J., Chupin, D., Pelot, J., 1975. Synchronisation of oestrus in heifers and cyclic cows with·progestagen and prostaglandin analogues alone or in combination. Ann. Biol. anim. Bioch. Biophys. 15 (2) 437-449.

FERTILITY CONTROL IN CATTLE BY DETERMINING PROGESTERONE IN MILK AND MILK-FAT

B. Hoffman[1]), E. Rattenberger[2]) and O. Günzler[2])
1) Institut für Physiologie, Südd. Versuchs- und Forschungsanstalt für Milchwirtschaft, Technische Universität München, D 805 Freising-Weihenstephan

2) Tiergesundheitsdienst Bayern e.V. 8011 Grub, FRG

1. INTRODUCTION

As early as 1968, Shemesh et al. suggested the measurement of progesterone levels in peripheral plasma of the cow as an indicator of early pregnancy. When applying this method we also showed in a distinct case (Hoffman, 1971), that the rather low fertility problem in this species was due to induction of early embryonic death following insemination performed during false oestrus. With the discovery that progesterone levels in cow's milk follow the same pattern as in plasma, and adequate methods being available to determine progesterone in milk (Heap et al., 1973; Hoffman and Hamburger, 1973), application of this control method seemed to be practicable on a larger scale, since there are no problems in obtaining and preserving the milk samples. A symposium, held under the chairmanship of Professor Laing in Potters Bar in March 1976 (Progesterone in Milk and Pregnancy Diagnosis; 1976) gave a good survey on the present status of application. During that symposium it also became obvious, that the method is applied for various reasons. While in most countries, especially in the United Kingdom, most emphasis is directed towards early pregnancy diagnosis, in Germany the trend is to achieve fertility control in general.

2. BACKGROUND AND PARAMETERS FOR PRACTICAL ASSAY APPLICATION

2.1. Background
Fertility of cows is an increasing problem. While only 4.3% of the cows were removed from herds due to infertility in 1950, this increased to 8.2% in 1974 or 32.3% of the total

number of animals removed (Zeddies, 1976); the corresponding
figure for 1975 is 37.4% (Schumann, 1975). The enormous economic
impact of fertility on cattle production was shown by Zeddies
(1976); in this respect attention was drawn to the fact that
more than 48% of the cows have a calving interval of more than
370 days.

The average number of cows per herd in the Federal Republic
of Germany (Arbeitsgemeinschaft Deutscher Rinderzüchter 1975)
is 9.4 and it is 8.9 in the State of Bavaria. These herds are
generally under good veterinary care; pregnancy diagnosis is
available, and widely done 6 - 8 weeks after insemination, by
rectal palpation. Thus, under our conditions a lack of large
scale pregnancy diagnosis seemed not to be a major factor
contributing to the increasing fertility problems. In consequenc
other factors must be responsible and a major one certainly
lies in the problem of oestrus detection. The phenomenon of
silent heat is well known and especially obvious when animals
are kept in stalls. In our opinion many problems related to
fertility can be related to the problem of heat-detection and
insemination at the wrong time. Thus, the idea was advanced,
to apply also the milk-progesterone-test for oestrus control.
Furthermore the test was offered as an aid to veterinary practic
for diagnosis and treatment of infertility (Günzler et al.,
1975).

2.2. Parameters for assay application

For practical assay application it was considered, that
progesterone concentrations in milk, besides depending on the
presence or absence of luteal tissue - are also directly
related to the milk-fat concentration (Hoffman and Hamburger
1973) and it was therefore suggested to use only the fat rich
'strippings' as a standardised milk sample for progesterone
analysis. Based on this and on the progesterone profile as
shown in milk during the cycle and early pregnancy (see Figure
1), the following criteria for test application were set up.

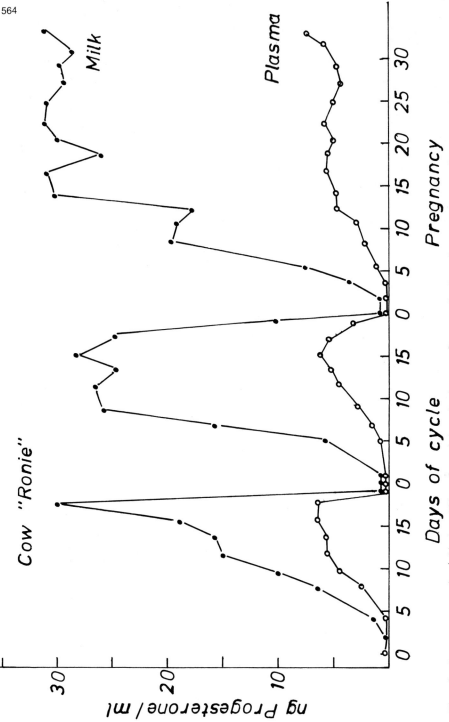

Fig. 1 Progesterone (ng/ml) in milk and peripheral plasma of a cow during the oestrous cycle and early pregnancy

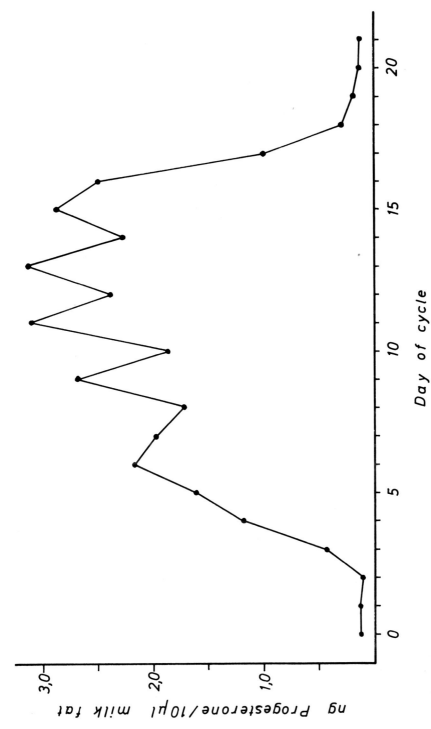

Fig. 2. Progesterone in milk fat (ng/10µl) of a cow during the oestrous cycle

a)   Control for oestrus:
     progesterone < 2.2 ng/ml milk=indicative of oestrus,
     progesterone   2.2 - 3.5 ng/ml milk = questionable,
     progesterone > 3.5 ng/ml milk = no oestrus:

Sample collection is necessary on the day of insemination
(day 0) or on the day of expected or apparently observed
oestrus.

b)   Control for early pregnancy:
     progesterone > 11 ng/ml milk = indicative of pregnancy
     progesterone 2.2 - 11 ng/ml milk = questionable,
     progesterone < 2.2 ng/ml milk = not pregnant:

Sample collection on day 20 - 22 after insemination

c)   Control of clinical situation; sample collection
according to special instructions, usually 5 samples at
2 day intervals.

2.3. Accuracy of methods

In spite of the high reliability of the radioimmunoassay
(RIA) for the determination of progesterone in milk, allowing
a measurement between 50 - 30 000 pg/ml, interpretation of the
results must be very careful.   There are only two situations
when the diagnosis based on the results obtained is practically
100% correct.   This is progesterone > 2.2 ng/ml = no oestrus
and progesterone < 2.2 ng/ml = not pregnant.   As far as we
could apply control checks no false diagnosis was obtained up
to the present (Hoffman et al., 1974, Günzler et al., 1975).   A
positive pregnancy diagnosis (day 20) is only correct in about
80% of cases and it can be expected that - due to the number of
cows per herd - the accuracy may vary from farm to farm, which
in many cases is not acceptable to the farmer.   On top of this
the samples diagnosed as questionable, range between 10 - 16%
(Hoffman et al., 1977).   Among the animals having progesterone
values < 2.2 ng/ml also, those suffering from follicular cysts

may be found. Such an animal exhibits an anovulatory oestrus and a diagnosis 'adequate oestrus for insemination' is wrong. But in cases like these the milk-progesterone-test has shown to be a highly valuable instrument in the hands of the veterinarian for setting up the correct diagnosis and treatment. However, a high precision of the RIA is necessary to monitor minute changes in ovarian function.

3. ROUTINE APPLICATION OF THE MILK-PROGESTERONE-TEST

The service system set up on a non-profit basis at the Bavarian Animal Health Service in Grub has now been working since the beginning of 1976. Table 1 shows the distribution of samples sent in up to June 1977 for progesterone analysis. It is obvious that there is an increase in samples from 1976 to 1977; the ratio is about 1 : 2 for samples provided for control of oestrus and control of early pregnancy and a substantial number of samples (38% and 19%) serves for control of clinical situations. In the latter connection the most common problems are: verification of cystic ovaries in respect of functional status of the cyst and subsequent therapeutic measures, control of oestrus synchronisation in special problem herds and monitoring of ovarian function in valuable, high yielding animals with 'subfertility'. The fact that the whole assay application is directed towards the problem-animal is also obvious from Table 2 which shows that 25% of the animals were not in oestrus when they were inseminated and that about 23% of the cows are not pregnant 20 days after insemination though they do not show symptoms of heat. These figures are by far greater than in a normal population where they range between 10 - 15% (Günzler et al., 1975; Hoffman et al., 1974) which indicates that the farmer - as it was hoped - makes use of the test system to check up on his problem animals particularly.

568

TABLE 1

DISTRIBUTION OF SAMPLES SUBMITTED FOR PROGESTERONE ANALYSIS TO THE
SERVICE LABORATORY IN GRUB

year	a Control for oestrus	b Control for early pregnancy	c Control of clinical situations	Total number of samples
1976	1388	3074	2750	7212
1977 Jan. - June	1888	3425	1237	6550

TABLE 2

RESULTS OF THE MILK-PROGESTERONE-TEST FOR 1976 AND THE FIRST HALF YEAR
OF 1977

Diagnosis	Control for oestrus	Control for early pregnancy
positive	71.3%	67.5%
negative	25.4%	22.6%
questionable	3.3%	9.8%

4. FURTHER METHODOLOGICAL DEVELOPMENTS

In view of the fact that the number of samples was increasing
it became necessary to simplify the method. However, due to
the specific nature of assay application no loss in accuracy
was acceptable; on the contrary it seemed necessary further to
increase the overall assay specificity in order to overcome the
problem of animals diagnosed 'questionable'. As we have reported
elsewhere, a constant source of error is the variable concentratic
of milk-fat and it was therefore attempted to measure progesterone
in milk fat directly (Hoffman et al., 1976). The results obtained
have shown that regardless of the stage of milking, constant
progesterone levels are being measured in milk-fat. Furthermore

TABLE 3

COMPARATIVE DETERMINATION OF PROGESTERONE IN MILK AND MILK FAT IN 177 SAMPLES COLLECTED ON DAY 20 FOR EARLY PREGNANCY TESTING*

Parameter	Milk fat-progesterone-test		Milk-progesterone-test		
	Pregnant	Non pregnant	Pregnant	Non pregnant	Questionable
Number of animals	124	53	105	42	30
%	70.1%	29.9%	59.3%	23.7%	16.9%
Pregnant at clinical examination	101	2	86	0	17
Agreement: laboratory clin. examination	81.4%	96.3%	81.9%	100%	–
Total number of animals diagnosed correctly	152=85.8%		128=72.3%		

* Taken from Hoffmann et al. (1977).

we could show in a previous study examining milk samples from day 20 (Hoffman et al., 1977) that the total number of animals diagnosed correctly rose from 72.3% to 85.8% (see Table 3); no more animals were falling in the questionable range (in this study 16.9% in the milk-progesterone test, which was performed simultaneously) when the following border lines were set up:

a) Control for oestrus;
   <0.3 ng/10 μl milk-fat; indicative of oestrus
   >0.3 ng/10 μl milk-fat; no oestrus

b) Diagnosis of early pregnancy;
   >1.0 ng/10 μl milk-fat; indicative of pregnancy
   <1.0 ng/10 μl milk-fat; not pregnant

Of course the cyclic variations can also be demonstrated by measuring progesterone in milk-fat (Figure 2).

The method involved includes a simple heating and freezing step to obtain a pipettable milk fat (5-10 μl are enough for one assay) and a separation of progesterone from milk fat by silica-gel column chromatography. While batchwise several hundred milk-samples can be heated and frozen simultaneously the efficiency of the method was initially rather limited due to the time-consuming column-chromatography. However, a semi-automated system was developed which permits parallel chromatograp of 25 samples within 20 minutes (75 samples per hour). As is shown in Figure 3 the system consists of three parts: collector, pump and column-block. The elution-profile identical for all columns is given in Figure 4 and it can be seen that almost 100% of the progesterone comes off the column in fraction 3 (2 ml of methanol). After taking the samples to dryness, the RIA is performed. Then simultaneous with this method, the milk-progesterone-test is applied to validate the procedure further.

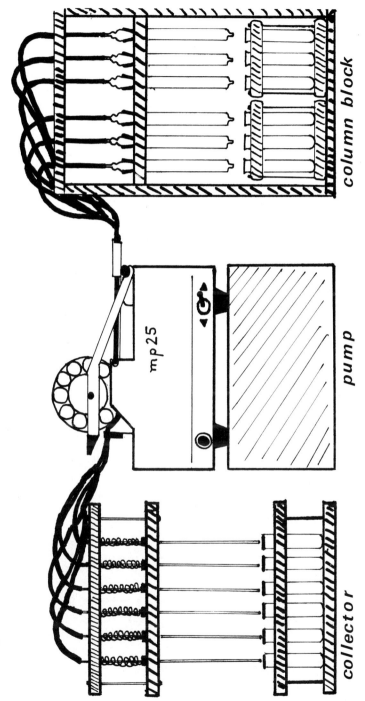

Fig. 3

Fig. 3. Scheme of the semi-automated system.

572

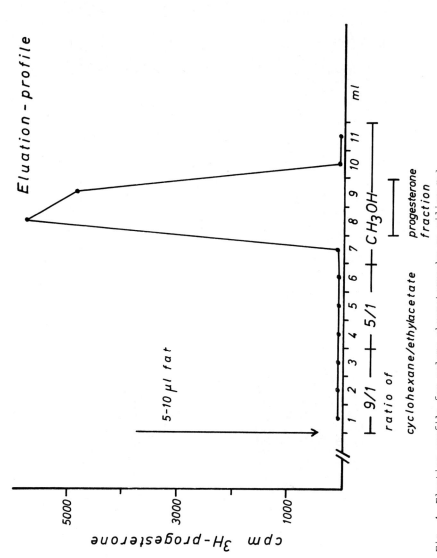

Fig. 4 Eluation-profile for column chromatography on silica-gel

5. CONCLUSIONS

The milk-progesterone-test has proved to be an adequate
tool to improve the fertility situation in cow-herd management.
This is mainly based on the fact that the farmer is enabled to
recognise his problem animals (for instance those showing
silent heat) early enough to overcome the situations leading
to economic losses. This is done either by paying special
attention to these cows and thus recognising true oestrus
(which is of course necessary for a successful insemination)
or to inform the veterinarian for biotechnical manipulations.
Application of this assay is increasing rather slowly and it
has become clear that for a more rapid progress a practical
feed-back system between the farmer and the laboratory is
absolutely necessary. Just transmitting the results is not
enough and under the conditions in Germany a very important
role must be given to the veterinary profession. The idea is
to promote this assay on a larger scale to check up on the
problem herds and problem cows and not on the regularly cycling
animal with good fertility. In part the problems may only be
management related (problem of oestrus detection) as we have
seen; in any case, but especially in those where the method is
applied for diagnosis of clinical situations, a very decisive
answer is necessary, and positive or negative as in pregnancy
diagnosis may not be sufficient. We therefore think, that the
latest methodological developments to measure progesterone in
milk fat directly may be a good way to maintain the accuracy
and high precision of the method and simultaneously increase
the practicability.

SUMMARY

The milk progesterone assay as developed 1973 was
established as a service method for routine fertility control
in cattle in Germany by the Bavarian Animal Health Service in
1975. Due to the average number of 8.9 cows per herd in the
Federal State of Bavaria and the fact, that pregnancy diagnosis

by rectal palpation is done rather routinely by the veterinary profession, emphasis for application of the assay was directed not so much for pregnancy-testing but rather toward the diagnosis of adequate oestrus at insemination (day 0) and for detection of eventual oestrus 20 days afterwards (day 20). In co-operation with the veterinarian control of the individual problem animals and special problem herds was sought. This is reflected by the distribution of samples; depending on the season, 20 - 50% were submitted for progesterone analysis to support clinical diagnosis and manipulations and for the rest of the samples the ratio of day 0 vs day 20 was 1 to 3.4. In order to meet the quality criteria of this highly specified assay application, progesterone was so far determined in the strippings, applying an extraction procedure which allowed measurement in the range of 50 to 30 000 pg/ml milk. With an increasing number of samples the method had to be simplified and a further simultaneous attempt was made to increase the specificity by working with separated milk and to determine progesterone in milk-fat directly. It has been established that due to the lipophilic properties of progesterone its concentration in milk fat is independent of the total concentration of milk fat and hence the time of sample collection. After determining progesterone in milk-fat the number of animals diagnosed correctly could be increased by 10%. The method consists of a simple heating and freezing step to obtain the pure milk fat and a simultaneous extraction and purification on a silica-gel column (5 - 10 µl milk fat are sufficient for one assay). For routine application semi-automated equipment was designed, allowing the simultaneous extraction and purification of 25 milk-fat samples within 20 minutes. After having evaporated the eluate containing the progesterone the RIA is performed.

REFERENCES

Arbeitsgemeinschaft deutscher Rinderzüchter e.v. Rinderproduktion 1975
Landwirtschaftsverlag GmbH, 4400 Münster-Hiltrup.

Günzler, O., Korndörfer, L., Lohoff, H., Hamburger, R. and Hoffman, B.
1975. Praktische Erfahrung mit der Progesteronbestimmung in der Milch
zur Erfassung des Fertilitätszustandes bei der Kugh Tierärztl. Umschau
30, 111-118.

Heap, R.B., Gwyn, M., Laing, J.A. and Walters, D.E. 1973. Pregnancy
diagnosis in cows; changes in milk progesterone concentration during
the oestrous cycle and pregnancy measured by a rapid radioimmunoassay
J. Agric. Sci. 81, 151-157.

Hoffman, B., 1971. Embryonaler Fruchttod bei einer Kuh im Anschluss an
eine Nachbesamung - nachgewiesen mit Hilfe der Progesteronbestimmung
im Blut. Zuchthyg. 6, 134-138.

Hoffman, B. and Hamburger, R. 1973. Progesteron in der Milch: Radioimmunologische Bestimmung, Beziehungen zur Gelbkörperfunktion und
Milchfettkonzentration Zuchthyg. 8, 154-166.

Hoffmann, B., Hamburger, R., Günzler, O., Korndörfer, L. and Lohoff, H.
1974. Determination of progesterone in milk applied for pregnancy
diagnosis in the cow. Theriogenol. 2, 21-28.

Hoffmann, B., Günzler, O., Hamburger, R. and Schmidt, W. 1976. Milk-
progesterone as a parameter for fertility control in cattle;
methodological approaches and present status of application in Germany.
Brit. vet. J. 132, 469-476.

Hoffmann, B., Hamburger, R. and Hollwich, W. 1977. Bestimmung von Progesteron
direkt im Milchfett als verbessertes Verfahren zur Fruchtbarkeitskontrolle
bei der Kuh. Zuchthyg. in press.

Laing, J.A. 1976. Special issue: Progesterone in milk and pregnancy
diagnosis. Brit. vet. J. 132, 443-550.

Shemesh, M., Ayalon, N. and Lindner, H.R. 1968. Early effect of conceptus
on plasma progesterone levels in the cow J. Reprod. Fert. 15, 161-164.

Schumann, H. 1975. Bestandsaufnahme und kritische Betrachtungen zur
Fruchtbarkeits-situation der bayrischen Kuhbestände. Seminar f. tier.
Prod. TU Weihenstephan am 16. 10.

Zeddies, J. 1976. Der wirtschaftliche Wert der Fruchtbarkeit 6.
Hülsenberger Gespräche Verlagsges. f. tierzücht. Nachrichten mbH
2000 Hamburg 54, 12-33.

USE OF PROGESTERONE CONCENTRATIONS IN PERIPHERAL PLASMA OR MILK IN CATTLE HERD MANAGEMENT

M. Thibier, M. Petit and P. Humblot

UNCEIA, 13, rue Jouët, 94700 Maisons-Alfort, and
INA Paris-Grignon (Centre de Grignon) 78850 Thiverval-Grignon
France

ABSTRACT

The determination of progesterone in peripheral plasma or milk in cows was used for three main purposes:

Early pregnancy diagnosis: this was studied from 2 746 females and accuracy was based on calving results. The negative accuracy decreased a little after more farmers applied for that service and was estimated from the new series of data as 94%. The mean positive accuracy was 69% but was higher in dairy heifers and suckling cows (74 to 75%) than in other groups (56 to 65%).

Ovarian activity diagnosis: this may well be measured with two samples 10 to 12 days apart. In connection with low dietary energy level, the percentage of cycling young heifers (12 - 18 months) was low (39%). Between 18 to 24 months, that activity was respectively, 92.85 and 65% in dairy, native and beef types (P <0.05). In suckling cows, the time of return to ovarian activity was very much delayed compared to dairy breeds.

Diagnosis of subfertile females: in a dairy herd, progesterone determinations and subsequent treatment when necessary, improved the interval from calving to conception by 10 days (96 vs 86 days). This kind of reproductive health programme seems very useful in that respect.

INTRODUCTION

In cattle herd management herd size and animal production are increasing whereas management practices are changing and particularly man hours per cow are decreasing. Yet, reproductive performance has to be maintained or improved to fulfill the potential for production. In that context, the availability of progesterone concentrations measurement techniques have given rise to a great deal of interest as a reproductive management aid. This may be classified into three main areas:

1) Early pregnancy diagnosis

It is almost 10 years since Shemesh et al. (1968) suggested that plasma progesterone concentrations could be used as an early pregnancy test in the bovine. Since then, the method has been improved either in plasma or more recently in milk (Laing and Heap, 1971; Heap et al., 1973). Commercial pregnancy testing services on a non-profit-making basis have been launched in France (Thibier, 1973) as in other countries (Booth and Holdsworth, 1976; Hoffmann et al., 1976).

2) Diagnosis of ovarian activity

Progesterone assays are also useful as an indicator of corpus luteum function (see review: Thibier et al., 1973$_b$) and hence with adequate samplings we have a convenient way of detecting ovarian activity in postpartum cows or in heifers whenever heat detection might not be reliable for some reason. With increasing demand by farmers for oestrous control techniques (Petit, 1977), it is essential to get data on ovarian activity or inactivity in order to choose the optimum treatment.

3) Diagnosis of subfertile females

In this area, progesterone determination may be useful as a basic tool in detecting accurate cow reproductive status in a reproductive health programme and hence give valuable information to the veterinarian to select adequate hormone treatment when necessary.

In the present paper, the method, accuracy and the benefits of those applications in field trials are described.

MATERIALS AND METHODS

1) Progesterone assay and samples collections

Progesterone was determined by radioimmunoassay as described previously (Thibier et al., 1973; Thibier, 1974; Thibier and Saumande, 1975). Blood (10 ml) was collected into heparinised vacutainers (Becton and Dickinson, Rungis-France) and centrifuged immediately (3 000g x 15 mn). Plasma was then decanted and stored at $-20^{\circ}C$ until analysed. It was shown (Thibier et al., 1976) that milk composition influences widely its progesterone concentration. It is therefore necessary to collect milk in strictly similar conditions. For convenience from the farmer's point of view, it was decided to collect milk from the morning milking and from pooled first milk from each healthy quarter.

2) Animals

2.1. Early pregnancy diagnosis (EPD)

Both blood and milk were sampled on the 21st day after artificial insemination. In order, first, to determine the magnitude of agreement between this diagnosis in blood and milk, both were sampled simultaneously from 114 dairy French Friesian cows on 16 farms.

The results were obtained from a series of 2 746 females. A first group of 472 animals were submitted to rectal examination at 60 - 90 days after insemination and accuracy of EPD was estimated when compared to the diagnosis of pregnancy by those examinations. The dates of calving of the 2 746 females were recorded and were checked in relation to the day of insemination. As indicated in Table 1, half of those animals were dairy heifers. They were all treated for oestrous control between December and March (1974 to 1976) either with a prostaglandin analogue (Estrumate ICI; n = 1 264) or with silastic coils (Abbott; n = 130). They were inseminated at the following oestrus. Beef heifers (mostly Charolais or crossbred) were also studied, they were also

previously treated with either a progestagen implant (Norgestomet - Searle) + PMSG (n = 26) or with Estrumate (n = 251) or
silastic coils + PMSG (n = 59) (Petit, 1977). Dairy cows (mainly
French Friesian breed) were diagnosed either from blood or milk
samples and suckling cows (beef breeds such as Charolais or native breeds such as Salers) were diagnosed from plasma sampling.
Oestrus in the cows was controlled either with Estrumate (n = 175)
or silastic coils (n = 174) in dairy cows and with Norgestomet
+ PMSG (n = 92) or silastic coils + PMSG (n = 109) in suckling
cows.

Finally, 466 dairy cows, routinely inseminated, were used as
controls and diagnosis was performed from milk samples. The
percentage of positive EPD is reported in Table 1.

TABLE 1

NUMBER OF FEMALES WHICH WERE SUBMITTED TO EARLY PREGNANCY DIAGNOSIS AND PER-
CENTAGE OF POSITIVE PREGNANCY DIAGNOSIS ACCORDING TO THEIR PHYSIOLOGICAL
STATUS

Groups of females	Total Number	Percentage of positive pregnancy diagnosis
Dairy heifers	1 394	80.4
Beef heifers	336	43.7
Dairy cows:		
Blood samples	189	61.3
Milk samples	160	66.2
Suckling cows	201	56.7
Control (dairy cows & milk samples)	466	86.7
Total	2 746	73.1

2.2. Ovarian activity diagnosis (OAD)

Ovarian activity was detected by two blood samples 10 or 12
days apart from animals not previously inseminated. It was estimated that two low progesterone levels from both samples would
indicate ovarian inactivity. This method will be referred to
as the double sampling method.

A first study on 156 dairy heifers was carried out to

determine the accuracy of OAD with only one blood sample when
compared to the usual double sampling method.

Also, 92 dairy heifers were submitted to rectal palpation in
order to determine the agreement between the results of such an
examination and those with the double blood sampling method.

In order to choose the most efficient oestrous synchronisation
treatment we tested the ovarian activity of 1 179 females according to physiological status, breed, age and environmental
conditions. From those, 709 heifers of different breeds (French
Friesian, Montbéliard, Normande, Charolais, crossbred) in private
farms were blood sampled at various months of the year (December
and January for dairy and Salers heifers - March to May for beef
heifers). These heifers were classified into 2 groups according
to age: 12 - 18 months (n = 96) or 18 - 24 months (n = 613). In
spring (March to June), 355 Charolais suckling cows and 115 dairy
cows (French Friesian and Montbéliard) were also studied.

2.3 Diagnosis of subfertile females

This study was undertaken in one herd (Institut National
Agronomique experimental farm in Grignon). Sixty nine cows of
the French Friesian breed were classified into 3 groups. Group
1 included 28 cows seen in heat during the first 45 days after
calving; they were inseminated during this period or at the following oestrus. All other animals (41 cows) were not seen in
oestrus at that time. Sixteen of those (group 2) were anoestrous
control animals and no treatment was administered. The remaining
25 cows (group 3) were blood or milk sampled for progesterone
assay between 45 and 60 days postpartum and subsequently treated
according to the progesterone levels. In the case of high levels
(ie luteal phase) prostaglandin analogue (Estrumate) was injected
(0.50 mg im; n = 12) and insemination was performed at the following oestrus (3 to 4 days after treatment). If progesterone concentrations were low (<1 ng/ml in plasma or <2 ng/ml in milk;
n = 13) synthetic LRH* was injected im (1 mg). Among these 13
females, 3 did not come into heat during the following 30 days
and were re-treated with LRH. Finally, 3 females were first
* A generous gift from Dr. A. Constantin (Intervet - Angers, France)

treated with LRH and not seen in heat after that but because of a high progesterone level they were treated at an average of 60 days after the first injection with Estrumate.

All cows were inseminated at each standing oestrus after the 45th day postpartum and early pregnancy diagnosis from milk was performed for each cow 21 days after AI. No cows were culled for infertility and all became pregnant by the end of this experiment. No evidence of metritis or other infections of the genital tract was seen.

RESULTS

1. Early pregnancy diagnosis (EPD)

1.1. Comparison in blood and milk samples

From Table 2, it can be seen that only 9 cows were diagnosed as pregnant from milk but not from blood samples and of those 9 cows one did not calve following that insemination. Diagnosis from milk and plasma was therefore in good agreement (90.4%) although milk diagnosis was slightly less accurate with respect to positive results.

TABLE 2

AGREEMENT BETWEEN EARLY PREGNANCY DIAGNOSED FROM BOTH PLASMA AND MILK SAMPLES

	Plasma	Milk	Number of cows
Agreement (90.4%)	+	+	65
	−	−	38
Disagreement (9.6%)	−	+	9
	+	−	2
Total			114

1.2. Accuracy of EPD based upon rectal examination

As shown in Table 3, the mean negative diagnosis accuracy was only 90% because of the significantly lower accuracy (P <0.05) in beef and native breeds of heifers and cows. Respectively 9 out of 50 and 17 out of 55 were considered pregnant by rectal examination whereas they were diagnosed non-pregnant on the 21st day after insemination. In contrast, the positive diagnosis accuracy (71.7%) was similar according to groups of females.

TABLE 3

EARLY PREGNANCY DIAGNOSIS (EPD) ACCURACY WHEN COMPARED TO RECTAL EXAMINATION AT 60 - 90 DAYS AFTER INSEMINATION

Groups of females	Positive EPD	Negative EPD
Dairy heifers	71/101	54/56
Beef heifers	46/64	41/50 *)
Dairy cows +)	26/36	21/21
Suckling cows	65/89	48/55 *)
Total	208/290 (71.7%)	164/182 (90%)

*) Significantly lower than other groups accuracy (P <0.05)
+) Milk samples

1.3. Accuracy of EPD based on calvings

The overall accuracy of both positive and negative pregnancy diagnosis compared to calving is reported in Table 4. No significant differences were seen between the negative accuracy according to groups of females or even within group according to any factor (breed, treatment, etc).

The mean accuracy of positive diagnosis was 69% but it varied significantly according to groups. It was higher in dairy heifers and suckling cows than in other groups of animals (P <0.001). In dairy heifers, neither oestrous control treatment (prostaglandin analogue vs progesterone silastic coils) nor the month of treatment have a significant influence on positive EPD accuracy. As far as the local effect (farm effect) is concerned, it must be noted that from 18 farms in which more than 8 heifers were simultaneously submitted to EPD, the range of accuracy was 43 to 100% but this was not significant (P >0.05). Neither the region nor the oestrous control treatment significantly affected the accuracy of EPD in beef heifers. In dairy cows, controls and

treated animals milk positive EPD was not significantly different
(P > 0.05), nor was it with the plasma diagnosis (P > 0.05)

TABLE 4
EARLY PREGNANCY DIAGNOSIS (EPD) ACCURACY WHEN COMPARED TO CALVINGS

Groups of females	Positive EPD	Negative EPD
Dairy heifers	74%* (1 122)	94% (272)
Beef heifers	65% (147)	93% (189)
Dairy cows Blood samples	62% (116)	96% (73)
Milk samples	56% (106)	91% (54)
Suckling cows	75%* (114)	98% (87)
Controls Milk samples	63% (404)	94% (62)
Total	69% (2009)	94% (737)

() Number of females;   * Significantly higher than in other groups (P<0.001)

2. OVARIAN ACTIVITY DIAGNOSIS (OAD)

2.1. Accuracy of OAD with only one plasma sample

It is shown (Table 5) that with only one progesterone determination, 66% of the 156 heifers were presumed to be cyclic
whereas with a second sampling 10 to 12 days later, 30 other
females were then diagnosed as cycling. This is an accuracy of
77% (104/134), with just one progesterone value.

2.2 Comparison of OAD by progesterone assay and rectal
palpation

From the 74 dairy heifers with ovarian activity as estimated
by the double sampling method, 67 (90.5%) had a detectable
corpus luteum on examination. Sixteen of 18 heifers with negative OAD by the progesterone method had no corpus luteum detected.

The accuracy based upon corpus luteum detection was therefore calculated at 90%.

TABLE 5

ACCURACY OF OVARIAN ACTIVITY DIAGNOSIS WITH ONLY ONE PLASMA SAMPLE COMPARED TO TWO, 10 DAYS APART, IN DAIRY HEIFERS

Groups of females	1st Sampling	2nd Sampling	Total
With high levels of progesterone (a)	104	30	134
With low levels of progesterone	52*	22	22

a) >1 ng/ml of plasma
* These 52 females were submitted 10 days later to another blood sampling

2.3. Ovarian activity in heifers

Based upon the double sampling method it was shown that a mean of 80% of the heifers showed ovarian activity. As illustrated in Table 6, significant differences of this activity occur between animals according to age or breed. Nutrition levels, in terms of energy intake, seem an important factor of variation of this activity (Table 7). Only one-third of the 12 - 18 months old heifers with low quantity of ordinary hay during winter, displayed luteal activity.

TABLE 6

MEAN CYCLING OVARIAN ACTIVITY IN HEIFERS ACCORDING TO BREEDS AND AGES

Breed / Age	12 - 18 months	18 - 24 months
Dairy breeds	79% (a) (77)	92% (b) (204)
Salers breeds	95% (c) (19)	85% (c) (158)
Beef breeds	-	65% (d) (251)

( ) Number of heifers.
a) VS (b) P <0.01   b) VS (c) P <0.02
c) VS (d) P <0.001   d) VS (d) P <0.001

TABLE 7

INFLUENCE OF NUTRITION LEVEL ON CYCLING OVARIAN ACTIVITY (MONTBELIARD HEIFERS)

Regime \ Age	12 - 18 months	18 - 24 months
Low - level*	39% (a) (19)	92% (b) (50)
High level **	92% (b) (13)	92% (b) (75)

```
*  Hay                        a)  VS (b)  Significantly different P <0.01
** Corn, silage.
```

2.4 Ovarian activity in cows

The cycling activity of dairy cows was well established
(Table 8) by 40 days postpartum (80%) and increased slightly
after this (95% after 90 days postpartum).

TABLE 8

OVARIAN ACTIVITY IN COWS ACCORDING TO BREEDS, SEASON AND POSTPARTUM INTERVAL

Breeds and season \ Postpartum interval	≤40 days	41 - 60 days	61 - 90 days	>90 days
Dairy cows	80% (125)	86% (28)	87% (23)	95% (39)
Charolais cows — March April	8% (71)	18% (129)	23% (70)	
May June	14% (56)	14% (29)	-	
Total	11% (a) (127)	17% (158)	23% (b) (70)	

```
()  Number of cows
a)  VS  (b)  P <0.05
```

In Charolais suckling cows, this activity was very low, until at least 90 days postpartum; it increased significantly
(P <0.05) from 11% (40 days postpartum) to 23% (60 - 90 days
postpartum). No significant difference was observed in this
study between the mean ovarian activity of Charolais females in
March - April vs May - June (P >0.05)

3. Diagnosis of subfertile females

As reported in Table 9, the interval from calving to conception in group 1 was short and significantly different to
those in other groups (P <O.05). Half of those cows were pregnant following their first insemination. The mean number of
services per conception was 1.75. Groups 2 and 3 conceived respectively 26 and 59 days later than group 1 although approximately
45% of those cows were pregnant following the first insemination
and the number of services per conception was 1.84 and 1.96. The
mean interval from calving to conception of groups 1 and 2 (no
treatment) was 96 days. By contrast that of groups 1 and 3 was
shorter and calculated as 86 days.

TABLE 9

MEAN ($\pm$ SD) INTERVALS BETWEEN CALVING AND CONCEPTION AND NUMBER OF SERVICES
PER CONCEPTION ACCORDING TO GROUPS (a)

Groups	Intervals between calving and conception (days)(b)	Number of services per conception	Number of cows served more than twice
Group 1 (n = 28)	74 $\pm$ 31	1.75	3
Group 2 (n = 16)	133 $\pm$ 55	1.96	5
Group 3 (n = 25)	100 $\pm$ 40	1.84	4

a) See text for definition of the 3 groups
b) Mean intervals between calving and conception are significantly
   different (P <O.05)
c) Conception assumption was based on early pregnancy diagnosis (21 days)

As far as the group 3 is concerned, the mean calving to conception interval was significantly lower than that of group 2
(P <O.05). From the 12 cows treated with prostaglandin analogue,
all came in heat in 3 to 4 days and all were inseminated. Six
were pregnant following the first insemination, 4 after the
second and 2 after the third one. This gave a mean interval
(Table 10) not significantly different to that of group 1. Following LRH, 10 females were seen in heat with a range of 16 to
53 days (mean interval 28 days) post treatment; only 4 conceived
at the first oestrus and one conceived after 7 inseminations.
The mean interval between calving and conception was 108 days.

Lastly, the 3 cows not seen in heat after LRH injection and with high progesterone levels (two were treated twice with LRH) were then treated with Estrumate. They were all inseminated 3 days later, two conceived on that insemination and one on the second. The mean calving conception interval (Table 10) was 138 days and the mean number of services per conception was 1.3.

TABLE 10

MEAN ($\pm$ SD) INTERVALS BETWEEN CALVING AND CONCEPTION AND NUMBER OF SERVICES PER CONCEPTION ACCORDING TO TREATMENT IN GROUP 3 COWS

Treatment	Calving conception intervals (days)	Number of services per con- ception
PG analogue * (n = 12)	85 $\pm$ 22	1.68
LRH ** (n = 10)	108 $\pm$ 50	2.2
LRH ** PG analogue (n = 3)	138 $\pm$ 43	1.3
Mean	100 $\pm$ 40	1.04

* Estrumate ICI
** Synthetic LRH - Intervet

DISCUSSION

1. Early pregnancy diagnosis

From blood and milk samples from the same cows good agreement in diagnosis of pregnancy status at day 21 was shown. Yet there was a slight non-significant overestimation of positive diagnosis from milk. However, because of the low magnitude of this difference together with a much more convenient sampling technique for the farmer, early pregnancy diagnosis may well be used in milking cows.

Accuracy of EPD is better estimated when based upon calvings as that is what matters from the farmer's point of view. The degree of accuracy of the negative diagnosis appears to be slightly lower than that reported in previous studies (Thibier,

1974; 1976) or by other investigators either in blood (Robertston
and Sarda, 1971; Wishart et al., 1975) or in milk (Heap et al.,
1976; Hoffmann et al., 1974 and 1976; Gunzler et al., 1975) but
similar to those reported by Dobson and Fitzpatrick (1976) or
in some cases by Pennington et al.(1976). The figure (94%) is
satisfactory for herd management because the probability of
error in a herd of normal size is very low. This figure slightly
decreased as the number of farmers applying for early pregnancy
diagnosis service increased and it might be postulated that
sampling was incorrectly carried out in a large number of cases.
It was shown (Thibier, 1973, unpublished data) that blood had to
be centrifuged very quickly and in less than one hour or else
progesterone concentration would drop rapidly according to the
environmental temperature. Mistakes in recording the number of
individual cows and mistranscription might also be taken into
account. Finally, Robertson and Sarda (1971), suggested that
some cows could have low blood plasma values although subsequently
shown to be pregnant. Very recently we found also from milk
samples taken every day over a long period, that pregnant cows
could have low progesterone levels on one or two days, either
with low plasma levels or without. In that case it might be
suggested that changes in milk composition could interfere with
low progesterone concentration as was previously shown (Heap et
al., 1976; Thibier et al., 1976) and therefore modify the negative accuracy.

The degree of accuracy of positive EPD is similar to that
found previously (Heap et al., 1976; Hoffmann et al., 1976;
Pennington et al., 1976; Pope et al., 1976$_a$). Among all the
factors which we tested and which could have modified the degree
of accuracy, we only found significant differences according to
the physiological status. This means that oestrus control, for
example, does not influence the accuracy of this test. Why the
accuracy in beef heifers and dairy cows (control and synchronised
oestrus females) should be lower is a matter for discussion. As
far as the latter are concerned it might be speculated that
abortion in connection with brucellosis might have occurred. This
would be in agreement with the fact that accuracy based upon

early rectal palpation is significantly higher. But no evidence
for such occurrence was recorded. The problem in beef heifers
might be related either to the use of PMSG when treated with
progestogens or to the random ovarian activity at the start of
Estrumate treatment as previously discussed by Deletang and
Petit (1976).

Nevertheless, it may be proposed that the farmers are to
expect a minimum of 94% accuracy in negative results and an average of 70% for positive diagnosis with some reservation either
in beef heifers or in dairy cows according to the herd health
status.

The comparison of accuracy in respect to rectal examination
or calving, shows that it is always preferable to rely on the
latter. As discussed above, the difference between positive
EPD and calving might be due to pathological events after the
date of rectal examination. We recently showed, (Thibier and
Rakotonanahary, 1977) that a cow population with a positive EPD
of 87% was also characterised by a non-return rate at 60 - 90
days of 66% and a calving rate of 53% from the same inseminations,
indicating a percentage loss to term (between 60 - 90 days and
calving) of over ten percent.

2. Ovarian activity diagnosis

Two samples, 10 to 12 days apart, seem necessary in order to
obtain a satisfactory accuracy of this diagnosis. Following the
use of only one progesterone determination, a quarter of cycling
animals are misdiagnosed. When compared to rectal palpation 10%
errors were found. Those might be related to either of the two
following observations reported by Dawson (1975): 1) non palpation of some active corpora lutea, 2) confusion of some
structures with a corpus luteum. In addition to that, as discussed above, active corpus luteum might be present together
with a low progesterone value under or at the conventionally
defined limit level.

In the 18 to 24 months old group the percentage of females

with cyclic activity was higher in dairy breeds than in native
and beef breeds. However, they were housed and fed different
nutrition levels; more precisely, the dairy heifers were better
fed than those of other breeds, particularly in late winter and
early spring. As already postulated (Chupin et al., 1977) underfeeding seems to block the ovarian activity of those females.
The situation is different in younger heifers, in that case
underfeeding prevents those animals from entering puberty (Laster
et al., 1972; Boyd, 1977) and this explains the low percentage
of ovarian activity in heifers with the low level of feeding we
observed.

The return of postpartum ovarian activity is much earlier
in dairy cows than in nursing cows and this agrees with reports
of numerous authors (see review: Wagner and Oxenreider, 1971).
This is due to both the suckling effect (Oxenreider, 1968; Short
et al., 1972) and the relatively low dietary energy levels (Wiltbank et al., 1964, Bellow et al., 1972) observed in the Charolais
cows in late winter or early spring before pasture. The percentage of postpartum ovarian activity in the Charolais breed is
similar to that reported by Chupin et al. (1976) although our
figures in the May-June period are slightly lower than that of
those authors.

It is therefore clearly established that progesterone determination from two samples is a useful method to characterise
the ovarian activity in both heifers and postpartum cows. Oestrous
control treatments may be selected according to the results.

3. Diagnosis of subfertile females

As reported by Attonaty et al. (1973); Britt, (1975); Pope
et al. (1976b); Manns and Richardson, (1976) an interval of 12
months between calvings results in optimum milk production. To
achieve this interval, cows must conceive about 70 to 80 days
postpartum. In most herds and in the one used here, this interval averaged 390 days before the experiment.

Shortening the interval from parturition to first insemination
to about 50 to 60 days postpartum was the first way used to
close the gap between this average and the optimum. The cows,

(group 1), first inseminated between 40 and 60 days postpartum, conceived before 80 days and averaged 50% conception rate at first insemination, similar to the report of Britt (1975). No more inseminations per conception were required than in groups 2 and 3 which were bred later. In groups 1 and 2 (ie, with no treatment) the mean interval between calving and conception (96 days) is relatively short compared to those indicated by several authors, (Britt and Ulberg, 1970; Barfoot et al., 1971; Galton et al., 1976). In groups 1 and 3, the mean interval between calving and conception was 86 days. The use of progesterone measurements and treatments permitted a gain of 10 days in the interval from calving to conception.

From the data of Attonaty et al., 1973, indicating that each day lost from calving to conception cost 5 F per cow, the gain of 10 days realised in the whole herd (69 cows) should bring an income of 3 540 F representing approximately the price of a cow. Moreover, we observed that, as reported by Galton et al., 1976, the percentage of animals culled for reproductive reasons decreased with the appliance of such reproductive investigations and treatments. This gain of 10 days is the result of our treatment of the large number of cows (25/69) not seen in oestrus by 45 days postpartum that we may consider subfertile animals as described earlier by Lamming and Bulman (1976).

Effectively in this herd, the most important problem was oestrous detection and behaviour. Unsuccessful inseminations were of a low influence and only a few cows needed more than 2 AI to conceive. The progesterone measurement allowed measurement of the level of ovarian activity and indicated that about 50% of cows (12/25) in group 3 had a high progesterone level and had started cycling some time before first observed oestrus. The remaining 13 cows were in true anoestrus. From this herd, the percentage of cycling cows at 45 days postpartum was low compared to the figures above or reported by Lamming and Bulman, (1976). Milk output in this herd is relatively high (>5 000 kg per lactation) and there is some indication of an antagonistic relationship between production and breeding efficiency

(Matsoukas and Fairchild, 1975)it may therefore be argued that high production delays the return to oestrus.

The mean time of conception of animals treated with a $PgF_{2\alpha}$ analogue was not significantly different from group 1 showing good efficiency of treatment. Use of LRH in true anoestrus cows gave inconsistent results; late or no return to oestrus after treatment was observed. This treatment did not hasten the init- iation of cyclic activity early after calving as reported by Manns and Richardson, 1976.

Therefore, the reproductive performance and subsequent gain obtained after such a programme including progesterone measure- ments, proved that this method could be considered as a very useful tool in therapeutic treatment.

ACKNOWLEDGMENTS

The authors gratefully acknowledge the excellent technical assistance of Mrs. N. Jeanguyot and Miss D. Hadida.

REFERENCES

Attonaty, J.M., Gastinel, P.L., Jalles, E. and Thibier, M. 1973 Consequences économiques des troubles de la fécondité - Compte-rendue des Journées d'information ITEB-UNCEIA PARIS - p. 16-53 ITEB Ed., (Paris)

Barfoot, L.W., Cote, J.F., Stone, J.B. and Wright, P.A. 1971 An economic appraisal of a preventive medecine programme for a dairy herd health management. Can. Vet. 12, 2-10

Bellows, R.A., Wagner, L.W., Short, R.E. and Panish, D.F. 1972 Gestation feed levels calf birth weight and calving difficulties. J. Anim. Sci. 35, 185-186

Booth, J.M. and Holdsworth, R.J. 1976 The establishment and operation of a central laboratory for pregnancy testing in cows. Br. Vet. J. 132: 518-528

Boyd, H. 1977 Anoestrus in cattle.. Vet. Rec. 100, 150-153

Britt, J.H. and Ulberg, L.C. 1970 Changes in reproductive performances in dairy herds using the herd reproductive status system. J. dairy Sci. 53, 752-756

Britt, J.H., 1975 Early postpartum breeding in dairy cows: a review. J. dairy Sci. 58, 266-271

Chupin, D., Pelot, J., Alonso de Miguel, M. and Thimonnier, J. 1976 Progesterone assay for study of ovarian activity during postpartum anoestrus in the cow. VIIIth Intern. Congress of Anim. Reprod. and Art. Insem. Cracow, p. 346-349

Chupin, D., Pelot, J. and Petit, M. 1977 Le point sur la maitrise des cycles sexuel chez les bovins. Bull. Techn. Insem. Artif. 5, 2-17 (Revue Tech. de l'Inséminateur, Rambouillet, France).

Dawson, F.L.M. 1975 Accuracy of rectal palpation - The diagnostic of ovarian function in the cow. Vet. Rec. 96, 218-220

Deletang, F. and Petit, M. 1976 Five years of field trials on control of oestrus cycle in cattle: Fertility following systematic artificial inseminations on oestrus induced by progestagen and prostaglandin $PgF_{2\alpha}$ analogue. VIIIth Intern. Congress of Anim. Reprod. and Art. Insem. Cracow, p. 457-460

Dobson, H. and Fitzpatrick, R.J. 1976 Clinical application of the progesterone in milk tests. Br. Vet. J. 132, 538-542

Galton, B.M., Barr, H.L. and Heider, L.E. 1976 Effects of a herd health programme on reproductive performances of dairy cows. J. dairy Sci. 60, 117-124

594

Gunzler, O., Korndorfer, L., Lossoff, H., Hamburger, R. and Hoffmann, D. 1975 Praktische Erfarhunger mit der Progesteronbestimmung in der Milch zur Erfassung des Fertilitätszustandes bei der Kuh. Tierarztl. Umsch. 30, 111-126

Heap, R.B., Gwyn, M., Laing, J.A. and Walters, D.E. 1973 Pregnancy diagnosis in cows; changes in milk progesterone concentration during the oestrous cycle and pregnancy measured by a rapid radioimmunoassay. J. agric. Sci. (Camb) 81, 151-157

Heap, R.B., Holdsworth, R.J., Gadsby, J.E., Laing, J.A. and Walters, D.E. 1976 Pregnancy diagnosis in the cow from milk progesterone concentration. Br. Vet. J. 132, 445-464

Hoffmann, B., Hamburger, R., Gunzler, O., Korndorfer, L. and Lohoff, H. 1974 Determination of progesterone in milk applied for pregnancy diagnosis in the cow. Theriogenology, 2, 21-28

Hoffmann, B., Gunzler, O., Hamburger, R. and Schmidt, W. 1976 Milk progesterone as a parameter for fertility control in cattle; methodological approaches and present status of application in Germany. Br. Vet. J. 132, 469-476

Laing, J.A. and Heap, R.B. 1971 The concentration of progesterone in the milk of cows during the reproductive cycles. Br. Vet. J. 127, XIX-XXII

Lamming, G.E., Bulman, D.C. 1976 The use of milk progesterone radioimmunoassay in the diagnosis and treatment of subfertility in dairy cows. Br. Vet. J. 132, 507-517

Laster, D.B., Glimp, H.A. and Gregory, K.E. 1972 Age and weight at puberty and conception in different breeds and breed-crosses of beef heifers. J. Anim Sci. 34, 1031-1036

Manns, J.G. and Richardson, G. 1976 Induction of cyclic activity in the early postpartum dairy cows. Cam. J. Anim. Sci. 56, 467-473

Matsoukas, J. and Fairchild, T.P. 1975 Effects of various factors on reproductive efficiency. J. dairy Sci. 58, 540-544

Oxenreider, S.L. 1968 Effects of suckling and ovarian function on postpartum reproductive activity in beef cows. Am. J. Vet. Res. 29, 2099-2102

Pennington, J.A., Spahr, S.L. and Lodge, J.R. 1976 Factors affecting progesterone in milk for pregnancy diagnosis in dairy cattle. Br. Vet. J. 132, 487-496

Petit, M. 1977 Maitrise des cycles sexuels chez les bovins. In Rapport d'activité des Services Techniques de l'UNCEIA. Elev. & Insem. 161, (in press).

Pope, G.S., Majzlik, I., Ball, P.J.H. and Leaver, J.D. 1976$_a$ Use of progesterone concentrations in plasma and milk in the diagnosis of pregnancy in domestic cattle. Br. Vet. J. 132, 497-506

Pope, G.S., Leaver, J.D., Majzlik, I. and Ball, P.J.H. 1976$_b$ Fertility of dairy cattle. ARC Research review 2, 49-54

Robertson, H.A. and Sarda, I.R. 1971 A very early pregnancy test for mammals and its application to the cow, ewe and sow. J. Endocrin. 49,407-413

Shemesh, M., Ayalon, N. and Lindner, H.R. 1968 Early effect of conceptus on plasma progesterone level in the cow. J. Reprod. Fert. 15, 161-164

Short, R.E., Bellows, R.A., Moody, E.L. and Howland, B.E. 1972 Effects of suckling and mastectomy on bovine postpartum reproduction. J. Anim. Sci. 34, 70-74

Thibier, M. 1973 Diagnostic précoce de gestation. In Rapport d'activité des Services Techniques UNCEIA p. 145-146 (UNCEIA Public. Paris)

Thibier, M. 1974 La progestérone dans le lait de vache. Diagnostic précoce de gestation. Elev. & Insém. 144, 27-32

Thibier, M. 1976 Diagnostic précoce de non gestation. In Rapport d'Activité des Services Techniques de l'UNCEIA. Elev. & Insém. 154, 19-21

Thibier, M., Castanier, M., Tea, N.T. and Scholler, R. 1973$_a$ Concentrations plasmatiques de la 17α-hydroxyprogesterone au cours du cycle de la vache. C.R. Acad. Sci. (Paris) 276, 3049-3052

Thibier, M., Craplet,C. and Parez, M. 1973$_b$ Les progestogènes naturels chez la vache. I. Etude physiologique. Rec. Méd. Vét. 149, 1181-1203

Thibier, M. and Saumande, J. 1975 Oestradiol - 17β, progesterone and 17α-hydroxyprogesterone concentrations in jugular venous plasma in cows prior to and during oestrus. J. steroid Biochem. 6, 1433-1437

Thibier, M., Fourbet, J.F. and Parez, M. 1976 Relationship between milk progesterone concentration and milk yield, fat and total nitrogen content. Br. Vet. J. 132, 477-486

Thibier, M. and Rakotonanahary, A. 1977 Concentrations de la progésterone plasmatique lors de l'insémination artificielle et taux de fertilité chez la vache laitière. Elev. & Insém. 159, 3-10

Wagner, W.C. and Oxenreider, S.L. 1971 Endocrine physiology following parturition. J. Anim. Sci. 32, Suppl. I, 1-16

Wishart, D.F., Head, V.A. and Horth, C.E. 1975 Early pregnancy diagnosis in cattle. Vet. Rec. 96, 34-38

Wiltbank, J.N., Rowden, W.W., Ingalls, J.E. and Zimmerman, D.R. 1964 Influence of postpartum energy level on reproductive performance of Hereford cows restricted in energy prior to calving. J. Anim. Sci. 23,1049-1053

OESTROUS CYCLE CONTROL BY MEANS OF PROSTAGLANDINS AND THE PROGESTERONE LEVELS IN MILK IN DAIRY CATTLE

C.H.J. Kalis and S.J. Dieleman.
Clinic of Veterinary Obstetrics Gynaecology and AI,
State University Utrecht, Yalelaan 7, de Uithof, Utrecht,
The Netherlands.

ABSTRACT

In this experiment the relationship between progesterone levels in milk from dairy cows and the fertility rate after oestrous cycle control with prostaglandins was investigated.

The cows were divided in three groups:

The first group of 41 cows received two injections of 0.5 mg cloprostenol, with an eleven-day-interval, followed by insemination 72 and 96 h after the last injection.

The second group of 32 cows was injected once with 0.5 mg cloprostenol when a corpus luteum was present. Insemination was performed at the standing heat following the injection.

The last group of 35 cows served as a control group. Insemination was performed at spontaneous oestrus.

The pregnancy rate of the three groups was respectively 44, 59 and 66%. Based on progesterone concentration in milk 25% of the cows from group I did not undergo luteolysis after the second PG injection. In fact these cows were inseminated while not in oestrus, explaining mainly the low pregnancy rate in this group. When these cows were excluded the pregnancy rate of group I was normal (58%).

Another result of this experiment was that both in the prostaglandin treated groups and in the control group a significant correlation existed between progesterone levels at 4 and 8 days after insemination and the fertility results.

The preliminary conclusion of this experiment is that a delayed development of the corpus luteum causes failure of both synchronisation and conception.

INTRODUCTION

The mean interval between parturition and conception in Dutch dairy cows is about 105 days. In order to obtain a one-year calving interval it has to be 85 days. Oestrous cycle control by means of prostaglandins could be used to shorten this interval. However, it has been demonstrated that pregnancy rate after double prostaglandin treatment in Dutch dairy cows is very low (de Kruif et al., 1976).

The aim of this investigation was to find an explanation for the low pregnancy rate caused by double prostaglandin treatment by monitoring the corpus luteum function by means of the progesterone levels in the milk.

MATERIAL AND METHODS

Cows

All cows belonged to the Dutch Friesian breed and had had a normal parturition and a normal puerperium. The cows were examined before treatment and only cows with a normal involuted uterus and with cyclic ovaries were used. The interval from parturition to treatment was between 40 and 80 days (average 61 days). The cows varied in age between 2 and 8 years (average 4.2).

Herds

The cows belonged to three experimental farms of moderate size (80 - 100 cows). During the winter period the cows were housed in cubicles. The herds were free from venereal infections, brucellosis and IBR. The fertility of the herds was good and there were no problems with retained placentae or abnormal discharge.

Treatments

Group I (41 cows) received two injections of 0.5 mg cloprostenol, eleven days apart. The cows were inseminated 72 and 96 h after the last injection.

Group II (32 cows) received one injection with 0.5 mg
cloprostenol. A well developed corpus luteum
could be palpated rectally. The cows were
inseminated at the time of oestrus 3 - 5 days
after treatment.

Group III (35 cows) received no treatment and was
inseminated at the first spontaneous oestrus.

Progesterone determination

Progesterone levels were determined in milk. The levels
were quantified by a direct Radio Immuno Assay, similar to the
procedure as reported by Heap et al., 1973. Antiserum* was
raised in rabbits against 11α hydroxy-progesterone haemisuccinate.
BSA conjugate was used with $/\bar{\ }1, 2, 6, 7\_/$ ^{3}H-progesterone as
tracer. Separation of bound and free progesterone was performed
with dextran-T 70 coated charcoal.

Milk sampling

Milk samples were taken from the cows on the day of
treatment and on the day of insemination. In order to follow
the development of the corpus luteum samples were taken also
on 5, 8 and 12 days after insemination. Cows were palpated
rectally on the days of sampling. The milk samples were taken
during the afternoon milking from the total milk and were
stored in the freezer.

RESULTS

Group I

Eighteen out of 41 cows became pregnant (44%).

A palpable and functional corpus luteum (progesterone
level >3 ng/ml) was present in 33 out of 41 cows at the moment
of the first PG treatment (T 0) (Table 1). At T 3 the

* Antiserum 5 R, PR 4 as described by de Jong et al. (1974) was kindly
donated by Prof. Dr. H.J. van der Molen, Erasmus University, Rotterdam.

progesterone levels were below 3 ng/ml in all 33 cows and
increased to normal levels in 25 animals at T 11. Due to the
second PG treatment (T 11) the values of these 25 cows dropped
to levels below 3 ng/ml (average 1.2 ng/ml); 14 became
pregnant.

In the remaining 8 animals the progesterone levels showed
a very slight increase from T 3 (av. 1.86 ng/ml) to T11 (av.
2.31 ng/ml); a more rapid increase took place thereafter (T 14
av. 7.3). None became pregnant. The progesterone levels in
8 out of 41 cows were below 3 ng/ml at the time of the first
PG treatment (T 0). In 6 of these animals a steady increase
of the progesterone levels occurred up to a maximum (av. 8.18
ng/ml) on T 11. The levels in all of these animals dropped
after the second PG treatment (av. 1.16 at T 14); 4 became
pregnant. In the 2 other animals the levels remained below
3 ng/ml until T 11. After T 11 a slight increase was shown
(av. 4.31). None became pregnant.

Eighteen of the 31 cows that showed luteolysis after the
second treatment became pregnant (58%).

Progesterone levels after insemination in pregnant and
non pregnant cows did not differ significantly from those
after insemination after a single treatment and were combined
in Table 2.

TABLE 1

PROGESTERONE LEVELS IN MILK FROM THE COWS ON SEVERAL DAYS DURING DOUBLE
PG TREATMENT

Cows responding to:	Number of cows	Progesterone concentration milk ng/ml					Pregnancy rate
		T 0	T 3	T 8	T 11	T 14	
1st and 2nd treatment	25	6.49	1.1	3.34	7.96	1.2	56% (14 cows)
1st treatment only	8	6.85	1.86	2.00	2.31	7.3	0%
2nd treatment only	6	1.47	4.01	7.45	8.18	1.16	67% (4 cows)
no treatment at all	2	1.28	0.85	1.60	2.97	4.31	0%

600

TABLE 2

PROGESTERONE LEVELS IN MILK AFTER CLOPROSTENOL TREATMENT

Day of sampling	prog. conc. in milk from cows which did not become pregnant (23) ng/ml ± SD (n)	prog. conc. in milk from cows which did become pregnancy (35) ng/ml ± SD (n)	Difference?
T O (PG inj.)	6.94 ± 3.13 (23)	8.54 ± 3.01 (34)	no diff.
T 3 (ins.)	1.06 ± 0.85 (22)	1.70 ± 1.35 (35)	no diff.
T 8	1.14 ± 0.85 (11)	6.03 ± 4.03 (18)	sign. diff.*
T 11	4.09 ± 3.36 (22)	9.49 ± 4.19 (34)	sign. diff.
T 14	8.20 ± 3.8 (8)	9.99 ± 3.51 (20)	no diff.

* Significancy tested by Student T-test with $P < 0.005$.

Group II

Nineteen of the 32 cows which had a functional corpus luteum at the time of treatment and which were inseminated at standing heat (3 - 5 days after treatment) became pregnant (59%).

Progesterone levels in the milk of those cows and of the 31 cows of group I that responded after the second treatment were combined in Table 2. At the time of treatment and three days later a slight, non-significant difference was found in the progesterone concentration in the milk between pregnant and non pregnant cows. On T 14 the difference between progesterone concentration in pregnant and non pregnant cows is not significant any more.

Group III

Twenty-three of the 33 cows of the control group became pregnant (66%). Progesterone levels in the milk are presented in Table 3. On the day of insemination and 12 days later no significant difference exists between pregnant and non pregnant cows. Seven days after insemination however this difference is significant.

TABLE 3

PROGESTERONE LEVELS IN MILK FROM COWS AFTER INSEMINATION AT A
SPONTANEOUS OESTRUS

Day of sampling	prog. conc. in milk from cows (12) which did not become pregnant ng/ml $\pm$ SD (n)	prog. conc. in milk from cows (23) which did become pregnant ng/ml $\pm$ SD (n)	Difference
D 0=ins.	1.50 $\pm$ 1.26 (12)	1.44 $\pm$ 1.35 (23)	no diff.
D 7	4.43 $\pm$ 2.95 (12)	8.20 $\pm$ 3.79 (23)	sign. diff*
D 12	8.51 $\pm$ 4.40 (12)	9.55 $\pm$ 3.42 (21)	no diff.

* Significancy tested by Student T-test with p 0.005

DISCUSSION

Pregnancy rates of the three groups are respectively 44,
59 and 66%. The results after a double prostaglandin treatment
are disappointing but in agreement with the results of de Kruif
et al. (1976). The low results can be explained by the fact
that 10 (25%) of the treated cows did not have a well developed
functional corpus luteum, sensitive to a prostaglandin treatment
at the time of second injection. As a consequence they were
inseminated at an inappropriate time.

Most probably sensitivity of a corpus luteum to prostaglandin
treatments is not determined by the age but by the functional
stage of the corpus luteum. Our findings suggest that in a
number of cows the functioning of the corpus luteum is delayed
after a PG administration and after a spontaneous oestrus as
well. This delay was strongly correlated with a failure to
conceive.

602

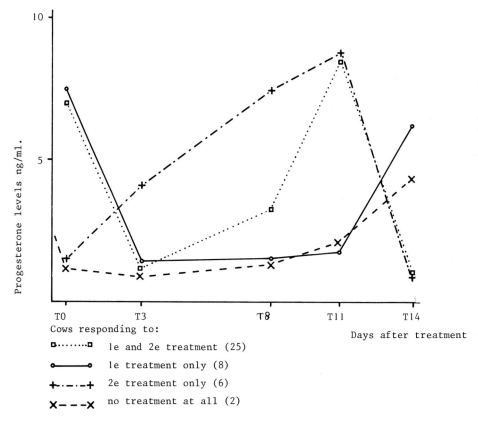

Fig. 1. Progesterone concentration in milk from cows, treated with two
         injections of prostaglandins, at days of treatment and at days
         of expected oestrus.

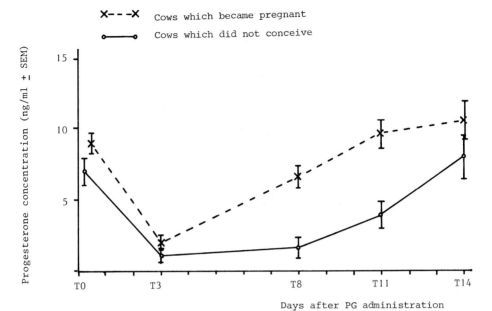

Fig. 2. Progesterone levels in milk (ng/ml $\pm$ SEM) after administration of cloprostenol.

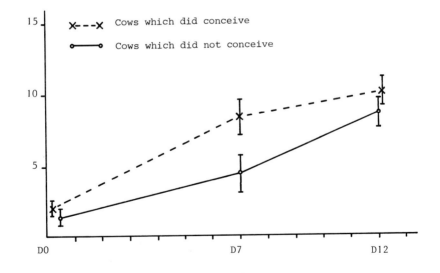

Fig. 3. Progesterone levels in milk (ng/ml $\pm$ SEM) after insemination at spontaneous oestrus.

REFERENCES

Folman, Y., Rosenberg, M., Herz, Z. and Davidson, M. 1973. The relationship between plasma-progesterone concentration and conception in post partum dairy cows maintained on two levels of nutrition. J. Reprod. Fert., 34, 267-278.

Heap, R.B., Gwyn, M., Laing, J.A. and Walters, D.E. 1973. Pregnancy diagnosis in cows; changes in milk progesterone concentration during the oestrous cycle and pregnancy measured by a rapid radioimmuno-assay. J. Agric. Sci. Camb. 81, 151-157.

Jong, F.H. de, Baird, D. and Molen, H.J. van der. 1974. Ovarian secretion rates of oestrogens, androgens and progesterone in normal women and in women with persistent ovarian follicles. Acta endocr. (Kbh), 77, 575.

Kruif, A. de and Brand, A. 1976. Use of prostaglandins in suboestrous cows. Tijdschr. Diergeneesk. 101, 9, 491-493.

Kruif, A. de, Zikken, A, Kommerij, R. and de Bois, C.H.W. 1976. Results obtained after synchronisation of oestrus by prostaglandins in dairy cows. Tijdschr. Diergeneesk. 101,22, 1257-1262.

DISCUSSION

J.M. Sreenan *(Ireland)*

And now the paper by Bridget Drew is open for discussion. Who will start?

K. O'Farrell *(Ireland)*

I would like to make one or two comments. First of all, eliminating oestrus detection with synchronisation, I don't think that is quite correct because there are a certain proportion of animals that will show oestrus after being inseminated. I think if they are not re-inseminated at this observed oestrus then you can expect them not to conceive. Perhaps you have some information on that.

The other comment I would like to make is on the calving to conception interval. I presume it refers to cows which conceive but it doesn't refer to cows which are culled from treated and controlled groups. In other words, if you have fewer cows which conceive on one side versus the other, then the figures can look better when you are referring to calving to conception intervals. So I think the culling rate for 'not in calf' is an important addition to that information.

Have you any information on your oestrous response rate following treatment with the PRID to your anoestrous cows. If not, can you give me any information on the return intervals?

You referred to treating cows 60 days post partum. Have you compared the effects of this with controls on the calving pattern?

B. Drew *(UK)*

Starting with the first question, you asked about the proportion of the animals which came into oestrus after fixed time insemination. This was the point that I tried to make in the last slide. With suckler cows and dairy heifers this is why I suggested that bulls should be turned in with the cows as soon as

the really active oestrus has subsided to cover the ones which
come later. Obviously, dairy cows should be re-inseminated.
We have found that normally about 8 - 10% of cows come between
Days 5 and 11.

Your second point is fair comment. The calving to conception interval data were only on the ones that conceived. There
was a higher proportion of control animals which failed to conceive. The figures were 4% of the treated and 8.2% of the controls that failed to conceive.

On the question of oestrous response following treatment
with the PRID, I personally feel that the oestrous response with
the PRID is probably more synchronised than with the other products although this doesn't necessarily affect the fertility.
We do seem to have a very precise and quite intense period of
heat on the first day that the animals are inseminated. Usually
by the second day there is very little to be seen at all. I am
afraid that we didn't observe the oestrus at regular intervals
on the trial on which we studied the non-cyclical cows so I can
not tell you whether the non-cyclic cows showed oestrus as well
as the cyclic cows, but the return service intervals were normal.
Of the 19 non-cyclical cows, 7 conceived to fixed-time inseminations; another 8 conceived at the first return period between
Days 17 and 24; one conceived at Day 38 and she lost the coil
in fact, so she wasn't inseminated the first time; two conceived
at Day 48 and one at Day 80. So, in all those cows, apart from
the one that lost the coil, a normal pattern followed.

On your last point, in the last one we did we tried to
average the treated ones to serve them at Day 50 but we did get,
as you saw in the slide, a very much lower proportion of animals
pregnant. Providing that you can ensure good fertility by good
herd management, I feel it is quite all right to leave it until
Day 60. It depends upon the fertility to first service. In one
herd where all the cows were past Day 60 (and this was the herd
where we had the best fertility) there were 64 animals treated,
of which 50 calved to the synchronised service and 11 to the

first repeat. So we had 61 out of 64 animals pregnant in 24 days.

J.M. Sreenan

I think we must leave this discussion there because the time does not allow for any more questions now but we may find time later when we are discussing the general aspects of the session. Thank you very much, Dr. Drew.

May we turn to Dr. Mulvehill's paper.

M. Boland *(Ireland)*

In relation to the twinning of heifers, at what stage did you slaughter your heifers to recover the embryos, and did you see any effect of the site of ovulation in relation to ovaries and the survival of the embryos?

J.M. Sreenan

Can I briefly comment first. The distribution of ovulations was very even in terms of unilateral and bilateral ovulation. However, the numbers per treatment group as you saw were too small to allow definite conclusions to be drawn.

M. Thibier *(France)*

Perhaps I missed the point but did you check for the ovarian cyclicity before the treatments in your large-scale field trials? If so, do you have any idea of the percentage of induction of ovulation in any of the three treatments you have tested?

P. Mulvehill *(Ireland)*

We did not test for cyclicity in animals treated; we accepted animals for treatment provided they were presented as non-pregnant by the farmer and 35 days had elapsed since calving.

M. Thibier

Did you have any idea of the percentage of cyclicity before the treatment?

P. Mulvehill

No, we had no information on the cyclicity of the animals.

P. Jackson

I wonder whether with your GnRH treated heifers you are not defeating your object in that you are advancing your ovulation, yes, but are you ovulating immature follicles?

P. Mulvehill

Looking at the animals to determine the time of ovulation, the large follicles prior to ovulation seemed normal to us.

J.M. Sreenan

If there are no further questions we will follow on to questions on the paper from Dr. Mawhinney.

D. Wishart (UK)

I wonder what the rationale was for choosing such a low dose of progesterone to incorporate with the oestrogen at the start of treatment?

S. Mawhinney (Ireland)

These values were taken from earlier trials where we had compared different rates of progesterone. We have found, certainly in a recent experiment carried out with either 50, 100 or 250 mg of progesterone, the best results were obtained with 100 mg. Both the lower and higher doses gave a less concise onset of oestrus.

D. Wishart

I asked the question because we have done very carefully planned titration studies using oestradiol valerate and

norgestomet which have a potency of some 33 times that of progesterone and it is only when one gets up to a dose level of
something like 3 mg norgestomet that one is getting the effect
one is looking for. I wonder whether you are not, in fact, about
tenfold short in your dosage?

S. Mawhinney

Well, in the first instance we wanted to know whether we
needed extra progesterone with the intravaginal device because
of its fast rate of progesterone release at the beginning. This
point of giving 100 extra mg of progesterone made no difference.
I take your point but in our trials we just found that not giving
any progestersone was effective. I am not trying to excuse
myself but I do think that the dosage of progesterone at this
point is not relevant really.

S. Willadsen (UK)

So you end up with some 55% of your animals pregnant to
the first insemination. I wonder, have you any information on
the ones that don't get pregnant?

S. Mawhinney

This point has been brought up to us many times before and
we made a study of last year's farm trials. The farmers keep
results for us and they take note of which animals repeat within
21 days and then from their calving dates we work out fertility
and that, again, is around 55%, which we have found from our
control animals is the average fertility in the animals we are
working with.

C.H.J. Kalis (Netherlands)

I would like to ask whether you saw vaginitis after removal
of the PRIDs and did you seek a relationship with subsequent
fertility?

S. Mawhinney

Well, whenever we removed the PRIDs, I wouldn't call it
'vaginitis' as such but they definitely don't smell very nice!

However, we have had trials where we dusted the PRIDs with anti-
biotic before insertion and it has made absolutely no difference
to the retention rate. If the animals were suffering from
vaginitis there would be a higher rate of expulsion of the coils
from irritation as much as anything else but dusting the coils
with antibiotics made no difference. Before insertion we dip
the coils and the speculum into an antiseptic solution and we
wash the perineal area with antiseptic solution but apart from
that we take no other precautions and we have had no obvious
problems.

H. Coulthard (UK)

We have done quite a few of these ourselves; we have prepared 20 heifers as recipients in this way, with and without
PMS. We have also done a group of 52 cows and another group of
about 18 heifers. One thing that we have noticed is that we have
had a bigger loss of PRIDs in the heifers than in the cows. I
wonder whether you have found this because I don't know why that
should be. Our results have shown pretty well what yours have.
Of course, we didn't know when they ovulated; we could only take
the first time observed in oestrus. Those animals which were
given PMS, their first observed oestrus was 51 hours after removal for the cows, for those which weren't given PMS it was
slightly shorter - 53.7 hours, not significantly different but
for our recipients we got a very high percentage at about 48
hours, which was ideal for our purpose.

On our pregnancy results: to one insemination there were
14 out of 26: to two inseminations there were 16 out of 26. As
far as using recipients are concerned, we obtained 8 pregnancies
out of 14 used, which was fairly good because at that time we
were only getting about 40% pregnancy rate anaesthetising animals
as recipients.

S. Mawhinney

In different trials we have found out that the retention
rate is affected by the diameter of the PRID and we now use the
coil that gives us optimum retention rate, one with a diameter

of 4.6 cm. We have found exactly the opposite to you; we have
found that in dairy cows the retention rate, on average, is 95%
whereas in heifers it is 98% - slightly higher. On the time of
insemination we started by looking at when the average oestrus
was following coil removal and then by trial and error as much
as anything we got down to the 56 hour single insemination. We
were slightly surprised at the timing of the insemination being
so critical. There was a drop of 5 - 6% in fertility by inseminating them just four hours too soon. We had hoped to say to
the farmer that if he removes the coils one morning he could
inseminate in the early afternoon of the second day. Now what
we have to say is, 'Remove them at 8 o'clock in the morning and
inseminate as near to 4 pm in the afternoon of the second day
as you can possibly manage'.

K. Betteridge *(Canada)*

I understand that if you put a PRID in water it will release progesterone steadily. If that is so why does the progesterone stay up for a limited time only in these treated
animals?

S. Mawhinney

I don't know about that one at all - possibly Dr. Gosling
from Galway University could help you because he has done studies
on progesterone removal from the coil. I honestly don't know.
Why it should only stay up for a certain length of time - it is
that the rate of release will gradually drop off. These coils
can be used again and again. At the second time of use you get
the high release at the beginning again. If you use them a third
time you still can get a response but it is not as much.

J.M. Sreenan

Perhaps this is another paper that we can discuss in the
General Discussion but for the moment we must move on to the
paper given by Dr. Mauleon.

D. Wishart

Dr. Mauleon, I am surprised that with the efficiency of

your AI organisations a bull with a 29% fertility should be used
for commercial AI service.

P. Mauleon *(France)*

This happens sometimes.

D. Wishart

Our AI centres would certainly never admit to that dis-
crepancy of fertility. What information is gathered from the
beef farms which allows the AI centre to monitor pregnancy rates
following AI on beef cows?

P. Mauleon

Nobody really knows the fertility in the beef herd. As I
said before, the cows are in the pasture with free running bulls.
You can know how efficient the AI is if you record the precise
dates of calving in relation to the time of mating and nobody is
really doing this. It is always difficult to know these things
without controls but in our particular conditions there is no
control at all because there is no oestrus detected in these
females.

M. Thibier

I would just like to add something to Dr. Mauleon's reply
because I am working with the AI organisation in France. In the
case of this Charolais bull, its fertility is rather low but it
is a special bull of a very high standard as far as the meat is
concerned. The farmers realise that because they are progeny
tested according to the carcass but also according to the fertility. So the farmer either wants it or he doesn't but he knows
what is going to happen.

As far as pregnancy diagnosis is concerned following insemination in the beef area, I will be talking about this a little
later on but I must admit that it is very difficult in the beef
area and up until now we have had quite a lot of trouble.

R. Newcomb *(UK)*

I hope Dr. Mauleon will forgive me if I have not caught
all his points but we have been so bombarded this afternoon.
However, when you are treating with PMSG, do you notice a tremendous variation in the time to the induction of oestrus, when
you are using PMSG at a very low level? The reason I ask is that
when we have injected PMSG at 1 000 iu, if we happen to have one
ovulation generally they take about three days to come into heat
after prostaglandin injection, whereas if they have got multiple
ovulations they come in quite quickly, in about two days. If
you do get this variation is it going to interfere with your
synchronisation programme?

P. Mauleon

Really the difficulty in monitoring experiments with the
Charolais breed in our experimental centre is that we really do
not have Charolais comparable with those on the farms in terms
of nutritional levels. So, if we treat Charolais with PMS in our
conditions at the research centre, we obtain very frequently -
with a low dose, say, 1 000 iu for example - a very high percentage of multiple ovulations. However, in these particular conditions we have no exact reference to the timing of ovulation.

J.M. Sreenan

I would like to comment because we have also used PMSG
at the end of short term progestagen treatments. I would
suggest, from our data, that PMSG at low dose levels has a
different effect on oestrous response and onset of oestrus
than at superovulation levels. We recognise this from our
donors and recipients. PMSG at high levels for superovulation
gives us a very early oestrous onset whereas at lower dose
levels (and our dose levels are perhaps a little bit lower
than yours) we don't see any effect in terms of timing of
oestrous onset.

P. Mauleon

I can answer in terms of oestrous control in sheep. It is
very clear with these conditions after progestagen treatment we

have evidence of something like 6 hours advance in the timing of ovulation.

H. Hardie *(Belgium)*

I just wanted to ask how long did you flush the animals - I didn't catch that.

P. Mauleon

Nine days, only during the period of treatment.

D. Wishart

You say in your paper that prostagens create unfavourable situations in the genital tract. Are you saying that this is the case with short prostagen treatment?

P. Mauleon

I don't know. It is very well known in the sheep. What we know is that on comparisons we have done between heifers and prostagen treatments and prostaglandin treatments, we have monitored two AI or one AI. When we have one AI it is necessary to increase the number of sperm that we put in the genital tract to get the same fertility. In my opinion, it is quite normal to decrease even with a nine days treatment the local conditions of the genital tract.

D. Wishart

The information we have from our own experience is that when we collected eggs surgically, admittedly we used one ejaculate and one bull. We selected the bull as the high fertility bull from the Milk Marketing Board Stud. We could find no difference in either pregnancy rate or egg fertilisation rate between animals treated with cronolone, and certainly with norgestomet, and the controls.

P. Mauleon

We have not done any experiments to demonstrate this fact.

T. Greve
    I just want to come back to Table 11. I have studied it
with great interest and I would just comment that there might be
a difference between the treatment and the onset of heat. I am
very surprised that by increasing the dose with 300 iu, you get
an increase in twinning rate from 7.3 to 20.2 when you consider
that even if we vary our dose at a higher level, for example,
from 2 000 to 4 000 iu, we can hardly detect any significant
difference in the number of ovulations. I think that is something to think about, that 300 units can make such a great difference.

P. Mauleon
    Really it is a survey of the results we obtained with dif-
ferent dose levels in a field trial. Therefore it is not sure
that only this factor exists. What it means is that we must
decide whether to increase or remain at a medium level without
risk. So we are using no more than 500 iu for the heifers and
600 iu for the Charolais nursing cows.

J.M. Sreenan
    Perhaps I may comment on that. I suggested earlier that
PMSG at a low dose rate acts in a slightly different way from
PMSG at high dose rates. The mean and variance increases with
dose level. The data that we have shows this, that you get
this small increase at a lower dose rate that is not possible
at a high dose rate because you have a greater variance.

S. Willadsen
    I think Greve is assuming a completely linear relationship
in response to PMSG and I think there is every evidence that, in
fact, you don't get that.

J.M. Sreenan
    It may be linear possibly at very low dose levels

S. Willadsen
    Yes, that's right.

P. Mauleon

It could be a very interesting subject for discussion -
about the relationship between dose and the number of ovulations.
In our opinion it is not a linear response.

J.M. Sreenan

Thank you very much, Dr. Mauleon. We must continue with
another French paper, that of Dr. Chupin.

T. Greve

Were the cows used in this study, Dr. Chupin, subjected
to clinical examination, and, if so, were they gynaecologically
normal before you initiated the study?

D. Chupin *(France)*

No, our experiments were made in this trial without any
prior clinical examination. The only thing that was done in
certain experiments was blood sampling, for progesterone assay
before treatment.

K. O'Farrell

I was interested in your Table which showed the interval
for returns. It is the first data I have seen other than my own
which indicates that the return interval is quite prolonged. For
example, in this Table you show about 55% return in the normal
18 - 26 days and that is obviously leaving quite a sizeable
proportion which do not return in that interval. The figures I
have from farm trials would indicate that about 33% do not return
in this interval and I think this is one of the problems with
fixed time insemination. It shows how necessary it is to have
progesterone and milk assay performed on fixed time inseminated
animals, otherwise valuable time gained from synchronising will
be lost.

H. Coulthard

Can you tell us why, in a superovulation situation, we give
PMSG prior to prostaglandin and yet when we get to this situation
we do it the other way round?

The other point is that what you are trying to do, by and large, is to shorten the interval between a cow calving and getting her into calf. Surely it is better to use two shots of prostaglandin even if it takes four or five days for her to come into season after the first injection, (as it often does, I agree), because you then have two chances to get her in calf rather than put a coil in and then prostaglandin? If you do the latter you are actually lengthening the time you are going to take to give you a chance to get her in calf the first time.

D. Chupin

In France the main reason for all oestrous control treatments and the main interest of the farmer is to suppress all oestrous detection. They need a treatment they can use with their eyes closed and be sure that they can inseminate on a given day.

H. Coulthard

What about the question of the other way round? Why not give the PMSG on Day 7 and the prostaglandin on the day you take the implant out? What difference does it make giving it early?

D. Chupin

In this experiment we do not want twinning; we just hope to get a little stimulation of follicular growth; we try to avoid twinning.

M. Boland *(Ireland)*

In one of your slides you showed that you can increase the pregnancy rate from about 41% to 51% by delaying the interval from your two prostaglandin injections from 11 to 13 days. Would you say that this is due to the lack of luteolysis at the first PG injection or is there some other factor involved?

D. Chupin

I said that when we increase the interval between the two injections we have a better chance of getting luteolysis after the second injection. This is because after the first injection

you get luteolysis in perhaps 75% of cows but you have a variation
If you increase the interval these young CLs are five, six or
seven days old and they can be destroyed.

R. Newcomb

Dr. Sreenan has been called away and so you have me in the
Chair for the time being. May we discuss Dr. Hoffmann's paper?

T. Greve

Did you find any difference in those positive or negative
questionable results in larger herds versus smaller herds?

B. Hoffmann (West Germany)

The largest herd we have examined was about 20 cows and
the smallest herd four cows and we have seen all kinds of variations there.

T. Greve

What I am getting at there, as you probably know, we have
done a survey for the Danish Breeders' Association. We had samples from all over Denmark from various sizes of herds and then,
as a control, we had two larger herds. We did not find any notable difference which we would have expected because many people
are not acquainted with the sampling of milk and so on - that was
what I wanted to hear - what experience you had in that context.

R. Newcomb

Let us move on to Dr. Thibier's paper. Perhaps I can ask a
question as a matter of information.

When you talk about cyclicity on the slides you were showing, how do you estimate this? Is it purely by oestrous detection
or by rectal palpation?

M. Thibier

We didn't know anything about the number in oestrus. The
farmers told us, "OK, you want to have those? You have them!"
I can't tell you whether or not they were in oestrus because I

haven't looked for it. If we had a high level of progesterone and further along a low level, or low level and high level, they were all considered as having cycled. That's what we mean.

R. Newcomb

Now I have a question for Dr. Kalis; I wonder if you would like to speculate as to whether it is possible that what you are doing is picking up non-ovulation and that the delay in production of progesterone might possibly be luteinisation of an ovulated follicle?

C.H.J. Kalis *(Netherlands)*

That is the reason why I palpated the cows four days after the treatment when most cows had ovulated. Of the cows not responding to treatment, one out of eight did not show an ovulation. You could not palpate the development of the corpus luteum as well. Thus in a few cows there may have been failure of ovulation but most cows, however, did ovulate.

P. Jackson

I congratulate Dr. Kalis on a very interesting piece of work. There are a couple of questions I would like to ask.

Firstly, was it possible to make any estimate of the average cycle length of the cattle in this herd before the experiment started? An indication to this, of course, might be the length of time it took for all your control animals to show heat - did they show in 21 days?

The second question is really an amplification of Ray Newcomb's question. I wonder, with these 'lag' animals, whether it was a delay in the period of progesterone drop to ovulation or whether it was ovulation to the time that the corpus luteum is producing an appreciable amount of progesterone?

Thirdly, was it possible to make any correlation between the lagging animals and milk yield?

C.H.J. Kalis

In answer to your first question, we recorded the observed oestrus by the farmers of those cows. Personally, I saw no changes in cycle length, not even from those cows later found to have a delayed corpus luteum formation. They came in oestrus at an average of 21 days later; there was no difference as far as I could see.

On the second question, do you mean the luteolysis of treated cows or do you mean the luteolysis from control cows?

P. Jackson

I meant that the long progesterone level could have been because after the luteolysis there was a long delay before follicular development occurred; or perhaps follicular development occurred straightaway and there was delay in the corpus luteum becoming active in terms of producing appreciable progesterone levels.

C.H.J. Kalis

Yes, you may be right but in that case I would have expected the prolongation of the next cycle from the delayed formation of the corpus luteum and I did not see that. They just came back into oestrus 21 days later.

The third question was about milk production. They were treated early after parturition but excepting one cow, which may have failed to ovulate, that was a very high producing cow, in that cow I should say that milk production was related to this effect, but in the others I did not find that correlation.

P. Jackson

As a comment on this, we recognise in England (work at Nottingham) that about 10% of post-partum dairy cows are endocrinologically abnormal. This is split fairly well down the middle into animals which at the period you mention are not in true cyclicity and haven't been in calf, and the other 5% who have perhaps one or two cycles and then go into a type of anoestrous condition. Obviously these things are mediated by stress and nutritional management and I suggest that those two categories parallel rather interestingly your eight animals in the first group and in the second category, the two animals in your first group. It looks as though you are getting the same sort of thing in Holland as we get in England.

C.H.J. Kalis

Yes, I think you are right. I just took animals which had been shown to be cyclic before but it may be that without the treatment they had stopped cycling anyway.

P. Jackson

Can you , in fact, estimate cyclicity from one rectal palpation?

C.H.J. Kalis

No, I did it several times. I also had progesterone levels, especially from those two cows, otherwise I would not have presented that.

K. Betteridge

Has field use of prostaglandins turned up any problems with serological testing for brucellosis?

C.H.J. Kalis

I don't know.

P. Jackson

May I answer that one personally afterwards? The answer is 'no' but I would like to amplify it.

R. Newcomb

If there are no more questions, thank you very much, Dr. Kalis.

The papers which we have had this afternoon have dealt with synchronisation, mainly with progestagens and also with

prostaglandin; they have also dealt with progesterone analysis
as a means of examining problems in cattle. I wonder if we could
deal with the synchronisation story first of all and direct the
questions to the appropriate speakers.

K. Betteridge

I would like to ask Dr. Hoffmann to give us an example of
how the test is being used to vary the treatments for cystic
ovaries in cows.

B. Hoffmann

There are basically two forms of cystic ovary, those which
produce probably oestrogen and those which have low progesterone
production. Differentiation can easily be done by rectal palpation and in these cases you can go into the animal either with
GnRH or HCG. If you have a cyst which produces oestrogen you
treat the animal with prostglandin analogue and you get cyclicity
afterwards. We have found the method very useful also for animals which are very valuable because they have a high milk yield.
The farmers take great care of these animals, they want the
animals treated, they want all kinds of things done. So you treat
them with this and that and after the third treatment the vet
gets frustrated at what he is achieving with respect to ovarian
function because this animal has a high breeding value. In quite
a few cases the problems can be solved by varying the dosage etc.

K. Betteridge

Is the test, in fact, proving useful for varying the treatment between GnRH and prostaglandin?

B. Hoffmann

I would say yes.

M. Thibier

I would like to make a comment on this subject and show
two slides if I may. They are from a paper which we gave last
year. There were two cows which had cysts on the ovaries and
they were nymphomaniac as well. Figure 1 shows the oestradiol

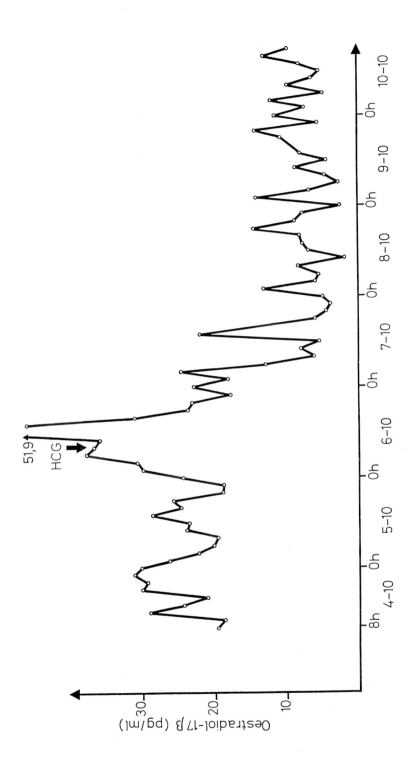

623

Fig. 1 cf: Saumande J. and Thibier, M. 1976 Hormonal levels in nymphomaniac cows. Proc. 9th Intern. Congr. on Cattle Diseases. Paris pp 875-883

624

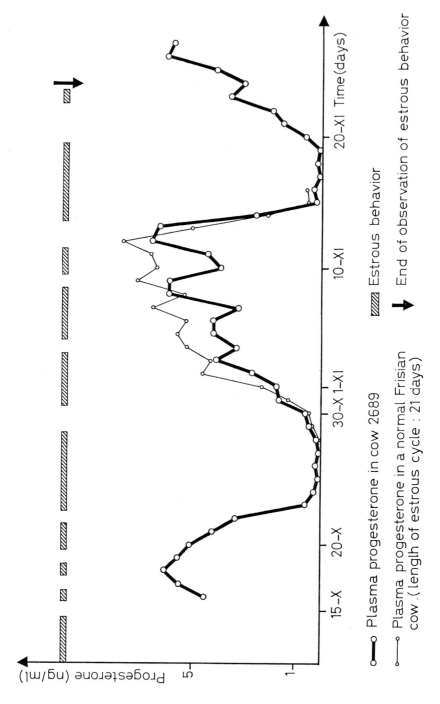

Fig. 2 cf: Saumande J. and Thibier, M. 1976 Hormonal levels in nymphomaniac cows. Proc. 9th Intern. Congr. on Cattle Diseases. Paris pp 875-883

concentration in the peripheral plasma and you can see that before any treatment the level was rather high, between 20 and 30
picograms per ml. After an HCG treatment the oestradiol level
dropped back to normal level.

If you look at the next slide, Figure 2, we have the other
situation. Here again there is a female with cysts and again a
nymphomaniac. All the periods of heat are reflected in the
dotted bar and you can see that in this cow the progesterone
level was quite normal, similar to the control. In other words,
it means that although there were some cysts and the cow was
definitely a nymphomaniac, yet there is a normal pattern of progesterone suggesting that the cyclicity was all right.

K. Betteridge
So it would be pretty hard to vary your treatment?

M. Thibier
That's the point, yes.

R. Newcomb
I would like to ask Dr. Thibier a question myself.

It was amazing to hear a Frenchman make some criticism of
the Charolais breed! He did mention that the Charolais has a
low cyclicity after calving, is this merely a feature of beef
suckler herds in general or is it a feature of the Charolais in
particular?

M. Thibier
There are two points here. The first is the breed effect
and the second is the breeding season. I believe that there is
probably an interaction between the two of them because definitely
a suckling cow has quite a difficulty to get into ovarian activity.
Three or four years ago we looked at the Limousin breed; their
breeding season was a bit later and we discovered that the ovarian
cyclicity returned very much quicker although they were nursing
cows. So we do not know that there is a breed effect.

T. Greve

On your last slide with the very nice progesterone profile, there are clinical signs of oestrus several times, was this cow examined clinically and what were the ovarian findings?

J. Saumande *(France)*

Once, I think in the middle of the luteal phase, we found a corpus luteum and a cyst on the same ovary.

T. Greve

But I think if you rely on the milk progesterone or blood progesterone in most cases and gave them prostaglandin treatment, the condition would cease.

M. Thibier

Well, we haven't done that so it is a matter of speculation but definitely you have quite a normal progesterone pattern in that case even though it was very quickly in oestrus as you can see.

J. Saumande

The point is that it was the first time that in a nymphomaniac cow we found progesterone and usually nymphomaniac cows are known to have a high level of oestradiol but this was not the case with this animal. There is a very low level of oestradiol and since we obtained these results we have taken a lot of plasma samples from nymphomaniac cows and in about 10 - 15% of them we found high levels of progesterone.

B. Hoffmann

I would like to put a general and perhaps provocative question to the audience here. I am very much interested in what is the practical need for oestrous synchronisation in cattle? I have had several discussions here with various people and there seem to be a number of views in this respect. Perhaps several people might give their opinions.

R. Newcomb

If nobody has got any comments on that it might be something to sleep on and discuss tomorrow.

J.M. Sreenan

This may well be a subject for an entire conference. There are two aspects to synchronisation: one of them is fertility and the factors which affect it, the other is the value of synchronisation to the farmer. If you start discussing this second aspect now it might well be tomorrow morning before you finish! That is not to say that it shouldn't be discussed but it is a large area on its own.

M. Thibier

There are very many reasons for oestrous synchronisation and I think one good example of why it is needed can be seen in the western part of France where there are large herds of Friesian heifers, 200 and sometimes even more. Until recently the farmers just concentrated on getting a reasonable daily gain and then put a bull in and that was it. However, as you will realise, in those conditions the genetic improvement was zero as far as these heifers were concerned because all the female calves were killed. The farmers in this practical situation were very pleased with the oestrous synchronisation because from the genetic point of view it enables them to have very good semen and after oestrous control to perform AI at a fixed time with a very high standard semen and thus get a rapid genetic improvement.

H. Hardie

One other aspect would be applicable to the situation that Dr. Hoffmann mentioned pertaining in Bavaria where there is a very long calving to calving interval. From some of the papers presented, the possibility of bringing down that interval by the use of some of the methods available would be valuable in such a situation.

T. Greve

I have one comment: I don't know whether it is the right

time to bring it up now but it concerns the residue by use of
various progesterone/progestagen treatments. What are the various regulations? I know we import a lot of Charolais cows for
meat in our country. I know that most progesterone treatments
are not allowed in Denmark and Sweden and certainly not in Nor-
way either. So what are the regulations?

R. Newcomb

I wonder whether one of our commercial representatives
would like to comment on that question.

H. Hardie

The situation is that progesterone is allowed in the countries you mention but progestagens are perhaps not allowed. As
far as the others are concerned I am not quite sure how long it
will take and how difficult it may be to overcome the problem
of residues with progestagens. With progesterone there is no
possibility of being able to decide which animal has been treated
and which has not. With the prostaglandins the residues are
particularly short lasting but I will leave that to Mr. Jackson,
I think - he may be able to say something more direct.

P. Jackson

Yes, well, I will speak only of our prostaglandin because
it is a very much smaller dose than the other one and I will
leave them to speak about their own. The milk residue situation
is that there is a zero withdrawal period in all other countries
where it has to date been registered and a statutory 24 hour
meat withdrawal period.

D. Wishart

As far as the norgestomet is concerned, I only have knowledge of the UK situation and there there are detectable residues
in milk and this has precluded the use of the combined treatment
of the injectible plus the implant in dairy cows but not in either
dairy heifers or beef cows, but there is a statutory holding
time.

SESSION 5

DISCUSSION ON CONFERENCE AND CONSIDERATION
OF FUTURE RESEARCH REQUIREMENTS

Chairman:

J.M. Sreenan

J.M. Sreenan *(Ireland)*

Firstly, I would like to thank all speakers and contributors for their participation over the past few days. The purpose of this morning's Session is to discuss in a general way, but particularly in relation to this conference and the results presented here, the area of reproduction in the cow and the various artificial attempts to control this. Perhaps arising from this is may be possible to recommend new approaches to some of the problems.

Briefly, I would like to go back and recollect the major topics of data presentation and discussion during the conference.

The major portion of the meeting was devoted to aspects of egg transfer and the major problem in this area was discussed in the opening session of the meeting, that is, the problem of a consistent source of viable fertilised eggs for use in transfer. The problem would seem to be two-fold: firstly, the PMSG preparation and, secondly, the individual animal response variation that has been shown in all the papers on this topic. For me, at least after this meeting, the problem is not so much the PMSG preparation, but that of the individual animal variation. We need, I feel, to understand a lot more of the conditions required to rescue the maximum number of follicles destined to become atretic. We need to understand much more of the hormonal interactions and ovarian status relative to superovulation response. Perhaps, also, we must look at alternatives to PMSG? Certainly, as the results from the rest of the meeting indicated, the limiting factor to the general use of egg transfer is at the moment the question of egg supply.

In the non-surgical ovum recovery which was the second part of the meeting, I think we have had some very exciting progress. I think this has been the one area in which there has been quite a lot of progress since the Cambridge meeting in 1975, which is

a very short time ago. In its own way this has opened up new
possibilities even with present superovulation methods where you
can repeat superovulate at low rates and achieve good embryo
recoveries.

On the non-surgical transfer side I think we have made some
progress; it has been a little bit slower, a little bit more
erratic and I am not sure that there is not a point where we
should stop some of the empirical experimentation and go back to
examine things a little more closely in relation to embryo survival and luteotrophic or antiluteolytic effects of the embryo.

In relation to embryo culture it would be nice if we could
culture the bovine embryo and I don't think we can. I am talking about the early stages, so that we would have more control
over egg supply and synchronisation between donor and recipient.
Egg storage, however, has made quite a lot of progress and it is
nice to think that since the Cambridge meeting progress has been
made here and in terms of pregnancies deep-freeze storage looks
quite a viable proposition at this time.

We have had a very detailed paper on embryonic sexing, and
this is also quite exciting. However, we must solve a lot of
problems before it can be useful in a practical sense.

On the question of synchronisation which was the second
part of the last day of the meeting, we have had a tremendous
amount of data produced. It would seem to me that there are a
lot of methods of synchronisation and that all of them work and
yet that all of them don't work - depending on the situation in
which you use them, particularly the management situation. I
think, perhaps, further evaluation is required here on an extensive basis as well as further refinements in treatment.

That is a very brief summary of what we have been talking
about over the past few days. I wonder whether anybody might
have a suggestion in terms of superovulation to start with.

Does anybody have any advice? Are there different approaches? Should we change our ideas in some way perhaps to study further the ovarian status at the time we are treating the animals?

L.E.A. Rowson (UK)

I think we have got to look at things rather radically. At the moment with our egg transfer/superovulation techniques, we are only getting probably about 7, 8 or 9 eggs per cow and this is not going to be adequate for any national scheme. It is all right for specialised schemes and so on. I think we have got to go back to the ovary itself; try to obtain follicles from the slaughtered animals, fertilise and culture these. This is a pretty big undertaking, to fertilise and culture these to a stage where they are capable of being transferred to suitable recipients. However, I think this is an aspect which the EEC ought to consider and more work should be carried out in this field.

R. Ortovant (France)

I am very much impressed, like you, by the results which have been reported at this meeting. I think that one of the principal problems is egg production. From this point of view, perhaps we need more basic research, particularly on the timing of follicular growth and on the control of folliculogenesis. We need not only the knowledge of the level of circulating hormones but also a study of the sensitivity of the various follicles to these hormones and especially of their receptors, in the cow. Professor Ryan has shown us that it is an important field. We had a communication in this meeting from Dr. Gosling on preliminary results on this subject. I think we have to explore this area very carefully.

A few weeks ago I heard a talk from Professor Friesen from Canada. Professor Friesen is working on producing receptors and he has purified these receptors from mammary glands and prepared specific antibodies against prolactin receptors. He has injected this antibody in his animals and observed that the ovulation rate

of the animal was increased from 6 to 11. It practically doubles
the ovulation rate just by injecting antibody against prolactin
receptors. Professor Friesen does not know why and I don't think
anybody else knows why, but it is a good illustration of the need
for knowledge in this field of receptors.

J. Gosling *(Ireland)*

I would like some comments on the following analysis: this
variability is occurring at two different stages, as I indicated
in my paper, and that is the variability in the number of follicles grown. That is the first point. The second point is the
variability in the success of the ovulation of the number of
follicles which are grown. I also found Professor Ryan's paper
to be very interesting on the first point. I think he has introduced a phrase which will be popular among us for the next year
or so, perhaps until we find a better one. The phrase is 'rescuing atretic follicles' - follicles which otherwise would be
atretic. Once we begin to think like this we begin to think
about eliminating the variability of this process. He also
brought in the idea that one way of increasing the response may
be to go back a little further. He mentioned two cycles - perhaps it is not necessary to go back this far, perhaps an improvement can be obtained by going back a little before the PMSG
treatment - to try to pave the way for it, or perhaps to the
previous cycle. I would appreciate comments on this.

L.E.A. Rowson

Whatever you do you are only going to increase the level
of ovulation by a matter of perhaps 50% - something of this order.
One is simply not going to get to grips with the problem as a
whole and this is why I suggest we have got to go back to the
ovary itself. One can obtain very large numbers of follicles
from the ovary from slaughtered animals and it is really a pretty
hard fundamental grind to get these fertilised, cultured and
transferred but otherwise we are only playing with the transplant
business. If you can average 11 ovulations per cow you are doing
extremely well.

J.M. Sreenan

Most of us would agree with that in the sense that ultimately if you want to make the proposition efficient this is what you have got to do. I agree with this. At the moment, however, the only possible method we have is superovulation so I think some resources must also be used to try to make superovulation more efficient and perhaps tie it down to repeated non-surgical embryo recovery. While the slaughtered animal may be very useful in some contexts, it will also be necessary to superovulate good genetic material, pure breeds, perhaps particular cows. We are part of the way with superovulation and while it is not going to be efficient ultimately I wonder at the moment whether there is any different approach we should have to superovulation first, and then go back and discuss Mr. Rowson's point.

R. Newcomb *(UK)*

May I just make a comment that in relation to the techniques of superovulation I think we have now got a very valuable tool in the non-surgical recovery methods. By repeated superovulation and repeated non-surgical recoveries, we may be able to examine more closely factors which are affecting superovulation. I think the French have done some very interesting work looking at breed differences; we have now got the chance to look at individual variation repeatedly and if we can mimic what is happening in a good responder then we might get to the sort of line that Dr. Gosling was suggesting.

The other comment which I might make is slightly contradicting Mr. Rowson, in that there is another way of looking at a supply of eggs and that is to use superovulation and slaughter and not just rely on recovering eggs from the ovary. I think there is a very important consideration when one thinks about egg supply, certainly in Britain, and I am sure it will apply to all the other continental countries, and that is that animal health regulations are going to be of tremendous importance. In a slaughter house situation, in the British context, it would be pretty certain that the supply of animals going in to a

slaughter house to supply ova, whether they be superovulated
ova or taken directly from the ovary, that supply of animals
will be closely screened by the Ministry of Agriculture and the
animals may have to come from approved premises, have approved
disease testing and so forth. So, I don't think we can assume
that the whole abattoir population is going to be at our dis-
posal. We are going to have to have a fairly closely screened
population. If one looks at the number of eggs which one can
recover from the ovary by evacuating follicles present in the
ovary, we have found it is something like 10 oocytes per cow.
I think it would be possible, under good circumstances, maybe
to obtain 10 superovulated, fertilised eggs per cow. Provided
the by-products of PMS and the very high oestrogen and progesterone levels didn't interfere with the meat for human consumption then we already have a very practical way of obtaining
a supply of eggs and probably we could supply as many eggs in
the immediate future by superovulation and slaughter as we would
be able to obtain by oocyte and fertilisation in vitro.

J.M. Sreenan

Or perhaps even more superovulation and repeated non-surgical recovery - I think more work on the intervals between
repeated superovulations in this area is needed.

W.W. Lampeter (West Germany)

Following that I would like to ask people here to consider
what we want the embryos for. It is not just that we need a
big number - we want to do something with the embryos. So, if
we go to the slaughter house, the material we will get there will
be just average, or even poor quality. If we are going for poor
quality embryos we are going the wrong way because if we use a
technique we have to achieve something and so we should have
top quality embryos and not slaughter house material.

J.M. Sreenan

There are two things involved here and I tried to make
this point earlier on, that in vitro slaughter house material
will not be suitable all the time because people will want to

use good genetic material.

R. Newcomb

There are two completely different problems. There is the harvesting of good genetic material from very valuable animals, genetically, and this is what Dr. Lampeter is concerned with at the moment. There is also the other aspect, which is what I am sure Mr. Rowson was referring to and which is one of the stated objectives of the European scheme, and that is to obtain twinning in beef cattle. In that situation I am sure that in every country in Europe there are a lot of very good quality cattle from the beef production point of view going to the slaughter house — there certainly are in the British context. If one simply inseminates those animals with semen from progeny tested beef bulls, the chances are one will get some jolly good beef calves. I think the two problems are quite different.

M. Boland *(Ireland)*

In conjunction with superovulation we should also be looking at the quality of the eggs — I think there has been evidence on this at this meeting. The French work, and some of our own work, would show that egg quality can be a factor of considerable importance and that even though you may start off with eight fertilised eggs per donor treated, when you take into consideration that 40% of your eggs can be abnormal, you end up with a fairly small supply of eggs from superovulation. This is one of the things that could be looked at — factors that are affecting the quality of the eggs when they have been in the uterine environment for a while.

A. Brand *(Netherlands)*

It is not only the quality of the egg we must consider but also the uterine environment. What we are doing is synchronising oestrus and, in my opinion, synchronisation of oestrus doesn't mean synchronisation of the activity of the endometria in recipient and donor animals. What is going on when we transfer an embryo into the uterine horn? At the moment we wish such an embryo 'good luck' and then we wait to see what happens during the subsequent two weeks. Some work has been done by Ras Lawson looking at embryos one, four and six days after transfer and I think we have to do that work in the cow as well to see which factors affect the survival rate. What is happening in the endometrium and what is happening to the embryo at this time is very important.

L.E.A. Rowson

We have got to be a little careful over this question of quality, that is, the quality of the egg. I remember that a number of years ago we looked at six eggs from an animal, they were morulae in fact, we considered that these were abnormal but we did transfer them and we produced six calves from them. Once you get to the morula stage it is extremely difficult to determine exactly which eggs are normal and which are abnormal.

J. Gosling

Dr. Baker showed a slide of a very interesting experiment and Dr. Willadsen commented on it and added some results of his own. This whole area of the synchronisation of events at the time of ovulation, if this is in any way abnormal then it can easily be understood that this would result in abnormal eggs - eggs which are either immature or over-mature. There seems to be a lot of agreement now in the literature that the interval between the onset of oestrus to the LH peak is related to the rate of ovulation, the success of the superovulation treatment. So, you have got not just factors of level and factors of sensitivity but, related with these, factors of timing. This experiment of Dr. Baker's shows in perhaps an extreme situation how crucial this can be.

J.M. Sreenan

I wonder if Dr. Baker would like to comment on that himself?

T.G. Baker (UK)

I didn't think anybody had noticed the significance of that experiment - thank you. However, that is absolutely right, one has got to be very careful when one uses foreign hormones like

PMSG and HCG to make sure that one does get the timing right.
One shouldn't be too surprised if there are some anomalies, like
some animals not responding. The only danger, as I see it, from
what was said earlier, is that if you are going to look at, genetically, the best animals, you must remember that they are not
necessarily the easiest animals to work with. Perhaps, with
tongue in cheek, one could quote the experience of the bull at
Cambridge that was a prize bull, and a very beautiful bull. The
only problem was that all its sperm heads were detached from
its sperm tails! If you are going to look at, genetically, the
best animals you must remember that their embryos may not be the
easiest to culture. Perhaps the point about using slaughter
house material is that although you may not have quite such
good animals they may culture a lot better.

W.W. Lampeter

Concerning the good and bad embryos, I think we must just
think in terms of normal and abnormal and say that in the normal
ones there are some stages we can handle more easily and some
stages that are more difficult. We have a considerable knowledge
of the stages of development of certain embryos; some of them
we can handle more easily as a morula or late trophoblast and
we get better results then. However, they may be just as good
at the early stages but we just can't handle them.

J.P. Renard (France)

On this question of embryo quality, I think we have to
move on from just the morphological observation of the embryos.
At the time of non-surgical recovery you can get great variation
in the morphological aspects of good embryos and so I think we
have to move on to much more fundamental studies with reference
to metabolic and biochemical aspects of cattle embryos, as distinct from other species, because the superovulated cattle embryo
has its own rules of development.

Y. Menezo (France)

A comment on this point of view is that if this way of
biochemical aspects is to be maintained it would be preferable

for everybody to work later because there is a very important difference from the metabolic point of view, a difference between a Day 10 embryo and a Day 13 embryo. The later you go, the better you can work with the embryo in this way.

G.E. Seidel *(USA)*

One other thing about the whole business is maybe we have been too hard on the superovulated embryo. If you just breed cows, normal cows, by oestrous synchronisation, you get probably 25% of the eggs degenerating - at least, you don't get calves - in the very, very best circumstances. A lot of the things we can learn will be important for less sophisticated agriculture - just plain getting cows pregnant.

Maybe we can make the greatest contribution, in the short run anyway, in trying to get every cow pregnant, just with one calf. We are in the fortunate situation of being able to collect superovulated and unsuperovulated eggs from the same population of donors and while the fertilisation rate is lower with super- ovulated eggs and the seeming ability of the reproductive tract to pick up the oocytes, get them fertilised, and developing, is different, once one gets fertilised eggs out of the reproductive tract, six, eight or twelve days after ovulation, while there are differences between superovulated and unsuperovulated embryos, these differences are not very great as I think has been shown with some data presented to this meeting. Maybe the pregnancy rate is 10% less with superovulated embryos but it is not more than that. Maybe the number of abnormalities is 10% higher with superovulated eggs but it is not more than that - at least to the point where the reproductive tract selected the embryos and that is the stage of embryo that we are working with. So, let's think about getting more basic information about the cow in a general way.

J.M. Sreenan

I would agree and that is our approach at the moment. I would suggest to the EEC that this is one area that they should be funding - to make the system in the normal situation more

efficient. One of the areas we are interested in is not the
abnormality rate of superovulated eggs but embryonic mortality
of non-superovulated cows. Perhaps we can use superovulation
techniques, mild or high and embryo transplantation techniques,
to study some of these things. I feel that we have had a little
too much emphasis on very intensive aspects of superovulation
and egg transfer without considering how best we could make up
this deficit of 25%.

I. Wilmut *(UK)*

I would simply echo your words. I think perhaps we can
develop your line of thought a little bit by saying that over
the last ten years various people have collected information on
hormone levels, follicle populations, uterine proteins, and so
on. We are now coming to a stage where it will be important,
not just to study these individual items, but to synthesise the
whole thing into a more general understanding with the hope that
either by improving our management systems, the environment that
the animals live in, improving their feed, improving our insemination techniques, or possibly by some selection procedures,
we may be able to reduce this 25% loss. This will depend not so
much on individual efforts in terms of endocrinology or follicle
populations or whatever, but much more on the kind of synthesis
that the people in Germany have been beginning to make between
hormones and follicle populations and this sort of thing. It is
important for us to begin to put together a much more complete
picture of reproduction in the cow using all the techniques that
we have.

G.E. Seidel

The problem, at least in the States, is that the fertility
rate with artificial insemination has actually declined over the
last several years. I don't know if the same thing is happening
here but I would guess that it probably is. As animal management
becomes more intensive it seems that we are creating new problems
which require new solutions. In a way we have gone backwards.

J.M. Sreenan

   This is perhaps relevant also to synchronisation - the
oestrous synchronisation techniques have served in many ways,
more to point out to us errors in management rather than to increase efficiency. We have got to go back and study these now
and then apply synchronisation techniques perhaps under modified
management.

M.T. Kane *(Ireland)*

   One thing that strikes me that the egg transfer techniques
might show up, although I am not sure exactly how, is how much
of the embryonic mortality is due to the egg and how much is due
to the uterine environment.

J.M. Sreenan

   My suggestion is that these are the kind of things we
could study in the normal situation as well as in the superovulation situation, and to use transfer techniques to study embryonic
mortality at various stages.

   On the question of embryos in terms of survival, are there
areas that we should explore more thoroughly in respect of non-
surgical transfer? This is an area in which progress has been
good but perhaps a little slower than non-surgical recovery. We
have carried out many careful transfer experiments and we have
an embryo survival level that we find it difficult to push up
much higher. Perhaps we haven't looked at transfer adequately
in terms of day of transfer. Does anybody have any ideas of
approaches we should have on non-surgical transfer?

J.P. Renard

   One thing that is important is the management of the recipient - to have a complete knowledge of the recipient in which
you put the embryo. The quality of the embryo used is also an
important aspect of non-surgical transfer.

J.M. Sreenan

   Yes, but I find a difficulty here because in our surgical

transfers we are able to get a very high conception rate. We are attempting to select our embryos in the same way for non-surgical work but we have difficulty in getting above about 30 to 40% embryo survival over large numbers.

T. Greve *(Denmark)*

May I suggest that some of the future work should be devoted to avoiding the cervical passage. I don't know, and I think very few people know, why we have that low survival rate but I would just throw out the suggestion of studying the para-cervical approach in one way or another.

J.M. Sreenan

Do you mean a minor surgical operation?

T. Greve

No. It would still be non-surgical but you would have to do away with passing the cervix because it certainly must do something wrong since recent data shows that it really doesn't matter exactly where you deposit the egg so other factors must influence it.

I agree with what Dr. Brand suggested a little while ago that we may have to study the location of the eggs - do we, in fact, get the eggs where we presume we put them, are they drawn back to the cervix by the Cassou gun because of some adherence of mucus? I think that perhaps infection has not been looked at closely enough and the microscopic changes we make in the endo-metrium by this method.

N. Rasbech *(Denmark)*

I want to come back to Greve's remarks. I think he is referring to the procedure and the instrument I mentioned the other day concerning para-cervical transfer of eggs. I think it is worthwhile to do more research there. I tried it a few years ago and gave it up but since then we have acquired better instruments and better techniques and I really think it is a question of having a suitable technique for going through the

vagina then it is very easy to place the eggs para-cervically.
So, in the near future I think we will do some comparisons between that route and the cervical route.

R. Newcomb

I think the only comment worth making with regard to trans-
cervical compared with para-cervical is that all the work that
has been done so far in by-passing the cervix has not resulted
in improved pregnancy rates after transfer. The French results
by trans-cervical transfer are, if anything, better than the
results they were having with the equipment which Testart developed. Likewise, Sugie's results were no better than the results
which everybody else is obtaining by going through the cervix.
So, I think we have to be a little cautious before we all go
diving off looking at para-cervical transfer.

G.E. Seidel

We dived off a couple of years ago! We transferred some
20 embryos that way. Unfortunately the first one got pregnant
and encouraged us - but the next 20 did not! It may be largely
a technical problem. It is very difficult to be sure that you
have got into the lumen with this approach. Maybe it just takes
a combination of ingenuity and practice. I rather think that we
did not get the embryo in the right place and that was most of
the problem - but I haven't been able to think of a simple way
of getting it to the right place. However, I don't think we
should say it is impossible; it may be the thing we are all doing
two years from now.

N. Rasbech

It is a pretty technical question.

J.M. Sreenan

Dr. Kane presented a review on egg culture in the cow, are
there any comments or suggestions on approaches other than those
he mentioned?

M.T. Kane

I could make a general comment. The whole area was more
egg manipulation. While I would not particularly like the culture medium, yet I think that the work that was done by Dr.
Willadsen and his people has made dramatic progress in terms of
embryo freezing, and that is something which we should not lose
sight of in discussing the problems of culture. Particularly
with the earlier stages of cattle eggs, the media that are available are very, very bad - most of them are just diluted forms
of serum. I would like to think that this is perhaps a general
biological problem and that possibly there are hormonal factors
controlling the early cleavage process in certain species, including the cow, but what factors I don't know.

K. Betteridge

Could I just get back to embryo transfer before we move on.
There have been odd comments about starting to use embryo transfer as a tool and the question of embryonic loss. Another one
which strikes me as important to Europe as an exporting group,
is disease transmission. As we progress with storing embryos
and import and export, there is going to be this problem of still
regarding the embryos as a possible source of infection for importing countries. Everyone who is importing embryos and animals
from Europe is going to be acutely aware of this possibility and
I would have thought it might be worth considering as a research
topic.

J.M. Sreenan

Yes, you are right. Does anybody wish to come in on this
question?

R. Newcomb

Merely to say, Mr. Chairman, that I wholeheartedly agree.
It is a completely neglected subject; we know very little about
it indeed and the more we know the better it will be from the
import/export situation.

D. Szollosi *(France)*

It is not known in bovine or ovine embryos but it is a
case in the mouse embryo that there is RNA type virus present
which is transmitted within species. We have not found one single
laboratory strain of mouse which did not contain this virus and
there is some evidence now that in the guinea pig, human, macaque,
there is a spontaneously occurring RNA type virus present in
embryos. Nobody knows the significance of these virus type particles and whether these are infective, but in the baboon and
human cases, and in some of the mouse cases as well, infective,
C-type, viruses have been found regularly. It does represent a
very serious problem and I don't think there are techniques
available at the moment to analyse or even attack this problem
systematically. If there is to be a regular export of embryos
from one country to another on a large scale, somebody will have
to tackle this problem totally from the beginning.

N. Rasbech

In this connection it might of interest for this discussion
if I just mention how the situation is in Denmark now. A year
ago the Government set up rules for importation of eggs just as,
for many years, we have had rules for importation of frozen semen.
I will briefly outline the main principles. For each importation
of eggs there must follow 20 ml of the flushing fluid used. All
eggs, which we expect will be frozen eggs in the future, are
quarantined in Copenhagen. The fluid is put into test heifers
in the same way as has been done for many years with imported
semen. The test heifers must be free of any disease and under
control: these test heifers are examined for a period of 50 days
for virus, bacteria, and by blood testing and uterine biopsies.
If there is no transfer of blue tongue or any other virus, the
eggs are released for use. However, the recipient cows must be
quarantined for a period of nine months plus one month after
calving. During this period various tests for brucellosis and
so on will be made

Those are the main principles. The most important thing
is to test the flushing medium in connection with importation

because if there is any type of virus transmission it should be possible to pick it up in the flushing medium.

P. Hartigan *(Ireland)*

As one who is not actually involved in the field of transplantation, I think I may offer some comment in relation to the RNA viruses. That sort of regimen would be totally inadequate to cope with the situation Dr. Szollosi has raised because of the incubation period with the C-type virus. For instance, in sheep the transfer of the bovine RNA virus of sheep has an incubation period of at least a year and a half to two years. So one could transmit a C-type virus in the ovum and this would not be covered by the regimen outlined by Professor Rasbech.

N. Rasbech

The regulations I mentioned apply to cows, not sheep.

P. Hartigan

Well, the same situation applies.

N. Rasbech

I agree, yes.

P. Hartigan

These are slow viruses which are endogenous and they may take many years and, in fact, we don't understand the circumstances which do activate them.

N. Rasbech

Yes, we have the same situation with leucosis in cattle. We may not be able to pick that up in such a short period.

P. Hartigan

This is particularly important in Ireland where we believe, although we are not absolutely sure of this, that we don't have leucosis virus and so we can't afford to take in ova from Denmark.

G.E. Seidel

It seems unlikely that one needs more stringent requirements

for an embryo that is in a test tube than one needs for an embryo
that is in a pregnant cow you have just bought. It seems quite
a reasonable process to apply the same regulations to the recipient as one would apply to a pregnant cow. We talk about C-type
viruses and RNA viruses that we know nothing about; we know even
less about situations which we know are dangerous, like blue
tongue. One of the things we have been doing in our lab is
studying virus embryo interactions in both the mouse and the cow
and looking at some of the known diseases to see if these are
transmitted that way.

   One other point, there will be meetings of the International
Embryo Transfer Society in Denver in January and we have a rep-
resentative from the Animal Health Inspection Service of USDA
at that meeting in an affort at least to set up some preliminary
guidelines for people at USDA to be thinking about, as far as
importing embryos is concerned. Some of you may be interested
in speaking with these people or providing input to setting up
these regulations in the United States.

D. Szollosi

   I don't want to be a 'Devil's Advocate' but I think the
problem is even more complex because that which you mentioned at
least can be picked up, cultured and handled by virologists. In
the mouse, which is certainly not a farm animal, there are other
types of viruses which are the A-type viruses which cannot be
cultured and which represent a very great problem to the virologists. No relationship between the A-type and C-type viruses
has been established until now but we find that in certain strains
of mice there is a high level of A-type virus. At the same time
those strains of mice develop spontaneous tumours at very high
rates, like the New Zealand Black, and so I think that at the
moment, when there is just a beginning in planning of examination,
these types of problems should be given serious consideration
even if they are virologically just on any level.

N. Rasbech

   I would just like to say that I don't think that leucosis
which Dr. Hartigan mentioned was a very good example because, of

course, the herd of origin of the donor must be under very strict
control for a period of one, two, three or four years - I don't
know - for leucosis. So, if you buy eggs from Denmark you can
be sure that they will come from herds that have been under control for years for leucosis. The same applies for semen. We
cannot make the regulations more strict for semen. We import a
lot of semen from Canada, the US and other countries and there
are established regulations regarding the herds of origin. So
we can control that. Concerning sheep, well we cannot import
sheep eggs; it was banned due to these long-lasting viruses.

P. Hartigan

Well, I just quoted the sheep, not because it is of any
great importance to reproduction but just to illustrate that a
recent experiment in transmission has shown that at a minimum
there is that risk. It is difficult to set up a minimum time
with cattle because the endemic viruses may already be there and
you are just activating them. This is not relevant to the point
of this meeting today; the only point was that it is important
to remember that in species like the mouse and so on, a lot of
these diseases are transmitted vertically from mother to offspring - presumably in the ovum.

J.M. Sreenan

I am sure it has a relevance. Does anybody want to talk
about the relevance of synchronisation?

K. O'Farrell (Ireland)

I would like to try to put together a few comments. I
think Dr. Hoffmann asked a very good question yesterday, what
are the possible advantages of synchronisation? Certainly if I
had been asked that question two years ago I would have given him
a full list. However, having done a series of trials and experiments, at this stage, I am not so sure that it has all that much
to offer, at least, at the present conception rates. I think
George Seidel mentioned that maybe we should be concentrating
more on the management factors which will improve individual cow
fertility. From experiments which we have done, comparing

synchronisation with control situations and using all the little management pieces that we can put together, we are showing that there is no advantage at all in synchronisation.

The best advantages that have been shown in synchronisation, I believe, are somewhere in the region of 10%. That means that if you treat 30 cows, which is about the average that will be available for treatment in this country, you will have three more cows calving in the first 30 days of your calving season. The cost of the synchronisation programme must then be put on those three additional animals. Working it out on very basic economics, that would mean a cost per three cows of £45. I don't think that that is an economic proposition. This is one of the problems facing synchronisation; the advantages will depend on the number of animals that you can submit to a synchronisation regime. So the farmer, who really needs synchronisation most of all is the one with a scattered calving pattern, but because of the scattered calving pattern he cannot submit many animals to the synchronisation treatment. Thus, what I think you have got to be looking at is improving the fertility to the synchronisation regimes, if this is to be an economic proposition.

M. Thibier

If I got your point correctly I don't agree completely with what you are saying. In France at least, there are two different cow populations, one of which is very suitable for oestrous synchronisation. That is the population in which the fertility is very good. It is obvious now that with the synchronisation or oestrous control treatment, it is no good trying to have an increasing calving or fertility rate. I would say that a good fertility rate prior to treatment is a prerequisite and then it might be a very good thing, in certain circumstances, to have oestrous synchronisation.

The second kind of cattle population is the one in which there are some herd management difficulties. As has already been said, management conditions are very important and in this respect we believe the approach which I tried to illustrate last

night is the one on which we ought to concentrate in order to
improve the fertility rates where necessary. So it is important
not to confuse the problems - at the beginning you have two
quite different situations.

J.M. Sreenan

I think Kevin O'Farrell is attempting to get synchronisation
in perspective by throwing out his comments rather than saying
it is of no use at all. I take it that you are trying to say
that we must examine the economics attaching to it.

K. O'Farrell

Yes. I think there are situations, particularly with the
heifer rearing programmes, where synchronisation is an obvious
advantage. Those are specific situations and I am referring
rather to the dairy situation where compact calving periods are
essential for maximising animal production, particularly in our
spring calving situation where our milk is produced totally from
grass. We have to look at the total economics.

J.M. Sreenan

Would you care to comment on this, Dr. Hoffmann?

B. Hoffmann

Well, it always comes back to the fertility problem and I
think we should mention the problem of early embryonic death
which is really an old story and has been widely discussed. It
has been the excuse for many things which went wrong. However,
we really don't know much about it. We don't know whether it is
a case of the embryo dying causing luteolysis or the other way
round. This is an area we should look into more closely. We
can monitor luteolysis with our new techniques but we cannot
really state the cause. This is something about which we haven't
talked at all during this meeting; what about the embryo in the
cow, how does it survive, what is its relationship to the mother
animal?

J.M. Sreenan

This is what I was suggesting earlier on relating to both superovulation and embryo survival in transfer and in normal breeding situations, and perhaps we can use embryo transfer techniques to study some of these things now.

B. Hoffmann

One major question is what is the luteotrophic factor of the embryo? Quite a few people are looking into it but to my knowledge it has not been identified, at least, not in the cow.

J. Gosling

The problem of inefficient luteolysis is an important contributory factor to the lack of success of synchronisation using the coil or sponge. I have a general principle when I am trying to think of ways of looking at things and that is to look at what is happening in the rat. The rat is a mammal and there may be something to be gained by just looking at what is happening in the rat. People who are studying the rat have got much greater numbers, better laboratory facilities, rats live in the laboratory and so on. One finding which has come out in recent years is that prolactin appears to protect against the inducement LH receptor loss by prostaglandin. So, if prostaglandin is not very efficient and if there is any problem of luteolytic breakdown, either directly induced by prostaglandin or indirectly in some way or another by oestrogen, perhaps if one were to minimise any possible contributory factor by prolactin, this might help to improve the efficiency of the treatment being used. Just to mention a drug that has been talked about recently, has anybody got any data to suggest that the simultaneous administration of Bromocryptine with a luteolytic treatment has any effect on the efficiency of the luteolytic treatment?

D. Schams *(West Germany)*

I would like to make a short comment about the prolactin story. I don't think we should be too optimistic about what prolactin does to the cow because we have done a lot of experiments and at the moment there are no indications that prolactin may have a luteotrophic effect, that it may stimulate the new cycle after parturition, or any other interaction in the ovary itself. So, I think there are really very strong species differences and I don't think the rat and other small animals are very good examples to compare with the cow. I think it is interesting from the scientific point of view but for use for superovulation I don't think there is any chance to get an improvement.

B. Hoffmann

We actually treated animals with a prolactin inhibitor during superovulation and it had no effect at all on the ovulatory rate.

J.M. Sreenan

Would any of the commercial people like to comment on synchronisation?

H. Hardie (Belgium)

The only comment I would add is that I think your summing up is a very good assessment of the situation, that under suitable circumstances the methods available are effective. Before any of them are put into operation the person applying them should be sure that the situation is suitable for the method being used. With those comments I think we have several effective methods which can be used to improve either the facility for getting AI to a group of animals where management makes that difficult, or, perhaps contrary to what Kevin O'Farrell was saying, the improvement in tightening up the calving season can be very considerable. It is not just a question of three animals calving down in the first month but it is possible to save quite a number of days per animal on calving to calving interval, as well as tightening up the season. In this way you can have a better annual production from each animal in the herd, not just from the three that he mentioned coming into the first month.

M.R. Woulfe (USA)

I would like to comment on Kevin O'Farrell's observation about usefulness or non-usefulness of synchronisation. I think that with the passage of time we will see regional advantages

654

develop within continental Europe in the same way that we have
seen within the United States, advantages developed on a geo-
graphical basis and certainly on a breed basis. The type of
animal in the Gulf States needs a different approach entirely
to the English Angus, Hereford, type of operation in the north
central prairie states. Again, we do have the climatic situation
where we need to coincide calf drop with the availability of
grass, availability of the onset of rain in certain areas, and
I am certain that within the United States we will see these
regional factors emerge in just the same way that we have seen
them emerge with the use of synchronisation in sheep. This is
going to happen by itself and not at the behest of private drug
companies or anybody else.

J.M. Sreenan

Has anybody else got any points to make on synchronisation
because we are running out of time and I would like to close the
session up at this stage unless there are further comments on
this or any other aspect?

D. Schams

I have one point on superovulation. We know that the profiles do not seem to be very different in the normal cycle but
there are no exact data about the timing of ovulation itself.
Is it only a few hours, or one day, or two days? Has anybody
any experience on this? If you know the time involved then it
is easier to manipulate. At the moment we just inject something
and after some days we are looking for something.

J.M. Sreenan

You are speaking about the time period over which all the
ovulations occur?

D. Schams

Yes.

J.M. Sreenan

I am not sure whether anybody has that data. We tried to

do this and produced some data showing ovulation occurring over
a 24 hour period, but, as I said the other day, we used repeated
laparascopy rather than laparotomy ovulation studies and we suspect quite strongly that the repeated laparoscopy affected the
time of ovulation. We have to look at it again. Has anybody
else data on timing in superovulated animals?

J. Mariana *(France)*

We have not spoken about planning experiments, how should
we conduct an experiment in order to have any conclusions! How
many animals do we require? We have not spoken about these problems, but they are important.

J.M. Sreenan

To conduct these types of experiments is always a problem
with us in terms of numbers of animals particularly. The big-
gest problem I see is that we try to do too many things with any
one group of animals. Perhaps we should select the things we
want to look at and increase the numbers of animals per treatment
group.

G.E. Seidel

Unfortunately with something like pregnancy rates one is
working with a binomial response and that is a terrible problem
to have to start with because the statistical techniques are so
unpowerful - one needs tremendous numbers of animals per group.
Having a hundred animals per treatment is still not adequate if
one is trying to find out the 10 or 15% difference. With a lot
of these things we are at a point where if we could establish
that one treatment is 15% better than another that would be quite
good and 100 animals per group is not enough to find that kind
of difference - on pregnancy rate.

J.M. Sreenan

This I agree with but I think we must continue with our
25, 30 or 40 animals per group because it is just not possible
to have 400 animals per group, we can, in this way, accumulate
trends or cumulative differences over groups.

M.T. Kane

There might be a possibility of putting data together at times from different labs, or from different countries even. It is possible that something which appears in one lab as no more than an interesting trend may appear in a number of labs. If it is reasonable to put the data together then maybe it should be put together and possibly even published together.

T. Greve

I would just like to say that the impression I have gained from private discussion as well as the public sessions is that there is a wide range of optimism and pessimism concerning embryo transfer work. From the Danish point of view, from our experience in the field both in collection, freezing and transfer, we are fairly optimistic and the farmers and co-ops we are dealing with are satisfied. I think that is our conclusion from the past 1½ - 2 years work.

J.M. Sreenan

Well, I suppose the views are as wide ranging as the responses are to treatments at times but I would agree that there is optimism there - enough to continue the work. I don't think it is pessimism that is there at times so much as people worrying about how to solve the problems, what new approaches to take. For me, the whole thing comes back to the fact that I think the Commission must start to fund not just for techniques of synchronisation and transfer but for the understanding of what goes on in the normal situation and then combine this with whatever new techniques are available.

I. Szollosi

I would like to come back to another point. On the first day of this seminar it became clear that we understand some of the hormonal effects on the graafian follicle and that there is very little understanding of what happens with the primordial follicle. I think if we ever hope to do follicular culture as a source of fertilisable eggs, as I think Dr. Baker and I hope to do, one may have to go back to a more fundamental study of

oogenesis, and determine whatever this mysterious factor is
which initiates time and again a new wave of primordial follicles
to develop. It will become of great importance to us to know
that in an ovary from which we may want to recover good follicles
for culture we have a fair number of good healthy follicles.
This subject was discussed on the first day and I would like to
come back to it because in my view our current data is very
promising that the culture system is becoming a usable tool. I
think we should re-examine the more fundamental process of
oogenesis.

J.M. Sreenan

Most people would agree with that. Does anybody want to
comment?

F. du Mesnil du Buisson *(France)*

I think in the future this will be the way of obtaining
eggs on an economic basis. It will be necessary to have very
large quantities of cheap eggs, not necessarily of a very high
genetic quality. For twinning, for example, it is necessary to
have cheap eggs and this method is the only one for the future.

T. Greve

I would just comment that I am working with dairy cows and
this question of oogenesis is very interesting but not particularly applicable so far. We have been doing histological studies
in superovulated dairy cows by ovarian biopsies - we are taking
ovarian biopsies through the superovulatory cycle and they seem
to be very encouraging.

J.M. Sreenan

Well, my proposal now is to close the session at this stage.
I am not going to attempt to summarise what has been said here.
In itself it will be taken as a summary of the meeting and per-
haps may be of some value in the planning of new experiments.

I would like to take this opportunity to thank everyone
who contributed to this discussion.

LIST OF PARTICIPANTS

Dr. W.R. Allen
TBA, Equine Fertility Unit
Animal Research Station,
307 Huntingdon Road,
Cambridge CB3 0JQ
UK

Dr. T.G. Baker
Dept. Obstetrics & Gynaecology
University of Edinburgh
Maternity Bldg.
Edinburgh EH3 9YW
UK

Dr. D. Beehan
Beef Breeding Section
Department of Agriculture
Agriculture House
Kildare Street
Dublin
Ireland

Dr. K. Betteridge
Animal Disease Research Institute
801 Fallowfield Road
Ottawa
Ontario K2H 8P9
Canada

Dr. C.H.W. de Bois
Kliniek Verloskunde
Gynaekologie en KI
Yalelaan 7
de Uithof-Utrecht
The Netherlands

Dr. M.P. Boland
Faculty of Agriculture
University College Dublin
Lyons Estate
Newcastle
Co. Dublin
Ireland

Dr. A. Brand
Kliniek Verloskunde
Gynaekologie en KI
Yalelaan 7
de Uithof-Utrecht
The Netherlands

Dr. A.J. Breeuwsma
Intervet International BV
PO Box 31
Boxmeer
The Netherlands

Dr. F. du Mesnil du Buisson
INRA
Station de Physiologie de la Reproduction
Centre de Recherches de Tours
Nouzilly 37380
France

Dr. D. Chupin

INRA
Station de Physiologie de la Reproduction
Centre de Recherches de Tours
Nouzilly 37380
France

Mr. H. Coulthard

Woodlea
Tattershall Road
Woodhall Spa
Lincolnshire
UK

Dr. R. Crinion

Department of Veterinary Medicine
University College Dublin
Ballsbridge
Dublin 4
Ireland

Mr. J. Cunningham

Galtee Cattle Breeding Station
Mitchelstown
Co. Cork
Ireland

Dr. Dhondt

Faculteit van der Diergeneeskunde
Rijks Universiteit
Casinoplein 24
B 9000
Ghent
Belgium

Dr. Bridget Drew

Ministry of Agriculture, Fisheries & Food
Agricultural Development & Advisory Service
Government Buildings
Christchurch Road
Winchester
Hampshire SO23 9SZ
UK

Dr. P. Fottrell

Dept. of Biochemistry
University College
Galway
Ireland

Dr. H. Frerking

Klinik für Geburtshilfe und Gynakologie
des Rindes
3000 Hannover 1
Bischofsholer Damm 15
W. Germany

Dr. W. Gehring

c/o Ambulatorische u. Geburtshilfliche
Veterinärklinik
Der Justus Liebig-Universität
6300 Giessen
Frankfurter Strasse 106
W. Germany

Dr. Catherine Godard-Siour	SERSIA 43 Rue de Naples 75008 Paris France
Dr. I. Gordon	Faculty of Agriculture University College Dublin Lyons Estate Newcastle Co. Dublin Ireland
Dr. A. Gorlach	Artificial Insemination Centre 8530 Neustadt a.d. Aisch Birkenfelder Strasse 21-27 Postfach 1220 W. Germany
Dr. J. Gosling	Biochemistry Department University College Galway Ireland
Dr. T. Greve	Institute for Animal Reproduction Royal Veterinary & Agricultural University Copenhagen Denmark

Dr. M.C. Gueur	Laboratoire d'Embryologie et
	    d'Anatomie Comparée
	5 Place Croix de Sud
	    B-1348 Louvain-la-Neuve
	Belgium

Dr. J. Hahn	Klinik für Geburtshilfe und Gynakologie
	    des Rindes
	im Richard-Götze-Haus der Tierärtzlichen
	    Hochschule
	3000 Hannover
	Bischofsholer Damm 15
	W. Germany

Dr. R. Hahn	Artificial Insemination Centre 8530 Neustadt a.d. Aisch Birkenfelder Strasse 21-27 Postfach 1220 W. Germany
Dr. H. Breth Hansen	'Transbova' East Jutland Animal Hospital Laurbjerg 8870 Langa Denmark

Dr. H. Hardie

Area Veterinary Adviser
(Abbot SA
        NV)
Parc Scientifique
Rue du Bosquet 2
1348 Louvain-la-Neuve
Belgium

Dr. Pat Hartigan

Dept. of Physiology
University of Dublin
Trinity College
Dublin 2
Ireland

Dr. S. Hanrahan

Agricultural Institute
Belclare
Tuam
Co. Galway
Ireland

Dr. W.C.D. Hare

Animal Pathology Division
Health of Animals Branch
Agriculture Canada
Animal Diseases Research Institute
Box 11300
Station H
Ottawa
Ontario K2H 8P9
Canada

Dr. Y. Heyman

INRA
Station Centrale de Physiologie Animale
78350 Jouy-en-Josas
France

Dr. L. Henriet

Université Catholique de Louvain
Sciences Vétérinaires
3 Place de la Croix du Sud
1348 Louvain-la-Neuve
Belgium

Dr. B. Hoffman

Institut für Physiologie
Südd. Versuchs-und Forschungsanstalt für
    Milchwirtschaft
Weihenstephan
Tech. Universität Munchen
8050 Freising
W. Germany

Mr. P. Jackson

Animal Health & Husbandry Department
Pharmaceutical Division
ICI Ltd.
Alderley Park
Macclesfield
Cheshire
UK

Dr. J. Jennings

Agricultural Institute
Grange
Dunsany
Co. Meath
Ireland

Dr. C.H.J. Kalis

Kliniek Verloskunde
Gynaekologie en KI
Yalelaan 7
de Uithof-Utrecht
The Netherlands

Dr. M.T. Kane

Physiology Department
University College
Galway
Ireland

Professor H. Krausslich

Institut für Tierzucht
Verebungs und Konstitutions forschung
    der Universität Munchen
D-8000 München 22
Veterinärstrasse 13
W. Germany

Dr. A. Kruip

Kliniek Verloskunde
Gynaekologie en KI
Yalelaan 7
de Uithof-Utrecht
The Netherlands

Dr. W.W. Lampeter

Institut für Tierzucht und Tierhygiene
    der Universität München
Lehrstuhl für Tierzucht
D-8000 München
Königinstrasse 13
W. Germany

Dr. O. Langley

Agricultural Institute
Moorepark
Fermoy
Co. Cork
Ireland

Dr. H. Lehn-Jensen

Institute for Animal Reproduction
Royal Veterinary and Agricultural University
Copenhagen
Denmark

Dr. J.C. Mariana

INRA
Station de Physiologie de la Reproduction
Centre de Recherches de Tours
Nouzilly - 37380
France

Dr. P. Mauleon	INRA Station de Physiologie de la Reproduction Centre de Recherches de Tours Nouzilly - 37380 France
Miss Sheelagh Mawhinney	Agricultural Institute Grange Dunsany Co. Meath Ireland
Dr. Y. Menezo	INRA Laboratoire de Biologie 406 20 Avenue Albert Einstein F-69621 Villeurbanne France
Dr. C. Menzer	Sudd. Versuchs und Forschungsanstalt für Milchwirtschaft Weihenstephan Tech. Universität München 8050 Freising W. Germany
Mrs. Patricia Morgan	Biochemistry Department University College Galway Ireland
Dr. Moustafa	3000 Hannover 1 Bischofsholer Damm 15 W. Germany
Dr. P. Mulvehill	Carcase Grading Section Department of Agriculture Agriculture House Kildare Street Dublin Ireland
Dr. Mary Murphy	Department of Biochemistry Trinity College Dublin Dublin Ireland
Mr. R. Newcomb	ARC, Institute of Animal Physiology Animal Research Station 307 Huntingdon Road Cambridge CB3 0JQ UK
Dr. K. O'Farrell	Agricultural Institute Moorepark Fermoy Co. Cork Ireland

Dr. P. O'Reilly

Department of Agriculture
Veterinary Research Laboratory
Ballycoolin Road
Finglas
Dublin
Ireland

Dr. R. Ortovant

INRA
Station Centrale de Physiologie Animale
78350 - Jouy-en-Josas
France

Dr. J.P. Ozil

INRA
Station Centrale de Physiologie Animale
78350 - Jouy-en-Josas
France

Professor H. Papkoff

Hormone Research Laboratory
University of California
San Francisco
California 94143
USA

Dr. E.J. Passeron

Director
Research and Development Division
Laboratorio ELEA
Saladillo 2452/68
Buenos Aires
Argentina

Dr. J. Pelot

INRA
Station de Physiologie de la Reproduction
Centre de Recherches de Tours
Nouzilly 37380
France

Dr. J.S. Perry

Scientific Adviser
ARC
160 Great Portland Street
London W1N 5DT
UK

Dr. S. Quirke

Agricultural Institute
Belclare
Tuam
Co. Galway
Ireland

Professor N.O. Rasbech

Royal Veterinary and Agricultural University
Bulowsvej 13
1870 Copenhagen V
Denmark

Dr. E. Rattenberger

Tiergesundheitsdienst Bayern. V.
D-8011 Grub/München
W. Germany

Dr. J.P. Renard

INRA
Station Centrale de Physiologie Animale
78350 - Jouy-en-Josas
France

Dr. Riemensberger

Bundesministerium für Ernährung
Landwirtschaft und Forsten
D-53-Bonn Duisdorf
Bonnerstrasse
W. Germany

Dr. W. Romanowski

Fachtierärzt für Zuchthygiene und Besamung
2820 Bremen 77
Postfach 77 0260
W. Germany

Mr. L.E.A. Rowson

Officer I/C, ARC
Institute of Animal Physiology
Animal Research Station
307 Huntingdon Road
Cambridge CB3 OJQ
UK

Dr. R.J. Ryan

Department of Molecular Medicine
Mayo Clinic
Rochester
Minnesota 55901
USA

Mrs. Charmian A. Russell

Upjohn Ltd.
Agricultural Veterinary Division
Fleming Way
Crawley
Sussex RH10 2WJ
UK

Dr. J. Saumande

INRA
Station de Physiologie de la Reproduction
Centre de Recherches de Tours
Nouzilly - 37380
France

Dr. D. Schams

Institut für Physiologie
Sudd. Versuchs und Forschungsanstalt für
    Milchwirtschaft
Weihenstephan
Tech. Universität München
8050 Freising
W. Germany

Dr. U. Schneider

Klinik für Geburtshilfe Gynakologie des
    Rindes
im Richard-Götze-Haus der Tierärztlichen
    Hochschule
3000 Hannover
Bischofsholer Damm 15
W. Germany

Dr. G.E. Seidel Jr.

Animal Reproduction Laboratory
Colorado State University
Fort Collins 80523
Colorado
USA

Dr. J.M. Sreenan

Agricultural Institute
Belclare
Tuam
Co. Galway
Ireland

Dr. D. Szöllösi

INRA
Station de Physiologie Animale
78350 - Jouy-en-Josas
France

Dr. S. Willadsen

ARC
Institute of Animal Physiology
Animal Research Station
307 Huntingdon Road
Cambridge CB3 0JQ
UK

Dr. I. Wilmut

ARC
Animal Breeding Research Organisation
Field Laboratory
Roslin
Midlothian EH25 9PS
UK

Dr. D. Wishart

Beechams Pharmaceutical Unit
Research Division
Walton Oaks
Dorking Road
Tadworth
Surrey
UK

Mr. M.R. Woulfe

G.D. Searle & Co.
Agricultural Regulatory Affairs
Box 5110
Chicago
Illinois 60680
USA

EEC PERSONNEL

Dr. P. L'Hermite   Commission of the European Communities
                              V1-E-4
Mr. J. Kuyl   200 Rue de la Loi
                              B-1040 Brussels,
                              Belgium

Mr. G.J. Breslin   Commission of the European Communities
                              European Centre
                              Kirchberg
                              Luxembourg

RECORDING PERSONNEL

Mr. S.E.W. Hallam   Janssen Services
                              14 The Quay
Mrs. Molly Robins   Lower Thames Street
                              London EC3R 6BU
                              UK